SOCIOLOGY OF CRIME, LAW AND DEVIANCE

SOCIOLOGY OF CRIME, LAW AND DEVIANCE

Series Editor: Jeffery T. Ulmer

SOCIOLOGY OF CRIME, LAW AND DEVIANCE VOLUME 2

SOCIOLOGY OF CRIME, LAW AND DEVIANCE

EDITED BY

JEFFERY T. ULMER

Crime, Law, and Justice Program: Department of Sociology, Penn State University, University Park, PA 16802, USA

EDITORIAL ASSISTANT: ROBERT V. GRANTHAM

Emerald

United Kingdom – North America – Japan
India – Malaysia – China

Emerald Group Publishing Limited
Howard House, Wagon Lane, Bingley BD16 1WA, UK

British Library Cataloguing in Publication Data
A catalogue record for this book is available from the British Library

ISBN: 978-0-7623-0680-0
ISSN: 1521-6136 (Series)

Printed and bound by CPI Group (UK) Ltd, Croydon, CR0 4YY

CONTENTS

DISCURSIVE ANALYSES OF LEGAL AND SOCIAL CONTROL PROCESSES

CONTRIBUTIONS TO THEORY DEVELOPMENT

LIST OF CONTRIBUTORS

Bruce A. Arrigo	Institute of Psychology, Law, and Public Policy, California School of Professional Psychology-Fresno, 5130 E. Clinton Way, Fresno, CA 93727, USA
Stacy Burns	20523 Wells Drive, Woodland Hills, CA 91364, USA
Gray Cavender	School of Justice Studies, Arizona State University, Tempe, Arizona, USA
Miriam DeLone	Department of Criminal Justice, University of Nebraska at Omaha, Omaha, Nebraska, USA
Morgan Blake Ward Doran	Human Development and Social Policy, Northwestern University, Evanston, Illinois, USA
David O. Friedrichs	Department of Sociology/Criminal Justice, University of Scranton, Scranton, PA 18510-4605, USA
Rebecca S. Katz	Assistant Professor, Department of Sociology, Rader Hall 347, Morehead State University, Morehead, Kentucky 40353, USA
Min Sik Lee	Division of International Criminal Justice and International Relations, Korean Institute of Criminology, 142 Woomyun-Dong, Sochu-Gu, Seoul 137-715, South Korea.
Elizabeth McLin	818 Eighth Street, West Vancouver, BC V7L 1S2, Canada
Albert J. Meehan	Department of Sociology and Anthropology, Oakland University, Rochester, MI, USA

Barbara Perry Department of Criminal Justice, Northern Arizona University, P.O. Box 15005, Flagstaff, AZ 86001, USA

Pamela Wilcox Rountree Department of Sociology, Patterson Office Tower, University of Kentucky, Lexington, Kentucky 40506, USA

Cassia Spohn Department of Criminal Justice, University of Nebraska at Omaha, Omaha, Nebraska, USA

Jerry Van Hoy Department of Sociology and Anthropology, Purdue University, 1365 Stone Hall, West Lafayette, IN 47907, USA

Barbara D. Warner Department of Criminal Justice and Policing, 467 Stratton Hall, Eastern Kentucky University, Richmond, Kentucky 40475-3131, USA

Ronald Weitzer Department of Sociology, George Washington University, Washington, DC 20052, USA

Christopher Williams Institute of Psychology, Law, and Public Policy, California School of Professional Psychology-Fresno, 5130 E. Clinton Way, Fresno, CA 93727, USA

PREFACE

I am very proud to bring you Volume 2 of *Sociology of Crime, Law, and Deviance*. Not only does it include thirteen high quality contributions, but these contributions represent a wide variety of substantive topics, methodological approaches, and theoretical developments. As I described in my preface for Volume 1, I see such diversity as a strength, and I want the series to showcase excellent work that represents such diversity.

The volume is organized into four thematic sections. The first section presents three quantitative studies of prominent criminological issues: criminal sentencing, communities and crime, and fear of crime. Cassia Spohn and Miriam DeLone contribute to the literature on courts and sentencing, extending Spohn's other work (Spohn 1995; Spohn and Holleran 2000) and other studies (Steffensmeier, Ulmer & Kramer 1998) empirically demonstrating the conditional nature of the influence of race and other extra-legal factors in criminal sentencing decisions. Barbara Warner and Pamela Wilcox Rountree articulate a exciting and provocative new theoretical model for research on community social organization, community-level culture, and crime – the cultural attenuation model. They then empirically test – and find substantial support for – the cultural attenuation model. Third, Min Sik Lee presents a very unique investigation of fear of crime among a population not often studied by criminologists, Korean Americans. Lee applies and extends Ferraro's (1995) symbolic interactionist-based risk assessment model of fear, and uses it to investigate whether potentially problematic relations with police in the Korean American community may exacerbate fear of crime.

Next, three excellent ethnographies examine legal and formal social control institutions. The first two ethnographies describe key transformations in law and policing. Jerry Van Hoy combines insights from the sociology of work and professions with a political economy perspective to illuminate the transformation of the status and work of lawyers near the bottom of the legal profession – in particular, those in large 'McDonaldized' franchise law firms. Albert J. Meehan, one of the most careful and insightful ethnographers of policing in American sociology, brings us an examination of how the proliferation of information technologies transforms the police subculture. Then, Elizabeth McLin's unique and sensitive ethnography presents a narrative-style thick description of sociological and ethical problematics surrounding the death penalty, as these confront a correctional officer whose work mandated his close and direct participation in executions on a routine basis.

The third section presents three contributions that focus in one way or another on discourse and symbols surrounding legal processes or criminal justice. Morgan Blake Ward Doran and Gray Cavender's discourse analysis draws from social constructionist perspectives on social problems, along with recent conceptions of framing activity in social movements, to track the careers of newspaper stories, frames, and logics surrounding anti-abortion violence. Bruce Arrigo and Christopher Williams bring theoretical conceptualizations of organizational symbolism and metaphors to bear to identify eight metaphors of prisons and how they structure correctional theory, research, and practice. Stacy Burns then develops the concept of 'impeachment work' in trials, and analyzes dynamics of impeachment work and its resistance in the context of the famous Menendez brothers murder trial.

The final four pieces are bold attempts to develop new directions in theory and/or reconceptualize existing theoretical formulations. Ronald Weitzer literally reconceptualizes an entire area of inquiry in the field of deviance – sex work. David O. Friedrichs uses the Clinton impeachment process and earlier scandals to delineate a new integrated, interdisciplinary theory of elite crime. Barbara Perry's contribution seeks a more complex and nuanced sociology of violence between oppressed minorities. Rebecca Katz's piece draws from and builds upon Gregg Barak's methods of theoretical integration in criminology, and she uses these methods to articulate a new integrated model of male violence and substance abuse.

I am also enthusiastic about a couple of changes for the series. First, starting with this volume the series will move to an annual, rather than biannual, format. Even numbered volumes will be "open" ones that publish contributions on a variety of topics, like Volume 1 and this current volume. Odd numbered volumes will be special thematic volumes, and these will usually be guest edited. For example, Volume 3, scheduled for 2001, will be a thematic volume on sociological studies of the legal profession, guest edited by Jerry Van Hoy. Another future special thematic volume planned focuses on crime and justice in Latin America, currently being developed by guest editor Juan Mario Fandino of the Federal University of Brazil, Rio Grande do Sul. And of course, I will be soliciting and welcoming a wide variety of high quality contributions for the next open volume, Volume 4 (scheduled for 2002).

In addition, this volume marks a transition of publishers. JAI Press was bought by Elsevier Science, an international publisher of books and journals in the natural and social sciences. I really think this transition will be a beneficial one for the series, and will help the series reach a wider audience and attract an even broader range of contributors. In particular, since the social sciences division of Elsevier is based in and marketed from Oxford, I hope the series gains a wider audience and pool of contributors in the United Kingdom and Europe. While I am enthusiastic about the future and working with our new publishing editor in Oxford, Ann Marie Davenport, I also want to express my deep gratitude to the series' previous executive editor, Sue Oppenheim. *Sociology of Crime, Law, and Deviance* was actually her idea, and it will always bear the mark of her guidance. Thank you, Sue. I hope the series continues to make you proud.

Jeffery T. Ulmer

Editor, *Sociology of Crime, Law, and Deviance*
Crime, Law, and Justice Program
Department of Sociology
211 Oswald Tower
Penn State University
University Park, PA 16802
U.S.A.

THANKS TO REVEWERS

I am grateful to the following scholars for serving as external reviewers for manuscripts submitted for consideration in this volume. Their careful reviews provided valuable guidance for me and for the authors.

Celesta Albonetti
David Altheide
Kevin Anderson
Tammy Anderson
Lonnie Athens
Gregg Barak
Spencer Cahill
Liqun Cao
Gray Cavender
Robert Grantham
Richard Hogan
Valerie Jenness
John Johnson
Min Sik Lee
Donileen Loseke
Kathleen Lowney
Paul Mazzerolle
Gale Miller
JoAnn Miller
Robert Perrucci
Alex Piquero
Martin Schwartz
J. William Spencer
Javier Treviño

QUANTITATIVE STUDIES OF SANCTIONS, CRIME, AND FEAR

WHEN DOES RACE MATTER? AN ANALYSIS OF THE CONDITIONS UNDER WHICH RACE AFFECTS SENTENCE SEVERITY

Cassia Spohn and Miriam DeLone

ABSTRACT

Research investigating the relationship between offender race/ethnicity and sentence severity suggests the existence of contextual discrimination. That is, minority offenders are sentenced more harshly than white offenders in some jurisdictions, for some types of offenses, and under some circumstances. The findings of our study confirm this. We compare the sentences imposed on black, Hispanic and white felony offenders in three jurisdictions: Cook County (Chicago), Illinois; Dade County (Miami), Florida; and Jackson County (Kansas City), Missouri. Our results reveal that offender race/ethnicity has both overt and subtle effects on the decision to incarcerate or not. Although race had no effect on the likelihood of incarceration in Kansas City, both blacks and Hispanics were more likely than whites to be sentenced to prison in Chicago, and Hispanics (but not blacks) were more likely than whites to be incarcerated in Miami. Further analysis revealed that offender race interacted

* This manuscript is based on work supported by the National Science Foundation under Grant SBR–93321852. Points of view are those of the authors and do not necessarily represent the position of the National Science Foundation.

Sociology of Crime, Law and Deviance, Volume 2, pages 3–37.

ISBN: 0-7623-0680-7

(although not always in the predicted way) with other legal and extralegal predictors of sentence severity.

INTRODUCTION

Nearly half a century after *Brown* vs *The Board of Education*, the historic Supreme Court decision that outlawed racially segregated public schools, the issue of race relations in the United States continues to evoke controversy and spark debate. Policymakers at all levels of government struggle to address contentious issues such as racial gerrymandering, affirmative action, racial preferences in granting contracts, and multicultural census categories. Politicians encourage Americans to voice their opinions and to offer suggestions for bridging the gaps between racial minorities and whites. During his second term in office, for example, President Clinton announced an Initiative on Race designed to "lead the nation in a great and unprecedented conversation about race." In January of 1998 former Senator Bill Bradley and Los Angeles City Councilman Mark Ridley-Thomas launched "National Days of Dialogue on Race Relations" in cities throughout the United States (NY Times, 1997). And on Internet sites ranging from Oprah Online to America Online's special site dedicated to "race relations in black and white" Americans of all political persuasions plunged into "the foamy churn of debating about race" (NY Times, 1998).

On no issue is the debate more spirited or are opinions more polarized than the relationship between race, crime, and justice. Politicians and scholars offer competing explanations for the disproportionate number of blacks arrested, imprisoned, and on death row. Those on one side contend that the war on crime – and particularly the war on drugs – has "caused the ever harsher treatment of blacks by the criminal justice system" (Tonry, 1995: 52) and charge that the over-representation of blacks in arrest and imprisonment statistics reflects systematic racial discrimination (Mann, 1993). Those on the other side assert that these results can be attributed primarily to the disproportionate involvement of blacks in serious criminal activity (Blumstein, 1982, 1993) and argue that the idea of systematic discrimination within the criminal justice system is a 'myth' (Wilbanks, 1987).

Decades of research examining the treatment of racial minorities at various stages in the criminal justice process, including well over one hundred studies exploring the effect of race on sentencing, have not resolved this debate. Although a number of studies have shown that blacks are treated more harshly than whites (Holmes et al., 1996; Kramer & Ulmer, 1996; Petersilia, 1983; Spohn et al., 1981–82; Spohn & Holleran, 2000; Steffensmeier et al., 1998;

Zatz, 1984), others have found either that there are no significant racial differences (Klein et al., 1990) or that blacks are treated more leniently than whites (Bernstein et al., 1977; Gibson, 1978; Levin, 1972). Moreover, there are relatively few studies that examine the treatment of Hispanics, Native Americans, Asians, or other racial/ethnic groups. Thus, definitive answers to questions concerning differential treatment of racial/ethnic minorities remain elusive.

The purpose of this study is not simply to add another voice to the debate over the *existence* of racial discrimination within the criminal justice system. Although we do attempt to determine whether racial minorities are sentenced more harshly than whites, we believe that this is a theoretically unsophisticated and incomplete approach to a complex phenomenon. Like Hawkins (1987) and Zatz (1987), we believe that it is overly simplistic to assume that racial minorities will receive harsher sentences than whites regardless of the nature of the crime, the seriousness of the offense, or the culpability of the defendant. Like Wonders (1996: 617), we believe that the more interesting question is "When does the particular social characteristic matter – under what circumstances, for whom, and in interaction with what other factors?" Our study addresses this question. We use data on black, white, and Hispanic offenders convicted of felonies in three large urban jurisdictions (Chicago, Miami, and Kansas City) to test for the direct effect of race/ethnicity on sentencing and, more importantly, to identify the conditions under which blacks and Hispanics are sentenced more harshly than whites. In doing so, we hope to inform the debate on race relations in the United States.

PREVIOUS RESEARCH

Research investigating the relationship between the defendant's race and sentence severity has not consistently supported the conflict perspective's contention that racial minorities will be sentenced more harshly than whites. This has led to conflicting conclusions. Some researchers (Hagan, 1974; Kleck, 1981; Pruitt & Wilson, 1983) assert that racial discrimination in sentencing has declined over time and contend that the predictive power of race, once relevant legal factors are taken into account, is quite low. Hagan (1974) and Kleck (1981), for example, have suggested that the direct effect of race on sentence severity would disappear in models that adequately controlled for the offender's prior criminal record. A recent review of 38 studies published since 1975, however, challenges this prediction; Chiricos & Crawford (1995) reported that many of the studies concluded that race had a direct effect on the decision to incarcerate or not, and that this effect remained even after the inclusion of controls for prior record and crime seriousness. Spohn's (2000) review of 32 state-level studies using post-1980 data produced a similar conclusion.

Other researchers (Klepper et al., 1983; Zatz, 1987) claim that discrimination has not declined or disappeared but simply has become more subtle and difficult to detect. While not discounting the possibility of a direct racial effect, these researchers argue that race influences sentence severity *indirectly* through its effect on variables such as bail status (LaFree, 1985b; Lizotte, 1978), type of attorney (Spohn et al., 1981–82) or type of disposition (LaFree, 1985a; Spohn, 1992; Uhlman & Walker, 1980), or that race *interacts* with other variables and affects sentence severity only in some types of cases (Barnett, 1985; Spohn & Cederblom, 1991), in some types of settings (Chiricos & Crawford, 1995; Hawkins, 1987; Kleck, 1981; Myers & Talarico, 1986), or for some types of defendants (Chiricos & Bales, 1991; LaFree, 1989; Peterson & Hagan, 1984; Spohn, 1994; Spohn & Holleran, 2000; Steffensmeier, 1998; Walsh, 1987).

A number of scholars argue that the inconsistent findings of research on race and sentencing reflect both specification error and an overly simplistic view of conflict theory. These scholars have called for research designed to delineate more precisely the conditions under which defendant race influences judges' sentencing decisions. Zatz (1987: 83), for example, contends that models of the relationship between race and sentencing that exclude indirect or interactive effects are mis-specified and, thus, "may erroneously conclude that discrimination does not exist when, in fact, it does."

Hawkins (1987: 721) presents an analogous but somewhat different argument. He argues that many of the so-called 'anomalies or inconsistencies' in sentencing research reflect 'oversimplification' of conflict theory. He contends that the work of early conflict theorists such as Quinney (1970) and Chambliss & Seidman (1971) does not support the proposition that "blacks or other non-whites will receive more severe punishment than whites for all crimes, under all conditions, and at similar levels of disproportion over time" (Hawkins, 1987: 724). Hawkins proposes a revision of the conflict perspective on race and sentencing to account for the possibility of interaction between defendant race and other predictors of sentence severity, and especially between defendant race, victim race, and the type of crime committed by the offender.

Researchers have begun to heed these recommendations. There is a growing body of literature demonstrating that the relationship between race and sentencing is nonlinear and non-additive. Researchers have shown, for instance, that the offender's race interacts with the offender's prior criminal record (Spohn & Cederblom, 1991; Spohn et al., 1998; Kramer & Ulmer, 1996; Zatz, 1984), the offender's gender and employment status (Chiricos & Bales, 1991; Holmes et al., 1996; Nobiling et al., 1998; Spohn & Holleran, 2000; Steffensmeier et al., 1998), and the seriousness of the offense (Baldus et al., 1985; Barnett, 1985; Spohn & Cederblom, 1991; Unnever & Hembroff, 1987). This research generally demonstrates that the offender's race more consistently affects sentence severity if the offender also has a relatively non-serious prior criminal record, if the offender is male and/or unemployed, or if the offense is less serious.

Limitations of Previous Research

The studies conducted thus far provide important insights into the sentencing process. They suggest that the criteria used to determine the severity of the sentence are not static, but vary depending upon the characteristics of the offender and the nature of the case. Even the more methodologically sophisticated studies, however, do not provide definitive assessments of the influence of race on sentencing. Many focused on only one type of crime – typically either homicide or sexual assault. There have been few systematic attempts to analyze the effect of defendant race on sentencing for a variety of crimes; and even those studies that did examine sentencing decisions for several types of crimes tended to focus on the more serious felonies. This is problematic, since, as noted above, researchers have shown that the effect of race is *not* constant across crime types and may be conditioned by the seriousness of the offense.

A second limitation of these studies is that most compared the sentences imposed on only black and white offenders. Many studies include only blacks and whites; all others are excluded from the analysis. Other studies examine the sentences imposed on whites and non-whites, with all non-whites lumped together. Both of these approaches are questionable, since a number of studies (e.g. Gruhl et al., 1984; Holmes et al., 1996; LaFree, 1985b; Zatz, 1984, 1985) have shown that sentencing outcomes may differ for blacks and Hispanics.

An additional limitation of the existing studies is that most analyzed data from the 1970s or early 1980s. The results of this research may not apply to the sentencing process in the 1990s. Sentencing reforms, and in particular the implementation of sentencing guidelines, may have reduced discretion, and thus the potential for discrimination, in the sentencing process. Studies have shown, for example, that the implementation of sentencing guidelines in Minnesota has reduced sentencing disparity, including disparities based on race and employment status (Miethe & Moore, 1985; Moore & Miethe, 1986;

Stolzenberg & D'Alessio, 1994; see Albonetti, 1997; Kramer & Ulmer, 1996; and Ulmer & Kramer, 1996 for contrasting results).

A final limitation of these studies is that most were single-jurisdiction studies that used different models and different statistical techniques to test for race effects. Some studies controlled for extralegal variables such as type of attorney, pretrial status, and type of disposition; others did not. Some researchers analyzed the decision to incarcerate and the length of the prison sentence separately; others either combined the two decisions into a single measure of sentence severity or analyzed only one of the two decisions. Some analyses of the length of the prison sentence corrected for sample selection bias in the earlier decision to incarcerate; others did not. Some studies tested for interaction between offender race and other predictors of sentence severity; others employed additive models only. These methodological differences make it difficult to compare the findings of existing research; they make it difficult to arrive at conclusions concerning the presence or absence of a *pattern* of racial discrimination in sentencing.

We address these limitations in this study. We use data on offenders convicted of felonies in 1993 and 1994 in three large urban jurisdictions: Dade County (Miami), Florida; Cook County (Chicago), Illinois; and Jackson County (Kansas City), Missouri. We use additive and interactive models to compare the sentences imposed on black, white, and Hispanic offenders convicted of violent crimes, property crimes, and drug offenses. We examine both the decision to incarcerate and the length of the sentence, and we incorporate a correction for sample selection bias in our analyses of sentence length.

Objectives and Hypotheses

The purpose of this study is to determine if racial minorities are sentenced more harshly, more leniently, or no differently than whites and "to address more specifically the question of the role played by various contingencies or mediating factors" (Hawkins, 1987: 721). Although we test for a direct race effect, we also explore the possibility that racial discrimination in sentencing is confined to certain kinds of offenses and to particular types of offenders. We probe for interactions between offender race/ethnicity and other legal and extralegal predictors of sentence severity.

Building on and extending previous research, we test the following hypotheses:

H1 Offender race/ethnicity will have a direct effect on sentence severity. Blacks and Hispanics will be more likely than whites to be incarcerated and the prison sentences imposed on blacks and Hispanics will be longer than the sentences imposed on whites (Kramer & Ulmer, 1996; Spohn et al., 1981–82; Ulmer & Kramer, 1996).

H1a Offender race/ethnicity will have a stronger and more consistent effect on the in/out decision than on the length of the prison sentence (Chiricos & Crawford, 1995).

H2 Offender race/ethnicity will interact with the seriousness of the offense (Baldus et al., 1985; Barnett, 1985; Spohn & Cederblom, 1991). Blacks and Hispanics convicted of property crimes and drug offenses will be sentenced more harshly than whites convicted of these offenses; there will be no racial differences in the sentences imposed on offenders convicted of violent crimes.

H3 Offender race/ethnicity will interact with the offender's prior criminal record (Ulmer & Kramer, 1996; Miethe & Moore, 1986; Spohn & Cederblom, 1991; Zatz, 1984). Among offenders without a prior prison sentence, blacks and Hispanics will be sentenced more harshly than whites. There will be no racial differences in the sentences imposed on offenders with a prior prison sentence.

H4 Offender race/ethnicity will interact with the employment status of the offender (Chiricos & Bales, 1991; Nobiling et al., 1998). Unemployed blacks and Hispanics will be sentenced more harshly than unemployed whites; there will be no racial differences in the sentences imposed on offenders who are employed.

RESEARCH DESIGN AND METHODS

The three jurisdictions included in this study are Cook County (Chicago), Illinois; Dade County (Miami), Florida; and Jackson County (Kansas City), Missouri. Chicago and Miami were chosen to represent northern and southern jurisdictions with large black and Hispanic populations and with crime rates well above the national average. Kansas City was chosen to represent a medium-sized midwestern jurisdiction with relatively small minority populations. Chicago was also chosen because it is one of only a few jurisdictions in the United States with more than a handful of black judges. An additional consideration was the type of statutory sentencing procedures in each jurisdiction. In Chicago[1] and Kansas City[2] judges impose determinate sentences; judges in Miami[3] sentence under sentencing guidelines.

Data Collection Procedures

We analyze data on 7,279 offenders convicted of felonies in the three jurisdictions. This includes 2,983 offenders in Chicago, 2,720 offenders in Miami, and 1,576 offenders in Kansas City. The data collection procedures varied somewhat in each jurisdiction.[4] In Chicago, we selected a random sample of all offenders convicted of felonies in 1993 from a list prepared by the Clerk of the Cook County Circuit Court. Data collectors read through the court file for each case included in the sample and recorded information about the offender and the case on an optical-scan form designed for the project. In Miami, we selected a random sample of all offenders convicted of felonies in 1993 and 1994; information concerning the case and the offender was provided by the Administrative Office of the Courts.[5] In Kansas City, we obtained data on all offenders convicted of felonies in 1993. The Department of Computer Services provided a printout listing the charges filed, the disposition of each charge, and other information about the case; information concerning the offender's background and prior criminal record was obtained from court files.

Dependent and Independent Variables

The dependent and independent variables, their codes, and their frequencies are displayed in Table 1. We present separate data for each of the three jurisdictions.

We include two dependent variables measuring sentence severity. The first variable measures whether the offender was sentenced to prison or not. The second measures the length of sentence (in months) imposed on offenders sentenced to prison.

In Miami, we analyze the decision to incarcerate (in jail or prison) as well as the prison/no prison variable. We include both measures of incarceration because of the obvious difference between a state prison sentence and a jail sentence and because of the large number of offenders (N = 1,172) who were sentenced to the Dade County Jail.[6] Under the Florida sentencing guidelines in effect in 1993 and 1994, many offenders convicted of property offenses and drug offenses did not qualify for a state prison sentence; instead, such offenders were sentenced to the Dade County Jail.

The independent variables included in the analysis are offender and case characteristics that have been shown to affect judges' sentencing decisions. We control for the offender's race, gender, age, and prior criminal record.[7] In Chicago and Miami, offender race is measured by three dummy variables –

black, Hispanic, and white.[8] Because there were only 47 Hispanic offenders in Kansas City, we could not analyze Hispanics separately; we therefore eliminated Hispanics from the data file and created two dummy variables (black offender, white offender) measuring offender race. White offenders are the reference category in all of the multivariate analyses.

The data file for this study included a number of measures of prior criminal record: number of prior arrests, number of prior felony arrests, number of prior felony convictions, number of prior prison terms of more than one year, whether the offender had previously been convicted of a drug offense, and whether the offender previously had been convicted of a violent offense. Some of these measures were highly correlated with one another;[9] they also differed in terms of their relationship to sentence severity. The variable we use – the number of times the offender previously had been sentenced to prison for more than one year – is the measure of prior record that had the most consistently strong (and statistically significant) relationship to these dependent variables.[10]

We control for three measures of offense seriousness – the most serious conviction charge, the class of the most serious conviction charge, and the number of current felony convictions.[11] The most serious conviction charge is a categorical variable; murder is the reference category. Because the classification of offenses differed in each jurisdiction, the variables measuring the class of the most serious conviction charge also differ. In Chicago, we differentiate between Class X, Class 1, Class 2, Class 3, and Class 4 felonies; Class X is the reference category. In Miami, we distinguish between life/1st degree felonies, 2nd degree felonies, and 3rd degree felonies; life/1st degree felony is the omitted category.[12] In Kansas City, we differentiate between Class A, Class B, Class C, and Class D felonies; Class A is the reference category.

We also control for characteristics of the offender's case that might influence the severity of the sentence imposed by the judge. We take into account whether the offender was on probation at the time of his/her arrest for the current offense, whether the offender was represented by a private attorney, whether the offender was released pending trial, and whether the offender pled guilty.[13]

Analytic Procedures

We analyze the data using both ordinary least squares (OLS) regression and logistic regression. Two different analytic procedures are required because of differences in the nature of the dependent variables. We use OLS regression to analyze the length of the prison sentence, which is an interval-level variable.

Table 1. Dependent and Independent Variables: Codes and Frequencies for Offenders in Chicago, Miami, and Kansas City

Variable	Code	CHICAGO (N = 2983) N	%	MIAMI (N = 2720) N	%	KANSAS CITY (N = 1576) N	%
Dependent Variables[a]							
Sentenced to Jail	1 = yes	NA		1861	68.4	NA	
or Prison	0 = no			859	31.6		
Sentenced to Prison	1 = yes	1822	61.1	689	25.3	646	41.0
	0 = no	1161	38.9	2031	74.7	930	59.0
Prison Sentence (Months)	Mean	57.01		61.08		81.52	
Independent Variables							
Offender's race[b]							
Black		2388	80.1	1619	59.5	1020	64.7
Hispanic		311	10.4	571	21.0	NA	
White		282	9.5	530	19.5	556	35.3
Offender's Gender	1 = male	2701	90.5	2471	90.8	1345	85.3
	0 = female	282	9.5	249	9.2	231	14.7
Offender's Age	Mean	27.15		31.15		28.91	
Most Serious Conviction Charge[c]							
Murder		60	2.0	35	1.3	62	3.9
Rape		34	1.1	NA		65	4.1
Robbery		246	8.2	156	5.7	94	6.0
Aggravated Assault		46	1.5	225	8.3	26	1.6
Burglary		366	12.3	500	18.4	161	10.2
Weapons Offenses		162	5.4	89	3.3	97	6.2
Larceny/Theft		149	5.0	318	11.7	274	17.4
Motor Vehicle Theft		189	6.3	NA		148	9.4
Possess Narcotics w/intent		775	26.0	985	36.2	309	19.6
Other Drug Offense		779	26.1	200	7.4	121	7.7
Other Property Offense		66	2.2	66	2.4	162	10.3
Other Felony		111	3.7	146	5.4	57	3.6
Class of Most Serious Conviction Charge[d]							
Chicago							
Class X		308	10.3	NA		NA	
Class 1		486	16.3				
Class 2		976	32.7				
Class 3		441	14.8				
Class 4		772	25.9				

Table 1. Continued

Variable	Code	CHICAGO (N=2983) N	%	MIAMI (N=2720) N	%	KANSAS CITY (N=1576) N	%
Miami							
Life or 1st Degree		NA		278	10.2	NA	
2nd Degree				833	30.6		
3rd Degree				1609	59.2		
Kansas City							
Class A		NA		NA		148	9.4
Class B						421	26.7
Class C						852	54.1
Class D						155	9.8
Mean No. of Current Felony Convictions		1.15		1.44		1.46	
Mean No. of Prior Prison Terms of > 1 Year		0.52		0.65		0.63	
Offender on Probation at Time of Offense	1 = yes	280	9.4	334	12.3	243	15.8
	0 = no	2703	90.6	2386	87.7	1293	84.2
Private Attorney	1 = yes	265	9.0	718	26.4	461	29.4
	0 = no	2673	91.0	2002	73.6	1107	70.6
Offender Released Prior to Trial	1 = yes	1104	37.0	1176	43.3	612	38.8
	0 = no	1879	63.0	1543	56.7	964	61.2
Offender Pled Guilty	1 = yes	2702	90.6	2674	98.3	1489	94.5
	0 = no	281	9.4	46	1.7	87	5.5
Offender Unemployed	1 = yes	1826	72.1	NA		999	69.6
	0 = no	707	27.9	436	30.4		

[a] In Miami we analyze the decision to incarcerate (in jail or prison) as well as the decision to incarcerate (in prison) because of the large number of offenders who were sentenced to the Dade County jail.

[b] There were only 47 Hispanic offenders in Kansas City, we therefore eliminated Hispanics from the data file.

[c] There were very few cases in the Miami sample in which the most serious conviction charge was rape (N = 16) or kidnaping (N = 27). Therefore, we eliminated these cases from the analysis. In addition, in Miami we could not distinguish between larceny and motor vehicle theft.

[d] Because each jurisdiction classifies offenses differently, the variables measuring the class of the most serious conviction charge also differ.

Because OLS regression is considered to be inappropriate for the analysis of dichotomous dependent variables, we use logistic regression to analyze the two incarceration decisions.

In analyzing the length of the sentence, we include a correction for sample selection bias (Berk, 1983; Berk & Ray, 1982). This type of bias results when some observations are systematically excluded from the sample being analyzed. Here offenders who were not sentenced to prison are excluded from the sentence length sample. Incarcerated offenders, in other words, are a selected population from the population of all convicted offenders; they were sentenced to prison because they exceeded some threshold of 'case seriousness'. Thus, the length of the sentence is a function not just of the usual linear combination of regressors (which suffices in the original population), but "[of] a hazard rate capturing the impact of the selection equation" (Berk & Ray, 1982: 369).

We use the procedures outlined by Heckman (1974) and Berk (1983) to correct for this problem. We use logistic regression to estimate the likelihood that the offender would be sentenced to prison. For each case the logistic regression model produced its predicted probability of exclusion from the sentence length sample – the hazard rate. We then include the hazard rate as a control in the regression equation for sentence length.

We perform separate analyses on the data for each jurisdiction and we use a two-stage analytic procedure to explore the relationship between race and sentencing. We first estimate the additive effects of the offender's race on sentencing outcomes, controlling for offender and case characteristics. We then test the remaining hypotheses, all of which focus on the possibility of interaction between the offender's race, other predictors of sentence severity (the seriousness of the offense, the offender's prior criminal record, and the offender's employment status), and the harshness of the sentence. In testing our hypothesis that race will affect sentence severity for property crimes and drugs offenses but not for violent crimes, for example, we perform a separate multivariate analysis on each type of conviction charge.

FINDINGS

Results of the Additive Analysis

Decision To Incarcerate or Not. We found some support for our hypothesis that offender race/ethnicity would have a direct effect on the judge's decision to incarcerate or not. As shown in Table 2, both blacks and Hispanics faced greater odds of imprisonment than whites in Chicago, while Hispanics (but not blacks) were imprisoned more often than whites in Miami. On the other hand, blacks were not more likely than whites to be sentenced to prison in Kansas City. Moreover, when we operationalized the in/out decision in Miami as the decision to sentence the offender to jail or prison, we found that neither blacks ($b = 0.20$; $SE = 0.11$) nor Hispanics ($b = 0.23$; $SE = 0.13$) were more likely than whites to be incarcerated.

The data presented in Table 2 reveal that legal variables are strong predictors of the decision to incarcerate or not in each of the three jurisdictions. As expected, the likelihood of a prison sentence was significantly affected by the number of conviction charges and by the seriousness of the (most serious) conviction charge. In each jurisdiction the likelihood of incarceration was also affected by the offender's prior criminal record and by whether the offender was on probation when he/she committed the current offense.

Several extralegal variables, in addition to race/ethnicity, are related to the decision to incarcerate or not. Males were significantly more likely than females to be incarcerated in Chicago and Kansas City, offenders who were released pending trial were less likely than those who were in custody prior to trial to be sentenced to prison in all three jurisdictions, and offenders who pled guilty were incarcerated less often than those who were tried in Miami and Kansas City. The offender's age and type of attorney, on the other hand, did not affect the decision to incarcerate in any of the three jurisdictions.

We used the logit coefficients presented in Table 2 to calculate the estimated probability of imprisonment for a 'typical' offender in each racial group (Hanushek & Jackson, 1977; Lichter, 1989). We calculated these probabilities for offenders with the following characteristics:

- male
- 30 years old
- convicted of one count of *either* burglary or possession of narcotics with intent
- convicted of a Class 3 (Chicago), 2nd Degree (Miami) or Class C (Kansas City) felony
- one prior prison term of more than one year
- not on probation at time of arrest
- represented by a public defender
- in custody prior to trial
- pled guilty

The formula used to calculate the probabilities was:

$$P_1 = \exp(Z_1)/1 + \exp(Z_1)$$

where

Table 2. The Effect of Race/Ethnicity on the Decision to Sentence the Offender to Prison

	CHICAGO			MIAMI			KANSAS CITY		
	b	SE	Odds[a]	b	SE	Odds	b	SE	Odds
Offender's Race[b]									
Black	0.38	0.19	1.47*	0.15	0.14		–0.06	0.14	
Hispanic	0.48	0.24	1.61*	0.38	0.16	1.46*	–	–	
Offender's Gender (Male = 1)	1.03	0.19	2.79*	–0.14	0.19		0.85	0.22	2.33*
Offender's Age	–0.001	0.005	–0.007	0.006	0.003	0.008			
Most Serious Conviction Charge[c]									
Rape	–1.76	1.18		–	–		–0.85	0.59	
Robbery	–2.28	0.85	0.10*	–1.04	0.49	0.35*	–0.36	0.61	
Aggravated Assault	–2.76	0.90	0.06*	–1.99	0.49	0.14*	0.51	0.99	
Burglary	–2.37	0.84	0.09*	–1.55	0.48	0.21*	–1.60	0.55	0.20*
Weapons Offenses	–1.28	0.86	0.09*	–1.81	0.49	0.16*	–1.56	0.63	0.21*
Larceny/Theft	–2.39	0.85	0.06*	–1.87	0.54	0.15*	–1.51	0.55	0.22*
Motor Vehicle Theft	–2.88	0.85	0.27*	–	–		–1.57	0.57	0.21*
Possess Narcotics w/intent	–2.25	0.83	0.10*	–2.41	0.47	0.09*	–1.57	0.54	0.21*
Other Drug Offense	–2.61	0.84	0.07*	–3.49	0.57	0.03*	–2.16	0.58	0.12*
Other Property Offense	–	–	–	–2.20	0.58	0.11*	–1.61	0.56	0.20*
Other Felony	–2.58	0.85	0.08*	–2.38	0.53	0.09*	–1.85	0.64	0.16*
Class of Most Serious Conviction Charged									
Chicago (Class X is reference category)									
Class 1	–1.00	0.31	0.30*						
Class 2	–1.52	0.31	0.17*						
Class 3	–1.84	0.36	0.13*						
Class 4	–1.84	0.36	0.13*						
Miami (Life or 1st Degree Felony is reference category)									
2nd Degree				–1.01	0.17	0.36*			
3rd Degree				–1.44	0.17	0.24*			

Table 2. Continued

	CHICAGO			MIAMI			KANSAS CITY		
	b	SE	Odds[a]	b	SE	Odds	b	SE	Odds
Kansas City (Class A Felony is reference category)									
Class B							–1.54	0.41	0.21*
Class C							–1.30	0.51	0.27*
Class D							–1.67	0.57	0.19*
No. of Current Convictions	0.48	0.16	1.61*	0.15	0.06	1.17*	0.31	0.09	1.37*
Number of Prior Prison Terms of > 1 Year	1.65	0.12	5.20*	0.60	0.04	1.82*	0.73	0.08	2.08*
Offender on Probation at Time of Offense	1.72	0.20	5.59*	0.78	0.14	2.18*	1.06	0.18	2.90*
Private Attorney	0.11	0.20		–0.15	0.12		–0.08	0.17	
Offender Released Prior to Trial	–2.36	0.11	0.09*	–0.41	0.11	0.66*	–1.49	0.16	0.22*
Offender Pled Guilty	0.15	0.19		–1.89	0.37	0.15*	–1.16	0.44	0.31*
Constant	2.87*	0.93		3.44*	0.65		2.29*	0.78	
N of Cases		2,938			2,718			1,519	
	X^2	df	P	X^2	df	P	X^2	df	P
–2 Log Likelihood	2253.96	2913	0.00	2425.73	2696	0.00	1413.58	1495	0.93
Goodness of Fit	31078.44	2913	0.00	2536.26	2696	0.00	1476.99	1495	0.62

[a] Odds ratios are presented only for independent variables that have a statistically significant effect ($P \leq 0.05$) on the dependent variable.
[b] Because of the small number of Hispanics in Kansas City (N = 47), we eliminated Hispanics from the data file and compared the sentences imposed on black offenders and white offenders. In Chicago and Miami, white offenders are the reference category.
[c] The most serious conviction charge is a categorical variable. Murder is the reference category.
* $P \leq 0.05$.

$$Z_1 = \sum_k B_k X_{ik}$$

As an illustration, the probability of incarceration for a black offender convicted of burglary in Chicago is computed as

$$\begin{aligned}\text{Prob}_{\text{black}} &= \exp(2.87 + 0.38 + 1.03 + (30)(-0.001) - 2.37 - 1.84 + 0.48 \\ &\quad + 1.65 + 0.15)/1 + \exp(2.87 + 0.38 + 1.03 + (30)(-0.001) - 2.37 \\ &\quad - 1.84 + 0.48 + 1.65 + 0.15) \\ &= \exp(2.32)/1 + \exp(2.32) \\ &= 10.176/11.176 \\ &= 0.9105\end{aligned}$$

The estimated probabilities, presented in Table 3, confirm that offender race had no effect on the likelihood of incarceration in Kansas City; in fact, the black/white differences in the probabilities of incarceration are less than two percentage points. Table 3 also reveals that the racial differences in the probabilities of incarceration in Chicago and Miami, while statistically significant, are not large. In Chicago there is about a four percentage point difference between white offenders and black offenders and between white offenders and Hispanic offenders for each of the two offenses. In Miami the differences between white offenders and Hispanic offenders are somewhat larger: eight percentage points for burglary and about five percentage points for possession of narcotics.

The Length of the Prison Sentence. We found no support for our hypothesis that offender race/ethnicity would have a direct effect on the length of the prison sentence. As shown in Table 4, there were no significant racial differences in the prison sentences imposed on offenders in any of the three jurisdictions. Consistent with the results for the in/out decision, the length of the prison sentence was determined primarily by case seriousness; in each jurisdiction, the best predictors of sentence length were the hazard rate, the number of conviction charges, and the most serious conviction charge. In fact, the only extralegal variable that consistently affected the sentence length was the type of disposition; offenders who pled guilty received shorter sentences than those who went to trial.

The results discussed thus far suggest that sentencing decisions in these three jurisdictions rest primarily on legally relevant considerations such as the seriousness of the offense and the offender's prior criminal record. Once these

Table 3. Probabilities of Incarceration for 'Typical'[a] White, Black and Hispanic Offenders

	Most Serious Conviction Charge	
	Burglary	Possess Narcotics with Intent
Probability of a Prison Sentence[b]		
Chicago		
White Offenders	0.874	0.885
Black Offenders	0.911	0.920
Hispanic Offenders	0.918	0.927
Miami		
White Offenders	0.261	0.130
Black Offenders	0.291	0.148
Hispanic Offenders	0.341	0.179
Kansas City		
White Offenders	0.553	0.413
Black Offenders	0.538	0.399
Probability of a Jail/Prison Sentence[c]		
Miami		
White Offenders	0.655	0.705
Black Offenders	0.698	0.745
Hispanic Offenders	0.705	0.750

[a] We calculated the probability of incarceration for white, black and Hispanic offenders with the following characteristics: male; age = 30 years; convicted of one count of either burglary or possession of narcotics with intent; class of the conviction charge = Class 3 Felony (Chicago), 2nd Degree Felony (Miami), Class C Felony (Kansas City); one prior prison term of more than one year; not on probation; represented by a public defender; in custody prior to trial; pled guilty. See p. 18 in the text for the formula used to calculate the probabilities.
[b] Calculated from logistic regresssion results presented in Table 2.
[c] Calculated from logistic regression results (not shown).

legal variables are taken into account, race/ethnicity has no effect on the length of the prison sentence in any of the three jurisdictions and does not affect the likelihood of incarceration in Kansas City. In Chicago, both blacks and Hispanics are more likely than whites to be incarcerated; in Miami, Hispanics, but not blacks, face greater odds of incarceration than whites.

Results of the Interactive Analyses

The fact that we found no direct race effects in Kansas City and limited effects in Miami and Chicago does not mean that the sentencing process in either

Table 4. The Effect of Race/Ethnicity on the Length of the Prison Sentence[a]

	CHICAGO			MIAMI			KANSAS CITY		
	b	Beta	T	b	Beta	T	b	Beta	T
Offender's race									
Black	0.30	0.002	0.07	–7.72	–0.06	1.07	6.72	0.04	1.44
Hispanic	5.04	0.02	0.95	–0.39	–0.002	0.47	–	–	–
Offender's Gender (Male = 1)	–6.50	–0.02	1.21	8.64	0.03	0.84	–8.54	–0.05	0.92
Offender's Age	–0.08	–0.01	0.58	–0.19	–0.02	0.65	–0.33	–0.04	1.28
Most Serious Conviction Charge[b]									
Murder	160.40	0.46	19.72*	–	–	–	–	–	–
Rape	55.59	0.12	5.77*	–	–	–	–39.20	–0.11	2.91*
Robbery	23.00	0.12	3.76*	–36.92	–0.18	2.60*	–57.70	–0.25	5.85*
Aggravated Assault	0.34	0.006	0.03	–42.89	–0.16	2.54*	–34.17	–0.09	2.51*
Burglary	–0.58	–0.003	0.10	–36.74	–0.23	2.62*	–123.73	–0.55	11.37*
Weapons Offenses	–27.98	–0.10	4.25*	–27.66	–0.07	1.39	–126.88	–0.31	8.73*
Larceny/Theft	–25.72	–0.09	3.49*	–39.40	–0.17	2.44*	–126.31	–0.61	11.59*
Motor Vehicle Theft	–1.60	–0.01	0.24	–	–	–	–124.63	–0.51	11.12*
Possess Narcotics w/intent	–9.14	–0.07	1.63	–27.81	–0.17	1.82	–92.62	–0.44	8.27*
Other Drug Offense	–29.27	–0.19	5.08*	–37.84	–0.07	1.50	–112.89	–0.30	7.80*
Other Property Offense	–	–	–	–40.40	–0.07	1.70	–131.74	–0.44	10.60*
Other Felony	–	–	-	–10.69	–0.04	0.53	–113.46	–0.20	6.09*
Class of Most Serious Conviction Charge[c]									
No. of Current Convictions	13.46	0.14	7.68*	–4.71	–0.06	1.81		11.37 .24	7.98*
Number of Prior Prison Terms of > 1 Year[c]									

Table 4. Continued

	CHICAGO			MIAMI			KANSAS CITY		
	b	Beta	T	b	Beta	T	b	Beta	T
Offender on Probation at Time of Offense	–9.47	–0.05	2.87*	–21.03	–0.12	3.07*	–7.76	–0.04	1.46
Private Attorney	3.25	0.01	0.78	15.31	0.09	2.50*	0.38	0.002	0.07
Offender Released Prior to Trial	8.41	0.05	1.94	–5.54	–0.03	0.92	6.38	0.03	0.83
Offender Pled Guilty	–9.99	–0.05	2.69*	–45.06	–0.13	3.42*	–35.48	–0.13	4.48*
Hazard Rate	39.11	0.15	5.46*	103.67	0.35	6.37*	46.43	0.17	3.44*
N of Cases		1790			685			598	
R^2		.51			.26			.59	

[a] The sentence imposed on offenders who were sentenced to prison. The sentence is measured in months.
[b] Because of multicollinearity among the dummy variables measuring the most serious conviction charge and the class of the conviction charge in Chicago, we were not able to model sentence length with murder as the reference category. Using 'other felony' as the reference category corrected for this problem.
[c] The hazard rate was strongly correlated with the class of the conviction charge and with the offender's prior criminal record. When the hazard rate was included in the model, the variance inflation factors (VIFs) for the dummy variables measuring the class of the conviction charge and the VIFs for prior record and the hazard rate all exceeded 5.0. We therefore eliminated the variables measuring the class of the conviction charge and the offender's prior record from the model.
* $P \leq 0.05$

jurisdiction is racially neutral. We noted above that Hawkins (1987), Zatz (1987) and others (Chiricos & Crawford, 1995; Spohn, 1994; Unnever & Hembroff, 1987) have called on researchers to abandon overly simplistic additive models of the relationship between race and sentencing. Asserting that conflict theory does not necessarily predict that racial minorities will be sentenced more harshly than whites for all crimes and under all types of circumstances, they have called for research designed to identify the conditions under which race/ethnicity influences sentencing decisions. As Chiricos & Crawford (1995: 301) note, "We are past the point of simply asking whether race makes a difference."

Building on past research, we hypothesized that offender race/ethnicity would interact with the seriousness of the conviction charge, the offender's prior criminal record, and the offender's employment status. To test these hypotheses, we ran separate analyses on offenders convicted of three types of crimes (violent crimes, property crimes, and drug offenses), on offenders with and without prior records, and on employed and unemployed offenders. We controlled for all of the independent variables listed in Table 1. The results of these analyses are presented in Tables 5 (decision to incarcerate or not) and 6 (length of the prison sentence). We present only the coefficients for the race/ethnicity variables. Because white offenders are the reference category, the coefficients reflect the differences in sentence severity for black (or Hispanic) offenders and white offenders.

The results of our analysis of the in/out decision (Table 5) confirm our finding that black offenders and white offenders faced similar odds of incarceration in Kansas city. In this jurisdiction, race had no effect on the likelihood of incarceration in *any* of the seven types of cases examined. In Chicago and Miami, on the other hand, the offender's race/ethnicity did affect the decision to incarcerate or not in some types of cases, but the patterns of results are not entirely consistent with our hypotheses.

In Chicago, for example, the offender's race/ethnicity interacted with the offender's prior record and employment status in the predicted way. Blacks and Hispanics without a prior prison sentence were significantly more likely than their white counterparts to be sentenced to prison for the current offense; there were no differences in the incarceration rates for white, black, and Hispanic offenders who had previously been sentenced to prison for a year or more. Similarly, unemployed Hispanics were more likely than unemployed whites to be incarcerated, but employed Hispanics did not face greater odds of incarceration than employed whites. Race/ethnicity, on the other hand, did not interact with the conviction charge in the predicted manner. Contrary to our hypothesis, in Chicago racial minorities convicted of property crimes or drug

Table 5. Effect of Race/Ethnicity on the Decision to Sentence the Offender to Prison in Various Types of Cases[a]

	CHICAGO				MIAMI				KANSAS CITY	
	Black		Hispanic		Black		Hispanic		Black	
TYPE OF CASE	b	SE	b	SE	b	SE	b	SE	b	SE
TYPE OF CONVICTION CHARGE										
Violent Crime	0.93	0.52	2.06*	0.82	0.21	0.29	0.29	0.35	–0.52	0.44
Property Crime	0.22	0.26	0.14	0.32	0.06	0.20	0.09	0.23	–0.17	0.18
Drug Offense	0.44	0.31	0.47	0.38	0.53	0.35	0.95*	0.38	0.32	0.33
PRIOR PRISON TERM OF MORE THAN 1 YEAR										
Prior Prison Term	–0.38	0.61	–0.40	0.78	0.12	0.21	0.60*	0.26	0.08	0.25
No Prior Prison Term	0.46*	0.20	0.56*	0.26	–0.02	0.21	0.20	0.23	–0.16	0.19
EMPLOYMENT STATUS										
Employed or Student	0.63	0.34	0.37	0.43	NA		NA		–0.12	0.28
Unemployed	0.40	0.29	0.80*	0.35					0.04	0.19

[a] We ran separate analyses on each type of case, controlling for offender race and for the other independent variables listed in Table 1. White offenders were the reference category in all of the analyses.

* $P \leq 0.05$

Table 6. The Effect of Race/Ethnicity on the Length of Prison Sentence for Various Types of Cases[a]

	CHICAGO				MIAMI				KANSAS CITY	
	Black		Hispanic		Black		Hispanic		Black	
TYPE OF CASE	b	Beta	b	Beta	b	Beta	b	Beta	b	Beta
TYPE OF CONVICTION CHARGE										
Violent Crime	29.32	0.03	−55.84	−0.04	13.46	0.08	8.32	0.04	−17.17	−0.07
Property Crime	77.64	0.10	65.39	0.06	−6.19	−0.04	−2.91	−0.02	6.57	0.12*
Drug Offense	9.74	0.02	74.49	0.09	−8.14	−0.07	8.08	0.07	14.09	0.20*
PRIOR PRISON TERM OF MORE THAN 1 YEAR										
Prior Prison Term	50.01	0.05	12.68	0.09	1.99	0.01	−1.30	−0.006	1.76	0.01
No Prior Prison Term	55.93	0.07	43.69	0.04	−10.00	−0.08	17.61	0.12	−1.36	−0.007
EMPLOYMENT STATUS										
Employed or Student	−25.62	−0.08	−31.21	−0.04	NA		NA		7.96	0.05
Unemployed	81.01	0.09*	89.11	0.05*					−0.40	−0.002

[a] We ran separate analyses on each type of case, controlling for offender race and for the other independent variables listed in Table 1. White offenders were the reference category in all of the analyses.

* $P \leq 0.05$

offenses were not more likely than whites convicted of these crimes to be incarcerated, but Hispanics convicted of violent crimes were more likely than whites convicted of violent crimes to be sentenced to prison.

The results for Miami are also inconsistent. There were no significant differences in incarceration between black offenders and white offenders for any of the five types of cases examined. We found a significant difference in incarceration between Hispanic offenders and white offenders for only two of the five types of cases examined, and one of these was the opposite of what we had predicted: Hispanics with a prior prison term faced higher odds of incarceration than whites with a prior prison term, but there were no differences for Hispanic and white offenders without a prior prison term. In fact, in Miami the only findings consistent with our hypotheses are that Hispanics and whites convicted of violent crimes faced similar odds of incarceration, while Hispanics convicted of drug offenses were more likely than their white counterparts to be incarcerated.

The results of the interactive analyses of the effect of race/ethnicity on the length of the sentence are presented in Table 6. Although the effect of the offender's race/ethnicity was not conditioned by the offender's prior criminal record in any of the three jurisdictions, it was conditioned by the type of conviction charge in Kansas City and by the offender's employment status in Chicago. In Kansas City, black offenders received significantly longer sentences than white offenders for property crimes and drug offenses; the difference was 14.09 months for drug offenses and 6.57 months for property crimes. Consistent with our hypothesis, black offenders convicted of violent crimes did not receive longer sentences than their white counterparts. In Chicago, unemployed blacks and Hispanics received substantially longer sentences than unemployed whites; the sentences imposed on employed offenders, on the other hand, did not vary by race/ethnicity.

DISCUSSION

The results of this study reveal that the relationship between race and sentence severity is complex. Offender race/ethnicity affected judges' sentencing decisions in some jurisdictions, for some types of offenses, and for some types of offenders. Although race had no effect on the likelihood of incarceration in Kansas City, both blacks and Hispanics were more likely than whites to be sentenced to prison in Chicago, and Hispanics (but not blacks) were more likely than whites to be incarcerated in Miami. Further analysis revealed that offender race interacted (although not always in the predicted way) with the type of conviction charge, the offender's prior criminal record, and the offender's employment status in Chicago and with the type of conviction charge and the offender's prior criminal record in Miami. In contrast, race/ethnicity had no effect on the length of the prison sentence in Miami and only affected sentence length in Kansas City if the conviction charge was a drug or property offense and in Chicago if the offender was unemployed.

Our finding that race/ethnicity had a direct effect on the decision to incarcerate or not but did not directly affect sentence length is consistent with previous research (Chiricos & Crawford, 1995; Kramer & Ulmer, 1996; Spohn et al., 1981–82). Chiricos & Crawford (1995: 297) reviewed 38 studies published since 1975 and found that "the frequency of significant racial disadvantage is approximately three times greater for in/out relationships as for those involving sentence length." Kramer & Ulmer's (1996) recent examination of departures from sentence guidelines in Pennsylvania similarly revealed that extralegal variables, including race, had a more pronounced effect on dispositional departures than on durational departures.

The fact that offender race/ethnicity influenced the decision to incarcerate or not (at least in Chicago and Miami) does not mean, of course, that this extralegal variable was the major determinant of sentencing. The race effects we found, while statistically significant, were rather modest. Moreover, legal factors were the primary determinants of sentence severity in all three jurisdictions. Both sentencing decisions were strongly influenced by the seriousness of the conviction charge, the number of conviction charges, and the offender's prior criminal record. This is not surprising; a long line of research has documented the effect of charge seriousness and prior record on sentencing decisions (Blumstein et al., 1983).

Our finding that race/ethnicity affected the decision to incarcerate or not even after the legal variables were taken into consideration, coupled with the fact that a number of other extralegal variables had significant effects, suggests that judges' sentencing decisions are not guided *exclusively* by factors of explicit legal relevance. Female offenders were substantially less likely than male offenders to be incarcerated in Chicago and Kansas City, offenders who pled guilty were sentenced less harshly than those who insisted on a trial in all three jurisdictions, and offenders who were detained prior to trial faced greater odds of incarceration than those who were released in all three sites. These findings add credence to assertions that sentencing decisions are an example of 'bounded rationality' (Albonetti, 1991) and reflect what Steffensmeier and his colleagues (1998) refer to as judges' 'focal concerns'. According to these perspectives, judges' sentencing decisions reflect their assessment of the blameworthiness or culpability of the offender, their desire to protect the community by incapacitating dangerous offenders or deterring potential offenders, and their concerns about the practical consequences, or social costs, of sentencing decisions. Faced with organizational constraints such as limited time in which to make decisions and limited information about offenders, judges may resort to stereotypes of dangerousness and culpability that rest, either explicitly or implicitly, on considerations of race, gender, pretrial status, and willingness to plead guilty.

With some exceptions, our findings are consistent with others published during what Zatz (1987) terms the 'fourth wave' of research on race and sentencing. Zatz (1987: 70) notes that studies analyzing data from jurisdictions using determinate sentencing have found "subtle, if no longer overt, bias against racial minorities." Although we did find evidence of overt discrimination against Hispanics in Miami and against both blacks and Hispanics in Chicago, we also discovered more subtle forms of bias in all three jurisdictions. As noted above, offender race/ethnicity interacted with the type of conviction charge, the offender's prior criminal record, and the offender's employment status.

Although these interaction effects attest to the validity of Hawkins' (1987: 721) assertion that "the conflict perspective must be revised to address more specifically the question of the role played by various contingencies or mediating factors," the fact that not all of them are consistent with our predictions makes their interpretation difficult. Our findings with respect to charge seriousness illustrate the problem. Building on previous research showing that race affects sentencing primarily in less serious or more ambiguous cases, we hypothesized that blacks and Hispanics convicted of less serious offenses would be sentenced more harshly than whites convicted of these offenses, but that race would not influence sentencing for more serious offenses. Consistent with our hypothesis, we found that race/ethnicity had no effect on sentence severity for violent crimes in Miami and Kansas City, that Hispanics (but not blacks) were more likely than whites to be sentenced to prison for drug offenses in Miami, and that blacks received longer sentences than whites for property crimes and drug offenses in Kansas City. In Chicago, on the other hand, there were no race effects for property or drug offenses, but Hispanics convicted of violent crimes were more likely than whites convicted of violent crimes to be sentenced to prison.

The fact that Hispanics convicted of drug offenses faced greater odds of incarceration in Miami comes as no surprise in light of theoretical discussions about the 'moral panic' in contemporary society concerning the war on drugs. Moral panic theorists argue that society is characterized by a variety of common sense perceptions about crime and drugs (Jenkins, 1994; Webb & DeLone, 1996) that result in community intolerance for such behaviors and increased pressure for punitive actions. Many theorists (see Chiricos & DeLone, 1992 for a review) argue that this moral panic can become ingrained in the judicial ideology of sentencing judges, resulting in more severe sentences for those believed to be responsible for such problems as crime and drugs. While it might be hypothesized that sentences for both black and Hispanic offenders would be affected by these beliefs, the fact that the effect was found only for Hispanics is not unexpected given the jurisdiction – Miami. Arguably, the most enduring perception about drug importation and distribution in Miami is that these activities are dominated by Hispanics of various nationalities.

This explanation is also applicable to our finding that blacks convicted of drug or property offenses received longer sentences than white drug or property offenders in Kansas City. The theories relating the impact of moral panics about drugs to community intolerance and judicial anxiety suggest that minorities who are perceived as having connections to drug use and drug distribution will be differentially punished. This linkage applies not only to drug offenses themselves but also to property offenses, which may be viewed as the means by which drug offenders obtain the money needed to buy drugs. Like Hispanics in Miami, blacks in Kansas City may be associated with drug use, drug distribution, and the property crimes that accompany a drug-involved lifestyle.

Our results concerning the interaction between prior record and race/ethnicity are similarly complicated. Previous research has shown that prior criminal record is a strong predictor of sentence severity and that, as a result, race comes into play primarily in cases involving offenders without serious prior convictions. The interaction we found between race/ethnicity, prior record, and the likelihood of incarceration in Chicago is consistent with this. Among offenders who previously had been sentenced to prison for a year or more, race/ethnicity had no effect on the decision to incarcerate or not; among offenders without a prior prison term, on the other hand, both blacks and Hispanics were more likely than whites to be sentenced to prison. These results suggest that the absence of a prior prison sentence enhances the judge's discretion and opens the door to considerations of extralegal factors such as race or ethnicity.

In Miami we found a race/ethnicity and prior record interaction that was contrary to our hypothesis. There were no race effects for offenders without a prior prison term, but Hispanics with a prior prison sentence were more likely than whites with a prior prison sentence to be incarcerated. This suggests that in Miami a prior criminal record – which, as noted above is a strong predictor of the decision to incarcerate or not – has a more detrimental effect on Hispanics than on whites.

These results are consistent with those reported by Zatz (1984), who found that prior record, which was an aggravating circumstance under California's determinate sentencing statute, had a significant effect on sentence severity only for Chicanos. Although prior record is formally factored into the score each defendant receives under the Florida sentencing guidelines, and thus should affect black, Hispanic and white offenders equally, there is nothing to preclude criminal justice officials from using prior record to aggravate or mitigate the sentence. The Florida sentencing guidelines do not eliminate judicial discretion in sentencing; in many cases the judge will be able to choose between prison, jail, or probation. Although we can only speculate, Miami judges and prosecutors may regard the presence of a prior prison sentence as a stronger indicator of dangerousness and recidivism for Hispanics than for whites. If this is the case, prior record should not be regarded as a racially neutral determinant of sentence severity.

A final comment concerns the interaction between race/ethnicity and employment status. We found some support for our hypothesis that race/ethnicity would affect sentence severity only among the unemployed. Consistent with our predictions, race/ethnicity did not affect either measure of sentence severity in either jurisdiction if the offender was employed. Contrary to our hypothesis, however, unemployed blacks did not receive harsher sentences than unemployed whites in Kansas City. In Chicago, on the other hand, unemployed Hispanics were more likely than unemployed whites to be sentenced to prison, and both unemployed blacks and unemployed Hispanics received longer sentences than unemployed whites. These latter findings are consistent with the labor surplus and punishment literature (see Chiricos & DeLone, 1992), which suggests that unemployed offenders will be seen as a threatening population that needs to be controlled and that one form of control is through incarceration (Greenberg, 1977; Melossi, 1985, 1989; Hale, 1989). The fact that we found this effect in only one of the two jurisdictions, however, suggests that we should exercise caution in drawing conclusions. Future research should continue to explore the race/unemployment/sentence severity relationship.

Our study does warrant some caveats. Although we controlled for the variables identified by prior research as significant predictors of sentence severity, we obviously were not able to control for all of the decision making criteria used by judges. We were unable, for example, to control for the offender's educational level, marital status, or responsibility for dependent children. Each of these factors has been shown to influence sentence severity and it is certainly possible that the race (and gender) effects we found might diminish if these variables were included in our models (see Daly & Tonry

(1997: 231–232) for a discussion of whether race- and/or gender-linked criteria are legitimate considerations at sentencing).

A second limitation concerns our inability to correct for sample selection bias in earlier case processing decisions (Klepper et al., 1983). We included a hazard rate reflecting the predicted probability of incarceration in our model of sentence length, but, because our data file included only offenders who had been convicted of felonies, we were not able to correct for selection bias in charging and plea bargaining decisions. If race/ethnicity affects these earlier decisions, the black, Hispanic and white offenders reaching the sentencing stage will not be representative of the population of cases entering the system. If, for example, prosecutors reject or divert a larger proportion of the less serious drug cases involving white offenders than black or Hispanic offenders, the remaining cases with white offenders will be more serious than those with black or Hispanic offenders. If these differences in case seriousness are not measurable (and thus not included as controls in the model of sentence severity), "then the biases induced by sample selection might mask the true extent of the discrimination" (Klepper et al., 1983: 65). Because of this potential problem, our finding that black offenders did not face higher odds of incarceration than white offenders in either Miami or Kansas City should be interpreted with caution.

CONCLUSION

The results of this multi-jurisdictional study of the sentencing process offer "clues to the contextual character of possible race effects" (Chiricos & Crawford, 1995: 284). There were small, but statistically significant, direct race effects in Chicago and Miami, more subtle effects in all three jurisdictions. Discrimination was directed primarily against Hispanics in Miami, but was directed against both blacks and Hispanics in Chicago. And race/ethnicity did not have consistent effects for all types of offenses or all categories of offenders. The patterns of results uncovered in this study attest to the importance of examining the effect of ethnicity as well as race and highlight the importance of testing interactive as well as additive models. Future research should continue to explore the complex interconnections among race/ethnicity, offender and case characteristics, and sentence severity.

Considered in conjunction with other research on determinate sentencing and sentence guidelines (Kramer & Ulmer, 1996; Tonry, 1996; Ulmer, 1997; Ulmer & Kramer, 1996; Walker, 1993), the results of our study suggest that sentencing reform cannot eliminate unwarranted disparity in the important decision regarding probation versus prison. We found significant race effects in one jurisdiction (Chicago) with determinate sentencing and in another jurisdiction (Miami) with sentence guidelines. Both sentencing reforms constrain the discretion of judges to impose excessively lenient or excessively severe prison sentences, but neither eliminates the judge's discretion with respect to the type of sentence. Although some offenses are non-probationable under determinate sentencing and a prison sentence is the presumptive sentence for offenders who accumulate a certain number of points under sentence guidelines, judges can "use the aggravating and mitigating circumstances loophole to alter the presumptive sentence" (Zatz, 1987: 79). As Tonry (1996: 180) notes "There is, unfortunately, no way around the dilemma that sentencing is inherently discretionary and that discretion leads to disparities."

NOTES

1. Judges in Illinois impose determinate sentences. Felonies are classified as either first degree murder, Class X, Class 1, Class 2, Class 3, or Class 4 felonies. Offenders convicted of first degree murder or attempted first degree murder, any Class X felony, or a number of other specified offenses (e.g., residential burglary or vehicular hijacking) cannot be sentenced to probation or to any other alternative to incarceration. For these offenses, the judge is required to impose at least the minimum prison sentence. The minimum and maximum terms of imprisonment for the six categories of felonies are:

First Degree Murder	20–60 years, life or death
Class X	6–30 years
Class 1	4–15 years
Class 2	3–7 years
Class 3	2–5 years
Class 4	1–3 years

The judge cannot impose a sentence that exceeds the maximum term of imprisonment unless he/she finds that at least one of the aggravating factors specified in the statute (e.g., the defendant received compensation for committing the crime or committed a crime against a person 60 years of age or older) was present.

2. Missouri has a determinate sentencing structure. Felonies are classified into four categories: Class A, Class B, Class C, or Class D. The terms of imprisonment associated with each category are:

Class A	10–30 years or life
Class B	5–15 years
Class C	a maximum of 7 years
Class D	a maximum of 5 years

For C and D felonies, the court can sentence the offender to one year or less in the county jail. The maximum terms for each category are increased if the offender is proved to be either a persistent offender or a dangerous offender. In this case, for example, the maximum term for Class B felonies is 30 years, the maximum term for Class C felonies is 20 years, and the maximum term for Class D felonies is 10 years.

If the offender is tried and found guilty by a jury, the jury determines the sentence unless the defendant requests in writing prior to *voir dire* that the court assess the punishment or the state pleads and proves the defendant as a prior offender, persistent offender, or dangerous offender. If the jury sentences the offender to prison, the judge cannot impose a harsher sentence (unless the term of years is less than the authorized lowest term for the offense); the judge can, on the other hand, impose a shorter sentence.

3. The state of Florida has had sentencing guidelines since 1983. The purpose of the guidelines is "to establish a uniform set of standards to guide the sentencing judge" and "to eliminate unwarranted variation in the sentencing process by reducing the subjectivity in interpreting specific offense-related and offender-related criteria." To meet these objectives, each offender is assigned a 'sentence score' based on the seriousness of the offense(s) and his/her prior criminal record. This score determines the recommended sentence. Judges retain some discretion under the guidelines. For example, if the total sentence points for a particular offender are less than 40, the presumptive sentence is a non-state prison sentence. In this situation, the judge has discretion to sentence the offender to county jail for a maximum term of 364 days or to impose probation or some other alternative to incarceration; the judge also has discretion to withhold adjudication. If the total points are greater than 40 but less than or equal to 52, the judge has discretion to sentence the offender to state prison or not. If the points total more than 52, the sentence must be a prison sentence, with the months in state prison calculated by subtracting 28 from the total sentence points; the judge can, however, increase or decrease the sentence length by 25% (without providing a written statement delineating the reasons for the departure) or more (with a written statement of the reasons for the departure).

4. In 1993 there were over 20,000 felony convictions in Dade County and over 30,000 in Cook County; there were only 1,600 felony convictions in Jackson County. We therefore selected a random sample of the felony convictions in Cook County and Dade County.

5. The original data file for Miami included 4,246 cases. This included 1,468 cases in which the sentence was 'adjudication withheld' rather than probation, jail, or prison. Because we did not have a comparable measure of sentence severity in the other two jurisdictions, we eliminated these cases from the analyses. (See Spohn et al., 1998 for a discussion of the effect of race/ethnicity on the decision to withhold adjudication. We found that race/ethnicity interacted with gender and with the offender's prior criminal record to affect the decision to withhold adjudication. Among offenders with no prior felony convictions, neither race/ethnicity nor gender was a significant predictor. Among repeat offenders, on the other hand, Hispanic males, and females in all three racial/ethnic groups, were more likely than black males to have adjudication withheld.) Because of the small number of cases in which the most serious conviction charge was either forcible rape ($N = 16$) or kidnaping ($N = 27$), we eliminated these cases from the analyses. We also eliminated 20 cases where the class of the most serious conviction charge was a capital felony or was missing.

6. Jail sentences were not common in the other two jurisdictions. Only 75 offenders were sentenced to jail in Chicago; 36 were sentenced to jail in Kansas City.

7. In Chicago and Kansas City we have data on the offender's employment status; we were not able to obtain this information for offenders in Miami. Because we believe it is important to use the same controls in each jurisdiction, we first model the sentencing

decisions without this variable; we then add a variable measuring whether the offender was unemployed (coded 1) or not (coded 0) to the models for Chicago and Kansas City. We also test our hypothesis concerning interaction between the offender's race and the offender's employment status in these two jurisdictions only.

8. We used different procedures to determine the offender's race/ethnicity in the three jurisdictions. In Chicago, all of the information was obtained from the court file. The offender's race was specified as white, black, Hispanic or other on the arrest report and on the information/indictment. In Kansas City, the computer printout of 1993 felony convictions included the offender's race, which was specified as either white, black, or Hispanic. In Miami, the data file provided by the Administrative Office of the Courts included two variables measuring race/ethnicity. Offenders were categorized as either white or black on the race variable and as either of Hispanic origin or not of Hispanic origin on the ethnicity variable. To create the three-category variable (white, black, Hispanic), we categorized as 'black' all offenders who were identified as black on the race variable. Offenders who were identified as white on the race variable and as non-Hispanic on the ethnicity variable were categorized as 'white'; those who were identified as white on the race variable and as Hispanic on the ethnicity variable were categorized as 'Hispanic'.

9 The correlation matrix for the measures of prior criminal record (for the Miami sample) is presented below:

	ARRS	FARRS	FCONS	INCARS	DCON	VCON
ARRS						
FARRS	0.80*					
FCONS	0.72*	0.91*				
INCARS	0.47*	0.64*	0.72*			
DCON	0.43*	0.51*	0.56*	0.26*		
VCON	0.36*	0.31*	0.30*	0.25*	0.15*	

ARRS = Number of prior arrests
FARRS = Number of prior felony arrests
FCONS = Number of prior felony convictions
INCARS = Number of prior prison terms of more than one year
DCON = Prior conviction for a drug offense
VCON = Prior conviction for a violent offense

10. The correlations between the measures of prior criminal record and the dependent variables (for the Miami sample) are presented below:

	INCARCERATE JAIL/PRISON	INCARCERATE PRISON	PRISON SENTENCE
ARRS	0.06*	0.07*	0.02
FARRS	0.05*	0.14*	0.05
FCONS	0.05*	0.16*	0.09*
INCARS	0.11*	0.33*	0.19*
DCON	–0.003	–0.06*	–0.09*
VCON	0.03	0.07*	0.05

ARRS = Number of prior arrests
FARRS = Number of prior felony arrests

FCONS = Number of prior felony convictions
INCARS = Number of prior prison terms of more than one year
DCON = Prior conviction for a drug offense
VCON = Prior conviction for a violent offense

We replicated the analysis using the number of prior felony convictions and whether the offender had previously been convicted of a violent offense. The results were identical.

11. We controlled for both the type of conviction charge and the class of the conviction charge because of the fact that each type of conviction charge includes various classes of felonies. For example, in Florida possession of cocaine with intent to deliver is a 2nd degree felony, while possession of cocaine with intent to deliver within 1,000 feet of a school is a 1st degree felony; both of these offenses are included in the broad category "possession of narcotics with intent." Including both variables as controls provides a more accurate and reliable measure of offense seriousness.

12. In Florida, felonies are classified as either capital crimes, life felonies, first degree felonies, second degree felonies, or third degree felonies. We eliminated 7 cases in which the most serious conviction charge was a capital crime. Because there were only 26 cases in which the most serious conviction was a life felony, we combined the life felonies and 1st degree felonies into a single category.

13. We could not control for whether the offender had a jury trial or a bench trial. The data obtained from the Dale County Court Administrator did not differentiate between jury trial and bench trial. Moreover, in Chicago there were only 21 offenders tried by jury; in Kansas city there were only 5 bench trials.

REFERENCES

Albonetti, C. (1997). Sentencing Under the Federal Sentencing Guidelines: Effects of Defendant Characteristics, Guilty Pleas, and Departures on Sentence Outcomes for Drug Offenses, 1991–1992. *Law & Society Review, 31*, 789–822.

Albonetti, C. (1991). An Integration of Theories to Explain Judicial Discretion. *Social Problems, 38*, 247–266.

Baldus, D. C., Woodworth, G., & Pulaski, C. (1985). Monitoring and Evaluating Contemporary Death Sentencing Systems: Lessons from Georgia. *U. C. Davis Law Review, 18*, 1375–1407.

Barnett, A. (1985). Some Distribution Patterns for the Georgia Death Sentence. *U. C. Davis Law Review, 18*, 1327–1374.

Berk, R. A. (1983). An Introduction to Sample Selection Bias in Sociological Data. *American Sociological Review, 48*, 386–398.

Berk, R. A., & Ray, S. C. (1982). Selection Biases in Sociological Data. *Social Science Research, 11*, 352–398.

Bernstein, I. N., Kelly, W. R., & Doyle, P. A. (1977). Societal Reaction to Deviants: The Case of Criminal Defendants. *American Sociological Review, 42*, 743–795.

Blumstein. A. (1982). On the Racial Disproportionality of United States

Prison Populations. *Journal of Criminal Law and Criminology, 73*, 1259–1281.

Prison Populations (1993). Racial Disproportionality of U.S. Prison Populations Revisited. *University of Colorado Law Review, 64*, 743–760.

Blumstein, A., Cohen, J., Martin, S. E., & Tonry, M. (Eds) (1983). *Research on Sentencing: The Search for Reform*, Volume I. Washington, DC: National Academy Press.

Chambliss, W. J., & Seidman, R. B. (1971). *Law, Order and Power.* Reading, MA: Addison-Wesley.

Chiricos, T. G., & Bales, W. D. (1991). Unemployment and Punishment: An Empirical Assessment. *Criminology, 29*, 701–724.

Chiricos, T. G., & Crawford, C. (1995). Race and Imprisonment: A Contextual Assessment of the Evidence. In: D. Hawkins (Ed.). *Ethnicity, Race and Crime*. Albany, NY: State University of New York Press.

Chiricos, T. G., & DeLone, M. (1992). Labor Surplus and Punishment: A Review and Assessment of Theory and Evidence. *Social Problems, 39*, 421–446.

Daly, K., & Tonry, M. (1997). Gender, Race, and Sentencing. In: M. Tonry (Ed.), *Crime and Justice: A Review of Research*, Vol 22. Chicago: The University of Chicago Press.

Gibson, J. L. (1978). Race as a Determinant of Criminal Sentences: A Methodological Critique and a Case Study. *Law & Society Review, 12*, 455–478.

Greenberg, D. (1977). The Dynamics of Oscillatory Punishment Processes. *Journal of Criminal Law and Criminology, 68*, 643–651.

Gruhl, J., Welch, S., & Spohn, C. (1984). Women as Criminal Defendants: A Test for Paternalism. *Western Political Quarterly, 37*, 456–467.

Hagan, J. (1974). Extra-Legal Attributes and Criminal Sentencing: An Assessment of a Sociological Viewpoint. *Law & Society Review, 8*, 357–383.

Hale, C. (1989). Economy, Punishment and Imprisonment. *Contemporary Crises, 13*, 327–350.

Hanushek, E. A., & Jackson, J. E. (1977). *Statistical Methods for Social Scientists.* New York: Academic Press.

Hawkins, D. F. (1987). Beyond Anomalies: Rethinking the Conflict Perspective on Race and Criminal Punishment. *Social Forces, 65*, 719–745.

Heckman, J. J. (1974). Shadow Prices, Market Wages, and Labor Supply. *Econometrics, 42*, 679–694.

Hembroff, L. A. (1982). Resolving Status Inconsistency: An Expectation States Theory and Test. *Social Forces, 61*, 183–205.

Holmes, M. D., Hosch, H. M., Daudistel, H. C., Perez, D. A., & Graves, J. B. (1996). Ethnicity, Legal Resources, and Felony Dispositions in Two Southwestern Jurisdictions. *Justice Quarterly, 13*, 11–30.

Jenkins, P. (1994). 'The Ice Age' The Social Construction of a Drug Panic. *Justice Quarterly , 11*, 7–31.

Kleck, G. C. (1981). Racial Discrimination in Sentencing: A Critical Evaluation of the Evidence with Additional Evidence on the Death Penalty. *American Sociological Review, 43*, 783–805.

Klein, S., Petersilia, J., & Turner, S. (1990). Race and Imprisonment Decisions in California. *Science, 247*, 812–816.

Klepper, S., Nagin, D., & Tierney, L.-J. (1983). Discrimination in the Criminal Justice System: A Critical Appraisal of the Literature. In: A. Blumstein, J. Cohen, S. E. Martin, & M. H. Tonry (Eds), *Research on Sentencing: A Search for Reform, 2*, 55–128.

Kramer, J. H., & Ulmer, J. T. (1996). Sentencing Disparity and Departures from Guidelines. *Justice Quarterly, 13*, 81–106.

LaFree, G. D. (1985a). Adversarial and Nonadversarial Justice: A Comparison of Guilty Pleas and Trials. *Criminology, 23*, 289–312

LaFree, G. D. (1985b). Official Reactions to Hispanic Defendants in the Southwest. *Journal of Research in Crime and Delinquency, 22*, 213–237.

LaFree, G. D. (1989). *Rape and Criminal Justice: The Social Construction of Sexual Assault.* Belmont, CA: Wadsworth.

Levin, M. A. (1972). Urban Politics and Policy Outcomes: The Criminal Courts. In: G. F. Cole (Ed.), *Criminal Justice: Law and Politics*. Belmont, CA: Wadsworth.

Lichter, D. T. (1989). Race, Employment Hardship, and Inequality in the American Non-metropolitan South. *American Sociological Review, 54*, 436–446.

Lizotte, A. J. (1978). Extra-legal Factors in Chicago's Criminal Courts: Testing the Conflict Model of Criminal Justice. *Social Problems, 25*, 564–580.

Mann, C. R. (1993). *Unequal Justice: A Question of Color.* Bloomington and Indianapolis, IN: Indiana University Press.

Melossi, D. (1985). Punishment and Social Action: Changing Vocabulary of Punitive Motive within a Political Business Cycle. *Current Perspectives in Social Theory, 6*, 169–197.

Melossi, D. (1989). An Introduction: Fifty Years Later. *Punishment and Social Structure in Contemporary Analysis. Contemporary Crises, 13*, 311–326.

Miethe, T. D., & Moor, C. A. (1985). Socioeconomic Disparities Under Determinate Sentencing Systems: A Comparison of Preguideline and Postguideline Practices in Minnesota. *Criminology, 23*, 337–363.

Miethe, T. D., & Moore, C. A. 1986. Racial Differences in Criminal Processing: The Consequences of Model Selection on Conclusions about Differential Treatment. *The Sociological Quarterly, 27*, 217–237.

Moore, C. A., & Mieth, T. D. (1986). Regulated and Unregulated Sentencing Decisions: An Analysis of First-Year Practices Under Minnesota's Felony Sentencing Guidelines. *Law & Society Review, 20*, 253–277.

Myers, M. A., & Talarico, S. (1986). The Social Contexts of Racial Discrimination in Sentencing. *Social Problems, 33*, 236–251.

New York Times, December 7, 1997.

New York Times, March 8, 1998.

Nobiling, T., Spohn, C., & DeLone, M. (1998). A Tale of Two Counties: Unemployment and Sentence Severity. *Justice Quarterly, 15*, 401–427.

Petersilia, J. (1983). *Racial Disparities in the Criminal Justice System.* Santa Monica, CA: Rand.

Peterson, R., & Hagan, J. (1984). Changing Conceptions of Race: Toward an Account of Anomalous Findings in Sentencing Research. *American Sociological Review, 49*, 56–70.

Pruitt, C. R., & Wilson, J. Q.(1983). A Longitudinal Study of the Effect of Race on Sentencing. *Law & Society Review, 7*, 613–635.

Quinney, R. (1970). *The Social Reality of Crime*. Boston: Little-Brown.

Quinney, R., & Spohn, C. (2000). Thirty Years of Sentencing Reform: The Quest for a Racially-Neutral Sentencing Process. NIJ Criminal Justice 2000, Vol. 3. Washington, DC: U.S. Department of Justice.

Spohn, C. (1994). Crime and the Social Control of Blacks: The Effect of Offender/Victim Race on Sentences for Violent Felonies. In: G. Bridges, & M. Myers (Eds), *Inequality, Crime and Social Control*. Boulder, CO: Westview Press.

Spohn, C. (1992). An Analysis of the 'Jury Trial Penalty' and Its Effect on Black and White Offenders. *The Justice Professional*, 93–112.

Spohn, C., & Cederblom, J. (1991). Race and Disparities in Sentencing: A Test of the Liberation Hypothesis. *Justice Quarterly, 8*, 305–327.

Spohn, C., DeLone, M., & Spears, J. (1998). Race/Ethnicity, Gender and Sentence Severity in Dade County: An Examination of the Decision To Withhold Adjudication. *Journal of Crime & Justice, 21*, 111–138.

Spohn, C., Gruhl, J., & Welch, S. (1981–82). The Effect of Race on Sentencing: A Re-Examination of an Unsettled Question. *Law & Society Review, 16*, 72–88.

Spohn, C., & Holleran, D. (2000). The Imprisonment Penalty Paid by Young, Unemployed Black and Hispanic Male Offenders. *Criminology, 38*, 281–306.

Steffensmeier, D., Ulmer, J., & Kramer, J. (1998). The Interaction of Race, Gender, and Age in Criminal Sentencing: The Punishment Cost of Being Young, Black, and Male. *Criminology, 36*, 763–797.

Stolzenberg, L., & Alessio, S. D. (1994). Sentencing and Unwarranted Disparity: An Empirical Assessment of the Long-Term Impact of Sentencing Guidelines in Minnesota. *Criminology, 32*, 301–310.

Tonry, M. (1995). *Malign Neglect: Race, Crime, and Punishment in America*. New York: Oxford University Press.

Tonry, M. (1996). *Sentencing Matters*. New York: Oxford University Press.

Uhlman, T. M., & Walker, N. D. (1980). 'He Takes Some of My Time, I Take Some of His': An Analysis of Judicial Sentencing Patterns in Jury Cases. *Law & Society Review, 14*, 323–341.

Ulmer, J. T. (1997). *Social Worlds of Sentencing: Court Communities Under Sentencing Guidelines*. Albany, NY: State University of New York Press.

Ulmer, J. T., & Kramer, J. H. (1996). Court Communities Under Sentencing Guidelines: Dilemmas of Formal Rationality and Sentencing Disparity. *Criminology, 34*, 383–408.

Unnever, J. D., & Hembroff, L. A. (1987). The Prediction of Racial/Ethnic Sentencing Disparities: An Expectation States Approach. *Journal of Research in Crime and Delinquency, 25*, 53–82.

Walker, S. (1993). *Taming the System: The Control of Discretion in Criminal Justice, 1950–1990*. New York: Oxford University Press.

Walsh, A. (1987). The Sexual Stratification Hypothesis and Sexual Assault in Light of the Changing Conceptions of Race. *Criminology, 25*, 153–173.

Webb, V. J., & DeLone, M. A. (1996). Drug Use Among a Misdemeanant Population: Exploration of a Legal Syllogism of the Drug War. *Crime, Law and Social Change, 24*, 241–255.

Wonders, N. A. (1996). Determinate Sentencing: A Feminist and Postmodern Story. *Justice Quarterly 13*, 611–648.

Zatz, M. S. (1984). Race, Ethnicity, and Determinate Sentencing: A New Dimension to an Old Controversy. *Criminology, 22*, 147–171.

Zatz, M. S. (1985). Pleas, Priors, and Prison: Racial/Ethnic Differences in Sentencing. *Social Science Research, 14*, 169–193.

Zatz, M. S. (1987). The Changing Forms of Racial/Ethnic Biases in Sentencing. *Journal of Research in Crime and Delinquency, 24*, 69–92.

IMPLICATIONS OF GHETTO-RELATED BEHAVIOR FOR A COMMUNITY AND CRIME MODEL: DEFINING THE PROCESS OF CULTURAL ATTENUATION

Barbara D. Warner and Pamela Wilcox Rountree

ABSTRACT

Present community-level crime models emphasize the impact of neighborhood structural features on the ability of residents to develop social ties and provide informal social control, thereby inhibiting crime. However, recent empirical evidence suggests that social ties do not sufficiently account for levels of informal control (Sampson, Raudenbush & Earls, 1997; Warner & Rountree, 1997). In addition, the effects of poverty and racial/ethnic isolation on crime rates are not well explained by variation in social ties. To address these gaps, we suggest that social control, and therefore crime rates, are affected not only by insufficient social ties, but also by the presence of 'ghetto-related' behaviors. The presence of behaviors in contrast to middle-class values can be viewed as diminishing social control through what Wilson (1996) refers to as cultural attenuation, or the loosening of social control due to diminished enforcement of conventional values. We hypothesize that poverty and racial/ethnic isolation affect the extent to which behaviors in contrast to

Sociology of Crime, Law and Deviance, Volume 2, pages 39–62.

ISBN: 0-7623-0680-7

conventional values appear in the community, which in turn, affect the control of criminal behavior. We empirically test these ideas using data from 100 Seattle neighborhoods and focusing on two 'ghetto-related' behaviors: rates of young, single motherhood and youth idleness. Our findings show that both social ties and 'ghetto-related' behavior arise from the structural conditions we examine. In turn, social ties decrease community rates of violence, and 'ghetto-related' behaviors increase violence rates.

INTRODUCTION

Blighted by high rates of poverty and racial segregation, many inner city neighborhoods became infested with high crime rates in the latter part of the twentieth century. Attempts to explain these variations in crime rates across neighborhoods have revolved around identifying community properties and processes that explain variations in informal social control. The reigning community and crime model, the systemic model, has relied on images of networks of community level social ties as providing the foundation for informal social control. This model argues that non-poor, stable and racially homogenous neighborhoods provide a favorable context for frequent and wide ranging friendship networks to develop. These networks are central to the systemic model as they provide the foundation for two basic processes to occur: the articulation of shared values across community members and, through this sense of shared values, a sense of support for behaviorally reinforcing those values through questioning strangers, supervising neighborhood children, and otherwise informally controlling undesired behavior.

While there are now several studies that provide encouraging evidence of the potential impact of local social ties for informally controlling crime and mediating the effects of residential mobility on crime rates (see e.g. Sampson & Groves, 1989; Warner & Rountree, 1997; Bellair, 1997; Elliot et al., 1996), there is also evidence that social ties, in and of themselves, do not adequately account for the variance in crime rates (Pattillo, 1998; Sampson et al., 1997) and do not affect crime rates in all types of neighborhoods (Warner & Rountree, 1997).[1] Neither do social ties mediate much, if any, of the effects of poverty and racial composition (Sampson & Groves, 1986; Warner & Rountree, 1997).

The above-cited work using a community-level systemic approach to social control is based upon traditional social control models and therefore assumes value consensus. In order for network ties to decrease crime, consensus regarding conventional values must seemingly be present. However, other recent social control models reject the assumption of value consensus. These models allow for cultural variation and suggest that social control is provided by groups with both conventional and oppositional values (see e.g. Heimer & Matsueda, 1994, and Felson et al., 1994).

Left unaddressed, however, have been ideas of cultural variation not in values themselves but in *value reinforcement*. Following ideas presented by Wilson (1987, 1996) we suggest a model which maintains the assumption of value consensus as in pure social control models, but allows for variation in the level of enforcement of those values. The present study attempts to move communities and crime studies forward by theoretically incorporating notions of attenuated culture into the contemporary social disorganization model and empirically examining several of the major propositions deduced from this model.

Just as structural conditions are posited to affect the emergence of social ties, structural conditions (particularly poverty) are presumed to cause certain 'ghetto-related' behaviors to 'take hold'. These behaviors arise directly from the level of poverty and racial isolation, not from diminished social ties. The emergence of high levels of such behavioral adaptations makes evident the inability of many community residents to carry out conventional values. The contradiction between articulated middle class values and the observed reality that large proportions of residents cannot, and do not, enact those values tempers the perceived appropriateness of residents enforcing those values. Thus, the presence of 'ghetto-related' behaviors leads to a loosening of community demand for conventional behaviors, thereby diminishing informal social control of more serious and more public breaches of conventional values, including crime.

While we do not have the data to test all aspects of the model we are proposing, we believe that to the extent our empirical findings support the major propositions presented, and to the extent that the proposed model is consistent with the assumptions of a social control model, it advances the development of a macro-level social control model for understanding the effects of community processes on crime. As a first step in this model-building and model-testing process, we review more fully the literature pertinent to the evolution of the community-control model proposed here.

THE COMMUNITY LEVEL SOCIAL CONTROL MODEL

Most discussions of contemporary community level crime models have been rooted in the systemic model of community attachment (Sampson, 1986; 1987; Bursik & Grasmick, 1993a, 1995).[2] The systemic model views community "as a complex system of friendship and kinship networks and formal and informal associational ties rooted in family life and on-going socialization processes" (Kasarda & Janowitz, 1974: 329). It is through local social ties and the processes that emerge from them that modern community and crime theorists explain the effects of community structural variables on crime rates.[3] Communities that have high rates of poverty, residential mobility, and are racially or ethnically heterogeneous are less likely to be able to support lasting and wide-ranging friendship networks. Neighborhood instability has been argued to make "the development of deep and lasting affective relational networks very difficult, [while] heterogeneity in the area limits the breadth of such networks" (Bursik & Grasmick, 1993a: 35). Although the theoretical role of poverty has been less clear in the systemic model it has nonetheless remained a part of the model. Bursik & Grasmick, (1993a) for example, suggest that poverty is important in that it affects mobility rates and levels of heterogeneity.

While neighborhood structure is predicted to affect the level of local social ties, these ties in turn are perceived as necessary to articulate shared values and implement informal sanctions, such as intervening in inappropriate behavior, when behavioral expectations are violated (Bursik & Grasmick, 1995). Therefore, communities with wider friendship and associational ties are predicted to have greater potential for informal social control. Although local social ties have been an important theoretical component in recent community and crime models, the empirical examination of their role remains sparse, and the results from the existing studies are mixed. While there were a handful of early studies attempting to examine social ties and informal social control, these studies had too few neighborhoods to provide more than speculative results (see for example, Macoby, Johnson & Church, 1958; Greenberg, Rohe & Williams, 1982; Simcha-Fagan & Schwartz, 1986). It is with more recent studies using sufficient numbers of neighborhoods that we have begun to develop an understanding of the effects of social ties.

Some support of the systemic model and the role of local social ties is reported by Sampson & Groves (1989). This study was the first to have a large enough sample to fully test the intervening effects of social ties. Sampson & Groves (1989) measure the extent of local ties with the average number of the respondent's friends who lived within a 15-minute walk of their home. Their results show that residential mobility significantly decreases the level of friendship networks, but socio-economic status and heterogeneity do not. The effect of local friendship networks on crime is also inconsistent. While they significantly decrease victimization in terms of mugging/street robbery, burglary and total victimization, they do not significantly affect victimization from stranger violence, auto theft or vandalism. In terms of self-reported delinquency, local friendship networks significantly decrease self-reported property offending rates, but not self-reports of personal violence. Sampson & Groves' (1989) findings do, however, suggest that local friendship networks mediate a substantial portion of the effects of mobility on some crimes.

In contrast, some studies suggest that social ties may not be an adequate explanation for the effects of community structure on informally controlling crime. First, findings from Warner & Pierce's (1993) study of Boston demonstrate that poor racially homogeneous neighborhoods have some of the highest crime rates. This finding calls into question whether racial homogeneity is a good predictor for social ties, and, if it is, whether social ties are the basis for informal social control (see Krivo & Peterson, 1996 for similar findings.) Further, Warner & Rountree's (1997) direct examination of the effects of social ties across different types of Seattle neighborhoods shows that social ties decrease informal social control of crime only in predominantly white neighborhoods. Their findings suggest that, particularly in minority neighborhoods, other neighborhood level variables must be examined to explain variance in controlling crime rates. Similarly, Pattillo's (1998) ethnographic study of a stable, middle-class black neighborhood in Chicago suggests that entrenchment of illicit networks, or criminal organizations, in such communities may render the effects of stability and social networks paradoxical, simultaneously *fostering* intervention behavior and *hampering* the removal of gangs and gang-related activity.

Other recent studies have also shown the limits of social ties for explaining informal social control. For example, Sampson et al. (1997: 923) found that while their measure of collective efficacy, or neighbors cohesiveness and willingness to intervene, is related to crime, friendship networks account for only about 25% of the variance in collective efficacy. Similarly Elliot et al. (1997) found that informal networks significantly decreased adolescent problem behavior in Denver, but not in Chicago. In short, while the idea that social ties serve as a foundation for effective informal social control is reasonable and has some empirical support, it is also apparent that social ties, in themselves, are not sufficient, particularly in certain contexts.

Other recent social control models have argued for incorporating cultural variation in values into a social control model. While a 'pure' social control model is theoretically inconsistent with cultural deviance theory because it assumes consensus in values and no variation in *motivation* to commit crime, Matsueda (1992) and Heimer & Matsueda (1994) present a 'differential social control' model that extends traditional control theory by including adolescent delinquent peer groups as an additional element of social control, along with families and institutions. This model, based in symbolic interactionism, argues that different groups (specifically adolescent peers vs. adult and institution based groups) exert control to act in *either* conventional or non-conventional ways. Specifically, Matsueda (1992) and Heimer & Matsueda (1994) argue that adolescent peer groups are a potential source of motivation for delinquent behavior through serving as generalized others who would reward or otherwise positively evaluate delinquent behavior. Thus groups are argued to provide controls for both conventional and delinquent behavior depending on the values of the groups and the extent to which those groups serve as generalized others.

While Matsueda (1992) and Heimer & Matsueda (1994)'s models are at the individual level, Felson et al. (1994) present a similar social control model, incorporating culture at a 'community' level. They examine the contextual effects of school-level attitudes toward aggression on interpersonal violence and delinquency, holding constant individual-level attitudes. Their study assumes that there is a singular pervasive attitude within schools that varies across schools, rather than different reference groups within schools. Their findings show that school-level attitudes significantly affect both interpersonal violence and delinquency, regardless of the individual's own values. They argue that rather than these school-level attitudes having a socialization effect, as traditional subculture of violence theories suggest, the effects are due to social control processes. That is, students perceive the general attitudes within their schools and then act to avoid punitive measures, rather than acting necessarily in accordance with their own beliefs or values regarding violence.

Both the Felson et al. (1994) and Heimer & Matsueda (1994) studies provide reason to speculate about the role of culture in community level social control models, yet some of their findings raise questions about the role of delinquent values. In our opinion, neither study provides strong evidence of specific *group* values. While there may indeed be *individual* variation in values, this is not equivalent to an actual group with shared values that are in contrast to conventional values. Felson et al. measure only individual level values and take the average of individual values to create the school value. They have no measure of students' perceived values within the school. Thus, they provide only a compositional rather than a contextual fact. Heimer & Matsueda (1994), on the other hand, provide only an individual level analysis, but do provide measures of perceptions of relevant others. While their findings show that having more delinquent friends increases the expectation of friends approval of delinquent behavior, friends perceived approval of delinquent behavior *does not* affect delinquent behavior. Further, perceived parental approval of delinquent behavior significantly *decreases* delinquent behavior. Thus it appears that families do not provide delinquent values and it is unclear as to whether peer reference groups actually provide motivation for delinquent behavior through expected appraisals of approval or disapproval.

While we remain skeptical about the evidence regarding differential value systems within or between communities as sources of motivation for crime, we do suggest that cultural variation between communities may exist in terms of reinforcement of community values. This issue, while clearly in line with traditional social control theory, has been surprisingly ignored in the social control literature. We draw on some of the literature from the study of the urban underclass to develop our theoretical basis of cultural attenuation.

GHETTO-RELATED BEHAVIORS AND CULTURAL ATTENUATION

Wilson (1987, 1996) argues that structural changes in inner city neighborhoods, such as the disappearance of manufacturing jobs, and out-migration of middle class minorities, have led to poor minority neighborhoods being socially isolated from middle class resources, value reinforcements, and role models. In these communities few residents are able to enact the middle class values in which they believe. Due to the impoverishment of these communities, behaviors not in line with middle class values appear, making obvious to other community members the diminished likelihood of acting out those values. Wilson (1996) argues that the presence of these 'ghetto-related' behaviors leads to the transmission of these behaviors by precept and a weakening of the demand to fully embrace conventional values. Wilson (1996) suggests that value differences across communities do not exist, only differential constraints on reaching those values, leading to more residents involved in behaviors in opposition to conventional values, and a transmission of those behaviors by precept.

This inability to enact conventional values does not necessarily reflect a commitment to oppositional or deviant values. Indeed, there is evidence that residents of even the most disadvantaged neighborhood articulate middle class values. For example, Wilson (1996: 67), citing data from the Urban Poverty and Family Life Study, states that "fewer than 3% of the black respondents from ghetto poverty census tracts denied the importance of plain hard work for getting ahead in society and 66% expressed the view that it is very important." Yet, given the constraints of every day life in these neighborhoods, many residents cannot and do not carry out the values that they articulate. As Wilson states (1996: 69) "They may strongly agree with mainstream judgments of unacceptable behavior and yet feel utterly constrained by their circumstances, forced sometimes to act in ways that violate mainstream norms." While 'ghetto-related' behaviors may, therefore, not reflect or affect internalized values, Wilson argues, they may nonetheless be cultural. "The more often certain behavior such as the pursuit of illegal income is manifested in a community, the greater will be the readiness on the part of some residents of the community to find that behavior 'not only convenient but also morally appropriate'" (p. 70). To the extent that non-normative behaviors are "manifested and tolerated in the overt behavior of their neighbors" (p. 70) community norms prohibiting those behaviors become weakened.

Ethnographic studies of disadvantage-related behavioral adaptations suggest the same ideas. For example, while in places Anderson (1991) seems to argue that high pregnancy rates among young women are due to cultural values encouraging pregnancy, in other places he clearly suggests that there is a strong attachment to middle class values, but in a context of limited opportunities and limited role models young women may nonetheless surrender to early pregnancy. It is clear in some of his statements, that indeed, the articulated values are middle class values – e.g. "Many of these young women have a dream of the good life: they want to be married to a good man ('someone who will treat me right') and settle down in a nice neighborhood" (p. 390). Despite the evidence of an adherence to middle-class values, the behaviors and attitudes accepting of the behaviors develop nonetheless, as other avenues diminish.

> Because a girl raised in the street culture may lack the outlook that would allow her to pursue options other than pregnancy, she often lacks the ability fully to appraise its consequences. She is so downtrodden and isolated from the wider society and its values that she may feel she is doing well by becoming pregnant and receiving welfare. Girls who have become single parents before her are often her role models.... These girls often have their independence, a welfare check, and food stamps (p. 389).

We follow the suggestion that mainstream values are pervasive, yet are sometimes unable to be manifested through conventional behaviors in some areas due to structural constraints. In communities where the failure to enact conventional values is common and easily observable to residents, the community level demand for conventional behaviors is diminished, thereby reducing the willingness to intervene in opposing this behavior or even more serious behaviors, and consequently decreasing the level of informal social control of crime. After all, even if residents view behaviors as inappropriate, if they also view few realistic alternatives, their expectations for appropriate behavior may be tempered by the reality of opportunities that exist in their community, and therefore they may be less likely to intervene. It is simply less likely that people will informally punish or denigrate people for inappropriate behavior if it is unrealistic to expect otherwise. Thus, we argue that communities are not different in the *values* they hold regarding what is right and wrong, but rather they are different in terms of the norms regarding the appropriateness of enforcing those values. Sampson & Wilson (1995: 50) also make this point. They state, ". . . in structurally disorganized slum communities it appears that a system of values emerges in which crime, disorder, and drug use *are less than fervently condemned* and hence expected as part of everyday life" (emphasis added).

Therefore, the perspective of attenuated culture falls within the traditional social control model, by assuming a *conventional* normative consensus and an absence of real cultural motivation toward crime, yet, at the same time, recognizing that the presence of behaviors contradicting conventional values varies across neighborhoods, *regardless of the level of social ties*. Extending these ideas further, we also argue that the presence of these behaviors then diminishes willingness of neighbors to intervene in more serious community behaviors.

Expanding the present community level crime model then, we suggest that informal social control of criminal behavior within a community is related not only to levels of social ties, but also relevant are community levels of non-conventional behavior that arise directly from community structural conditions and affect community level norms regarding the appropriateness of intervening behaviors. In communities where limited opportunities make behaviors in line with conventional goals difficult, the community-level demand (in terms of intervening behaviors) for compliance with conventional goals decreases. For example, in communities where jobs are scarce, dropping out of school and not having a job becomes more common. Likewise, in communities where there are few marriageable men, early pregnancies outside of marriage also become common. The commonality of these behaviors within the community reminds community residents of the diminished likelihood of community members enacting middle class norms and thereby undermines the general willingness of neighborhood residents to intervene in or punish inappropriate behaviors, including criminal behavior.

THE PRESENT STUDY

In the present study we extend existing community-and-crime research by presenting a community-level social control model of crime rates that includes measures relevant to informal social control, in the form of social ties *and* behavioral antecedents of attenuated culture. The theoretical statements and deductions comprising this perspective are as follows:

(1) Community structural conditions, particularly poverty, affect the level of behaviors not consistent with conventional values ('ghetto-related behaviors').
(2) The proportion of residents engaging in these non-conventional behaviors is not related to the level of social ties. Thus these behaviors *do not arise out of a lack of a foundation for intervention* (i.e. a lack of social ties).
(3) As the proportion of residents engaging in 'ghetto-related' behaviors increases, the norms regarding intervening in other community members behaviors decreases. (This is what we refer to as cultural attenuation.)
(4) As the norms regarding the appropriateness of intervening decrease, rates of criminal behaviors increase.

From 3 and 4 above we can deduce,

(5) As the proportion of residents engaging in 'ghetto-related' behaviors increases, rates of criminal behavior increase.

While we cannot empirically test all aspects of this model, we focus on the examination of statements 1, 2 and 5. To the extent that we are able to empirically support these propositions, the theoretical model we are suggesting is strengthened.

In this study we examine two 'ghetto-related' behaviors in relation to community structure, social ties, and violent crime rates: young single motherhood and youth idleness. Early childbearing has been discussed in the urban underclass literature as an example of behavior emerging from situations of disadvantage (Crane, 1991; Jencks, 1991; Anderson, 1991). Middle class values clearly emphasize postponing child bearing until after marriage. For young women especially, having a child outside of marriage presents a barrier to other accomplishments valued by society, such as completing one's education or finding a suitable marriage partner. Therefore, in communities with a strong middle class culture, girls growing up face strong statements from neighbors, friends, and family members about the importance of resisting adolescent pressures promoting sexual activity. To the extent that girls do not receive or observe others receiving these strong statements from adult authority figures in the community, however, they may succumb to the 'normality' of what they see around them and become sexually active and pregnant. As Anderson (1991: 393) states, "The wish for many is to go to college or land a job downtown. But sooner or later they make do with what they have at their disposal. In effect, they may settle for babies because there is 'nothing else to do.'"

The other behavior that we examine is youth idleness – in particular, joblessness among high-school-aged teens who have neither completed nor are enrolled in school. Middle class values emphasize the importance of education and work, and stress that young people should either be in school or working. While unemployment of young adults may be partially related to the unavailability of jobs, it is the fact that these young adults are not employed *and* not in school that we believe makes this an appropriate measure. In neighborhoods where the profits from education or employment are difficult for people to see, dropping out or not working may be accepted more easily, and statements against such behaviors may not be as strongly articulated among community members. Thus, higher levels of youth idleness would lead to less generalized intervention, or greater cultural attenuation.

Sample and Data Sources

This study uses data from 100 census tracts in Seattle. Census tracts are relatively small in size and often have homogenous populations, thus providing reasonable approximations for 'neighborhoods'. They are also a standard geographic unit for which much of the data needed for this study are available. In particular, the present study required data on ecological characteristics of neighborhoods, crime rates, social ties, and culturally relevant behavior. The ecological data and measures of culturally relevant behaviors come from 1990 U.S. Census data. Crime data are from 1989 and 1990 Seattle Police Department annual reports.

The measure of local social ties, on the other hand, was constructed by aggregating individual-level responses from a 1990 survey of 5,302 Seattle residents. The survey respondents were clustered within 300 city block pairs and 100 of the city's 121 census tracts. While the 100 sampled census tracts represent a random sample of the city's 114 stable census tracts (those tracts which had not changed boundaries for several decades), the three block-pairs sampled within each tract were non-randomly selected. One block in each pair contained a street address at which a burglary had been reported during the year preceding the survey (see Miethe & Meier, 1994; or Miethe & McDowall, 1993 for further discussion of the sample). A sample of housing units was then drawn from residents of the three city-block pairs within each tract, with a reverse telephone directory serving as the sampling frame. Within each of the 100 census tracts, approximately 50 adult respondents were interviewed.

Measures of Variables

As much of the public and political concern about crime focuses on violence in particular, we examine the effects of neighborhood ecological characteristics, social ties, and culturally relevant behavior on violent crime rates. Our

Table 1. Descriptive Statistics

Variables	Mean	S.D.	Min.	Max.
Dependent Variable				
Violent Crime Rate	27.10	33.95	0.00	140.30
Independent Variables				
Poverty	0.13	0.10	0.02	0.57
Proportion Black	0.10	0.16	0.00	0.75
Residential Stability	0.44	0.12	0.15	0.66
Downtown	0.04	0.20	0.00	1.00
Local Social Ties	0.58	0.09	0.33	0.77
Young Single Motherhood	0.11	0.16	0.00	1.05
Youth Idleness	0.08	0.10	0.00	0.49

[a] Descriptive statistics based upon N = 100

measure of violent crime is based upon aggravated assaults and robberies reported to police in Seattle for the years 1989 and 1990.[4] The counts for assault and robbery in both years were summed within each tract, divided by the total number of persons in the tract, and multiplied by 1000 in order to obtain a tract-level *two-year violent crime rate* (per 1000 persons). As the descriptive statistics provided in Table 1 suggest, violent crime varies tremendously across Seattle's neighborhoods. Furthermore, its distribution is heavily skewed. Consequently, the natural log of the violent crime rate is used as the dependent variable for analysis purposes. Because some neighborhoods had zero violent crimes, we added a constant (1) before logging.

The independent variables used in this study include neighborhood ecological characteristics, social ties, young single motherhood and youth idleness. More specifically, we focus on the exogenous ecological variables of poverty, proportion black population, and residential stability. The data for these three measures come from the 1990 U.S. Census STF-3 data tape. *Poverty* is measured as the proportion of the population in each Seattle census tract living below poverty according to the 1990 U.S. Census. *Proportion Black* is measured by the proportion of residents in each tract who reported their racial identity as black on the 1990 U.S. Census. While some studies estimating community and crime models use measures of racial or ethnic heterogeneity rather than measures of minority population size, we focus on the latter measure here since most of the theoretical work on attenuated culture suggests that 'ghetto-related' behaviors are most likely to arise in poor, minority neighborhoods (e.g. Wilson, 1996). We focus on the size of the black population in particular because blacks constitute the largest minority group living in U.S. ghettos and are thus most susceptible to the problems examined here (Jargowsky, 1997; Wilson, 1996). Further, since the data are from Seattle, a measurement of the proportion minority (rather than the proportion black) would be affected by a significant Asian population – a minority population *not* specifically addressed in the underclass literature. Finally, we measure *residential stability* using the 1990 Census-reported proportion of residents within each tract who had lived in the same house for at least five years.

Since the underclass literature suggests that the effects of poverty are greatest in minority neighborhoods – and, particularly, in African-American communities – we also include in our models an interaction between poverty and black population (*poverty* proportion black*). An indicator of local social ties is included in the analyses as a mediating variable. We use three questions from the Seattle survey that appear to address the extent of neighboring or *social ties* among residents. These questions concern whether respondents: (1) had borrowed tools or food from neighbors, (2) had lunch or dinner with neighbors, and (3) had helped neighbors with problems. Each of these questions was originally coded as a 'yes' (1) or 'no' (0) response. For each question, the number of residents responding 'yes' were summed and divided by the total number of survey respondents for each tract, producing the proportion of residents engaging in each individual neighboring activity per tract. These three aggregate measures were, in turn, summed and divided by three in order to compute a tract-level average proportion of persons engaging in activities suggestive of local social ties.

As discussed earlier, we examine two disadvantage-situated behavioral adaptations. Both measures were constructed from tract-level data from the 1990 U.S. Census STF-3 tape. Our first behavior, *young single motherhood*, is measured by constructing a birth rate for young unmarried women. Specifically, we divide the number of children born to unmarried women age 15–24 by the number of unmarried women 15–24. Our second culturally relevant behavior, *youth idleness*, is measured as the proportion of 16 to 19 year olds who are not in school and who are either unemployed or have not looked for work in the last four weeks.

It should be noted that we also control for census tracts within *Downtown* Seattle in the models presented herein. The small number and unique character of Downtown residents introduced some concern that such a population might affect results. Therefore, following the lead of previous studies examining Seattle census tracts (Crutchfield, 1989), we created a dummy variable indicating whether or not each tract was located Downtown. Our sample contained four census tracts within Downtown.

RESULTS

We begin our analysis by examining the extent to which our measures of social ties and 'ghetto-related' behavior are predicted by community structural characteristics. The systemic model of social disorganization theory suggests that social ties are most strongly affected by residential stability and racial composition.[5] In particular, we would hypothesize a positive relationship between residential stability and social ties but a negative relationship between proportion black and social ties. The systemic model does not make strong predictions regarding the effect of poverty on social ties. On the other hand, drawing upon the urban underclass literature, we would hypothesize a positive relationship between poverty and our measures of 'ghetto-related' behavior. This part of the model would also hypothesize an effect of racial isolation in conjunction with poverty. That is, these behaviors are predicted to be most prevalent in the most socially isolated communities, with the effects of poverty increasing with racial isolation, or alternatively, the effects of percent black occurring only in the context of high poverty. In contrast, little theoretical attention has been given to the relationship between residential stability and 'ghetto-related' behaviors. So, while we certainly expect stability to have a positive effect on social ties (as posited by the systemic model), we have no theoretical reason to hypothesize a strong relationship between stability and the behaviors considered here. Further, we expect that social ties will not be related to these behavior, i.e. we expect that these behaviors arise directly from poverty and the interaction between poverty and percent black, and not from a lack of social ties. Results from these models examining the effects of structural conditions on social ties and 'ghetto-related' behaviors are presented in Table 2.

As predicted, residential stability is positively and significantly related to social ties. Stable neighborhoods have higher levels of social ties. The effect of the proportion black on social ties is non-significant. While the literature addressing social ties focuses more on racial heterogeneity as a block to social ties, proportion black and heterogeneity are highly correlated in these data ($r = 0.74$). Poverty, both alone and in interaction with proportion black, is also non-significant in relation to social ties.

We next examine the effect of community structural characteristics on measures of youth idleness and young single motherhood. As expected, poverty is positively related to both. Young, single motherhood is positively affected by poverty, but is not affected by proportion black or the interaction of these two variables. Youth idleness is positively affected by both poverty and proportion black. The significant interaction term suggests that the effect of proportion

Table 2. OLS coefficients (and standard errors) for the effects of community ecological characteristics on social ties and ghetto-related behaviors

	SOCIAL TIES	YOUNG SINGLE MOTHERHOOD	YOUTH IDLENESS
Poverty	–0.19 (0.13)	0.84** (0.21)	0.21 (0.14)
Proportion Black	–0.09 (0.72)	0.10 (0.11)	0.11⁺ (0.07)
Residential Stability	0.35** (0.09)	0.45** (0.16)	–0.03 (0.10)
Poverty* Proportion Black	–0.04 (0.72)	1.62 (1.23)	1.66* (0.76)
Downtown	0.03 (0.04)	0.00 (0.07)	–0.08⁺ (0.04)
Social Ties	–	–0.19 (0.18)	–0.10 (0.11)
Constant	0.57	0.00	0.14
R^2	0.38	0.36	0.35

⁺ $p<0.10$; * $p<0.05$; ** $p<0.01$

black on youth idleness is greatest as the percentage of residents in the neighborhood living in poverty increases. For example when poverty is low (one standard deviation below the mean), the effect of proportion black on youth idleness is non-significant (s.m.e. = –0.05, s.e. = 0.12, t = –0.375). However, when poverty is high (one standard deviation above the mean) the simple main effect (s.m.e.) of proportion black is positive and significant (s.m.e. = 0.28, s.e. = 0.08, t = 3.5). Importantly, neither idle youth or young single motherhood is found to be significantly related to social ties. Finally, residential stability positively and significantly affects the pervasiveness of young out-of-wedlock births, but it has no effect on youth idleness.

Having examined the effects of community structure on the mediating variables, we now turn to an examination of the violent crime models. These results are presented in Table 3. In the first model we present the effects of just the structural variables on the violent crime rate. This model shows that poverty, percent black and being downtown significantly increase violent crime rates, while stability decreases them. The interaction term for percent black and poverty, while not quite significant ($p = 0.055$), suggests that the effect of percent black on violent crime decreases as poverty increases.

Table 3. OLS coefficients (and standard errors) for the effects of exogenous ecological variables, social ties, and ghetto-related behaviors on violent crime rates

	Violent Crime 1	Violent Crime 2	Violent Crime 3
Poverty	2.97**	2.54**	1.40
	(0.89)	(0.86)	(0.86)
Proportion Black	4.16**	3.97**	3.66**
	(0.45)	(0.44)	(0.41)
Residential Stability	–2.54**	–1.78**	–2.12**
	(0.60)	(0.62)	(0.60)
Poverty* Proportion Black	–9.81^{+}	–9.89*	–14.50**
	(5.05)	(4.82)	(4.61)
Downtown	1.39**	1.45**	1.60**
	(0.30)	(0.28)	(0.27)
Social Ties	–	–2.21**	–1.86**
		(0.69)	(0.65)
Young Single Motherhood	–	–	0.88*
			(0.38)
Youth Idleness	–	–	1.90**
			(0.62)
Constant	3.88	4.81	4.53
R^2	0.78	0.80	0.83

$^{+}$ $p<0.10$; * $p<0.05$; ** $p<0.01$

In the second column of table 3 we examine a typical systemic model. That is, we examine the effects of community structural variables along with a measure of social ties. This model shows that social ties decrease violence rates as expected, but social ties do little in terms of mediating the effects of the exogenous variables. Although the effects of poverty, proportion black, and stability are all somewhat smaller they remain statistically significant with social ties included in the model.

In the third model we add our measures of 'ghetto-related' behavior: the number of births to young unmarried women and youth idleness. As can be seen in this model, social ties, young single motherhood, and youth idleness all significantly affect informal social control as demonstrated by their significant effects on violent crime rates. Young, single motherhood and youth idleness positively and significantly increase vioence rates while social ties maintain a significant negative effect. Further, in this model, the main (i.e. non-conditional effects) effects of poverty on violence are mediated by the behavioral measures included in this model, although the conditional effects of poverty are increased. Significant direct effects of stability and percent black remain, suggesting that neither social ties nor ghetto-related behaviors do a good job of mediating the effects of these variables.

Nonetheless, there is some evidence that these non-conventional behaviors are operating in a manner consistent with the expectations of an underclass model. Specifically, we further examine the effect of the poverty percent black interaction term across the last two models presented in Table 3. For example, Table 4 shows that while the model which includes only social ties as a measure of informal social control (VIOLENT2) shows some differences in the effects of percent black in low poverty, poverty and ghetto neighborhoods (4.96 vs. 2.98 vs. 1.99), the model that includes young single motherhood and youth idleness (VIOLENT3) demonstrates that the effect of percent black on violent crime is substantially diminished in poverty communities and *completely* accounted for in ghetto neighborhoods (e.g. 5.11 vs. 2.21 vs. 0.76). This suggests that a good proportion of the effects of proportion black on violence are being accounted for by the 'ghetto-related' practices measured here.

In summary, our results suggest that both social ties and ghetto-related behaviors are affected by community structural conditions. While these measures are not entirely effective in mediating the effects of structural conditions, social ties among neighbors do appear to increase social control

Table 4. Conditional Effects of Proportion Black on Violence

	LOW POVERTY[a]	POVERTY[b]	GHETTO POVERTY[c]
Violent Crime 2			
Coefficient	4.96	2.98	1.99
standard error	(0.76)	(0.51)	(0.89)
t-ratio	6.48	5.81	2.22
Violent Crime 3			
Coefficient	5.11	2.21	0.76
standard error	(0.71)	(0.51)	(0.88)
t-ratio	7.20	4.33	0.86

[a] Defined as 1 standard deviation below the sample mean
[b] Defined as 1 standard deviation above the sample mean
[c] Defined as 2 standard deviations above the sample mean

while higher rates of behaviors inconsistent with conventional values diminish social control of crime.

CONCLUSIONS AND DISCUSSION

In this study we provide a theoretical framework for incorporating ideas of diminished value re-enforcement, or cultural attenuation, into a community-level model of informal social control. Further, we provide a partial empirical test of the propositions deduced from this framework. Culture refers to the value system, rules (norms) and moral evaluations shared by a society. In contrast to some recent adaptations of social control theory (e.g. Felson et al., 1994; Heimer & Matsueda, 1994), the framework presented here does not view oppositional values as motivational forces for criminal behavior, rather, it suggests that cultural norms regarding interventions in non-conventional behavior become attenuated with the increasing presence of certain behaviors not in line with conventional values.

Although recent theoretical and ethnographic works have talked about attenuated culture, discussions, until now, have not been presented in a clearly testable model. In focusing on two 'ghetto-related' behaviors posited to lead to cultural attenuation – young single motherhood and youth idleness – we find support for the idea that these behaviors (and the consequential weakening of norms regarding intervening in inappropriate behavior) stem from structural conditions and, in turn, allow violent behaviors to flourish.

Consider first the effects of exogenous community characteristics on the 'ghetto-related' behaviors examined here. These findings present some empirical support for work of scholars who suggest that non-conventional behaviors arise due to structural conditions found in poor minority neighborhoods. While only residential stability has a significant effect on the formation of social ties, poverty and proportion black increase youth idleness. Furthermore, the significant interaction between poverty and black population in estimating youth idleness suggests that the effects of poverty on youth idleness are exacerbated in black communities. This finding is consistent with the arguments of Wilson (1996: 66–67), who suggests that the social isolation of many poor blacks "contributes to the formation and crystallization of ghetto-related cultural behaviors," including ". . . an overt emphasis on sexuality, idleness, and public drinking."

Poverty was also found to increase out-of-wedlock childbearing among young women, as was residential stability. The effect of stability on youthful motherhood is certainly plausible in that low rates of mobility could allow for any particular manifestation of behaviors to 'take hold'. Nonetheless, while the effect is plausible it was unanticipated since little theoretical attention has been paid previously to the effects of stability/mobility on ghetto-related behaviors. Instead, most of the recent focus has been on understanding why ghetto-related behaviors prevail in underclass neighborhoods (Wilson 1987, 1996). Our findings regarding the structural predictors of young, single motherhood suggest, however, that this behavior is not tied to neighborhood racial composition. Despite the fact that the pattern of structural effects is not entirely consistent when considering young single motherhood versus youth idleness, it is nevertheless clear that structural conditions, in general, affect both behaviors.

In contrast, social ties do *not* affect either behavior, indicating that these behaviors do not arise from a state of weak social networks (see Table 2). This is important because it makes it difficult for one to argue that our measures of culturally relevant behavior simply result from rather than lead to diminished control. While these behaviors do not stem from loose networks, both young single motherhood and youth idleness are related to community violence rates. Although we are not able to directly test the hypothesized processes through which they work, we argue that high rates of these behaviors in communities lead to a weakening of norms regarding the appropriateness of intervention that are important in controlling criminal behavior. Thus, our models imply that inadequate social control may not only be preceded by weak social ties among neighbors but perhaps also by weakened demands for normative behavior. With the restraints of a strongly tied community that successfully *reinforces* middle-class values broken down by the presence of behaviors that make obvious the high rate of failure to achieve conventional values, communities are no longer able to control the more serious problems of violent crime, that ultimately also arise from the extreme disadvantage in these neighborhoods.

Though the 'ghetto-related' behaviors examined here and violent crime do ultimately stem from structural disadvantage, we believe there are important differences between these various behaviors that make the causal process outlined here feasible. It is our contention that violent crime represents more serious and more public norm-violating behavior than do young, single motherhood and youth idleness, which are essentially private behaviors. Nonetheless, the continual presence of more minor, private norm violations can reinforce a view of community in which few will achieve conventional values, and thus lead to an overall diminished requirement for intervention in inappropriate public behavior.

In addition to our findings regarding the effects of 'ghetto-related' behaviors on violence, our findings that these behaviors completely account for the effect of proportion black on crime at very high levels of poverty are important.

Specifically, in ghetto neighborhoods (proportion in poverty at 33% or more), the effect of proportion black population was non-significant when controlling for 'ghetto-related' behaviors, suggesting that it is the presence of these behaviors and not racial composition that is most relevant in understanding ghetto violence. In contrast, proportion black continues to have a significant positive effect on violence in non-ghetto communities, even when accounting for social ties, young teen motherhood, and youth idleness. The reasons for such a continued effect are not directly addressed in this study. However, there are several possible explanations. For instance, Patillo (1998), points to the spatial proximity of middle-class black neighborhoods to higher-crime, poorer neighborhoods and the inter-relationship between licit and illicit networks that sometimes exist in these communities. The implications from her study are twofold. First, ghetto-related behaviors may not be present in middle-class black neighborhoods but due to spatial proximity to high-crime neighborhoods, they themselves may experience higher levels of crime. Related to the spatial proximity issue, strong supervisory networks that manage crime in such areas are sometimes hampered by the entrenchment of criminal organizations.

While our findings open the door for further development of a macro social control model of crime, many issues remain. First, despite the fact that our measures of 'ghetto-related' behavior are related to violence rates, they do *not* do a good job of mediating the effects of proportion black (except at very high rates of poverty) or residential stability. In fact, both exogenous variables continue to have significant direct effects even after measures of both social ties and culturally relevant behaviors are added to the model. The significant direct effects of these exogenous variables on violence in the presence of social ties and ghetto-related behaviors are somewhat surprising, as community-level control theory suggests that these effects should be indirect, through indicators of informal social control. These findings suggest that perhaps other sources of control, such as external community ties and ties among criminal groups, need to be included in a comprehensive macro social control model (see e.g. Bursik & Grasmick's (1993) discussion of the role of public control, and Pattillo's (1998) discussion of ties among licit and illicit groups).

Second, we have been consistent with traditional social control models in maintaining that divergent value systems are not motivations for criminal behavior. However, new extensions of social control models, such as Matsueda's (1992; Heimer & Matsueda, 1994) differential social control theory, which suggest that ties to groups provide motivation to both conventional and non-conventional behaviors would require researchers to examine more complex measures of social ties than those examined here. For example, if there are extensive ties within a community, but those ties are among community members with deviant values, then such ties would not be predicted to decrease rates of crime. A model which measures the values of groups within neighborhoods would be necessary to determine whether social ties also exert motivations for criminal behavior, thereby diminishing the appropriateness of an assumption of value consensus.

Finally, we realize that our test of the model proposed here is limited. We posit that young, single motherhood and youth idleness are behaviors that go to the heart of a community's self-identity. In that these behaviors are clearly not in line with the conventional value system, their presence reveals the level of failure and disheartenment with which many community residents live on a daily basis. This constant reminder of the improbable odds of most community residents successfully living out conventional standards weakens the intervening behaviors that may otherwise naturally arise from social ties within the community. While we believe the logic of our argument is both sound and based on previous literature, we recognize that perceptions of appropriateness of intervening are not measured. Without more explicit measures of the implied process, we can not speak definitively to the validity of the framework proposed here.

Nevertheless, our framework and results do fit with recent findings by Sampson et al. (1997) who measure levels of non-intervention as part of a concept referred to as collective efficacy. Collective efficacy is "defined as social cohesion among neighbors combined with their willingness to intervene on behalf of the common good" (p. 918). Sampson et al.'s (1997) collective efficacy model is based on a perception of neighborhood *incapacity* to intervene due to a sense of powerlessness and alienation (p. 919). While their interpretation is somewhat different, their measure of willingness to intervene (combined with a measure of social cohesion within the neighborhood) is negatively related to violent crime, suggesting that intervention levels in communities do play an important role in crime control. In addition, collective efficacy is strongly (negatively) affected by concentrated disadvantage. Sampson et al. (1997) suggest that concentrated disadvantage results in a collective sense of powerlessness and alienation felt by residents of these neighborhoods. Sampson et al. do not test this aspect of their model. In summary while Sampson et al.'s findings can be viewed as consistent with our model, our model focuses on *cultural norms* of intervention, while theirs focuses on *structural capacity* for intervention. Both of these issues are undoubtedly important in explaining variations in social control and we would encourage other community and crime scholars to further develop an understanding of intervention through the collection of survey data allowing for more explicit measures of frequency of articulation of mainstream values, appropriateness of intervention, and intervening behaviors. Until such data are available, however, we view this study as an important step in theoretically developing a community-and-crime model incorporating subtle cultural cues relevant to norms of intervention.

NOTES

1. For example, Elliot et al. (1996) find that their measures of social integration and informal networks do not affect problem behavior in Chicago, and social integration does not affect problem behavior in Denver. Sampson et al. (1997) also find that social ties in terms of friendship networks account for only 25% of the variance in their measure of social control which they refer to as collective efficacy.

2. There is a closely related body of literature on urban crime rates that uses entire cities or MSAs as units of analysis, that is also sometimes discussed as community and crime research. This literature however focuses on much broader social factors such as de-industrialization, rather than the *neighborhood* level factors that have been the mainstay of community social control theories.

3. While we realize that the systemic model also discusses organizational participation and supervisory or surveillance behavior, we view local social ties in terms of neighboring as the most seminal aspect of the systemic model. For example, supervisory or surveillance behavior depends upon the establishment of local friendship ties to articulate what the shared values are that should be enforced in supervising neighborhood activites.

4. We realize that homicide and rape are also violent index crimes. However, these crimes are such rare events at the neighborhood level that few community-level studies examine them. For example, many of Seattle's census tracts (including 62 of the 100 included in our sample) had no homicides in either 1989 or 1990.

5. The systemic model also suggests that female-headed households may be an important variable in controlling crime. We therefore included a measure of female headed households in each of our original models. However, the measure never approached significance in any model. Because of our limited sample size we chose not to include variables that never approached significance.

REFERENCES

Anderson, E. (1991). Neighborhood effects on teenage pregnancy. In: C. Jencks & P. E. Peterson (Eds), *The Urban Underclass* (pp. 375–98). Washington, D.C.: The Brookings Institution.

Anderson, E. (1994). The code of the streets. *The Atlantic Monthly, 273*, 81–94.

Bellair, P. E. (1997). Social interaction and community crime: Examining the importance of neighbor networks. *Criminology, 35*, 677–703.

Bursik, R. J. Jr. (1988). Social disorganization and theories of crime and delinquency: Problems and prospects. *Criminology, 26*, 519–551.

Bursik, R. J. Jr., & Grasmick, H. G. (1993). *Neighborhoods and Crime: The Dimensions of Effective Community Control*. New York: Lexington Books.

Crane, J. (1991). The epidemic theory of ghettos and neighborhood effects on dropping out and teenage childbearing. *American Journal of Sociology, 96*, 1226–1259.

Cohen, A. K. (1955). *Delinquent Boys: The Culture of the Gang*. New York: The Free Press.

Conklin, J. E. (1971). Criminal environment and support for the law. *Law and Society Review, 6*, 247–259.

Elliott, D. S., Wilson, W. J., Huizinga, D., Sampson, R. J., Elliott, A., & Rankin, B. (1996). The Effects of Neighborhood Disadvantage on Adolescent Development. *Journal of Research in Crime and Delinquency, 33*, 389–426.

Felson, R. B., Liska, A. E., South, S. J., & McNulty, T. L. (1994). The Subculture of Violence and Delinquency: Individual vs. School Effects. *Social Forces, 73*, 155–173.

Hannerz, U. (1969). *Soulside: Inquiries into Ghetto Culture and Community*. New York: Columbia University Press.

Heimer, K., & Matsueda, R. L. (1994). Role-taking, role commitment, and delinquency: A theory of differential social control. *American Sociological Review, 59*, 365–390.

Jargowsky, P. A. (1997). *Poverty and Place: Ghettos, Barrios, and the American City*. New York: Russell Sage Foundation.

Jencks, C. (1991). Is the American underclass growing? In: C. Jencks & P. E. Peterson (Eds), *The Urban Underclass* (pp. 28–100). Washington, D.C.: The Brookings Institution.

Krivo, L. J., & Peterson, R. D. (1996). Extremely disadvantaged neighborhoods and urban crime. *Social Forces, 75*, 619–650.

Kornhauser, R. R. (1978). *Social Sources of Delinquency*. Chicago: University of Chicago Press.

Macoby , E. E., Johnson, J. P., & Church, R. M. (1958). Community integration and the social control of juvenile delinquency. *Journal of Social Issues, 14*, 38–51.

Matsueda, R. L. (1992). Reflected appraisals, parental labeling, and delinquency: Specifying a symbolic interactionist theory. *American Journal of Sociology, 97*, 1577–1611.

McNulty, T. (1995). Disorganization and crime in contemporary urban society. Paper presented at the annual meeting of the American Society of Criminology, Boston, MA.

Miller, W. B. (1958). Lower class culture as a generating milieu of gang delinquency. *Journal of Social Issues, 14*, 5–19.

Miethe, T. D., & McDowall, D. (1993). Contextual effects in models of criminal victimization. *Social Forces, 71*, 741–759.

Miethe, T. D., & Meier, R. F. (1994). *Crime and Its Social Context*. Albany, NY: SUNY Press.

Pattillo, M. E. (1998). Sweet mothers and gangbangers: Managing crime in a black middle-class neighborhood. *Social Forces, 76*, 747–774.

Rossi, P. H., Waite, E., Bose, C., & Berk, R. (1974). The seriousness of crime: normative structure and individual differences. *American Sociological Review, 39*, 224–237.

Sampson, R. J., & Groves, W. B. (1989). Community structure and crime: Testing social disorganization theory. *American Journal of Sociology, 94*, 774–802.

Sampson, R. J., Raudenbush, W. J., & Earls, F. (1997). Neighborhoods and Violent Crime: A Multilevel Study of Collective Efficacy. *Science, 277*, 918–924.

Sampson, R. J., & Wilson, W. J. (1995). Toward a theory of race, crime and urban inequality. In: J. Hagan & R. D. Peterson (Eds), *Crime and Inequality* Stanford, CA: Stanford University Press, pp. 37–54.

Shaw, C. R., & McKay, H. D. (1969). *Juvenile Delinquency and Urban Areas* (2nd ed.). Chicago: University of Chicago Press.

Shihadeh, E. S., & Flynn, N. (1996). Segregation and crime. The effect of black social isolation on the rates of black urban violence. *Social Forces, 74*, 1325–1352.

Taylor, R. B. (1997). Social order and disorder of street blocks and neighborhoods: Ecology, microecology, and the systemic model of social disorganization. *Journal of Research in Crime and Delinquency, 34*, 113–155.

Warner, B. D., & Pierce, G. L. (1993). Reexamining social disorganization theory using calls to the police as a measure of crime. *Criminology, 31*, 493–517.

Warner, B. D., & Wilcox Rountree, P. (1997). Local social ties in a community and crime model: Questioning the systemic nature of informal social control. *Social Problems, 44*, 520–536.

Wilson, W. J. (1987). *The Truly Disadvantaged: The Inner City, the Underclass, and Public Policy*. Chicago: The University of Chicago Press.

Wilson, W. J. (1996). *When Work Disappears: The World of the New Urban Poor*. New York: Alfred Knopf.

APPENDIX: BIVARIATE CORRELATIONS

	Poverty	%Black	Interaction term	Idle Youth	Young single Motherhood	Stability	Downtown	Log Violent Crime Rate
POVERTY	1.000	0.475**	0.472**	0.436**	0.510**	–0.563**	0.325**	0.731**
%BLACK	0.475**	1.000	0.605**	0.468**	0.457**	–0.016	–0.020	0.647**
Interaction term	0.472**	0.605**	1.000	0.489**	0.477**	0.015	–0.042	0.346**
IDLE YOUTH	0.436**	0.468**	0.489**	1.000	0.408**	–0.166	–0.094	0.488**
Young Single Motherhood	0.510**	0.457**	0.477**	0.408**	1.000	–0.025	0.101	0.488**
Stability	–0.563**	–0.016	0.015	–0.166	–0.025	1.000	–0.237(*)	–0.498**
Downtown	0.325**	–0.020	–0.042	–0.094	0.101	–0.237(*)	1.000	0.397**
Log Violent Crime Rate	0.731**	0.647**	0.346**	0.488**	0.488**	–0.498**	0.397**	1.000

** Correlation is significant at the 0.01 level (2-tailed).
* Correlation is significant at the 0.05 level (2-tailed).

POLICING, CULTURE, AND FEAR OF CRIME IN THE KOREAN AMERICAN COMMUNITY

Min Sik Lee

ABSTRACT

This research examines how the quality of local policing, measured in both personal- and community-levels, and cultural (or acculturation) factors influence fear of crime among Korean Americans, drawing on Poole & Pogrebin's (1990) ethnographic study of crime and law enforcement policy in the Korean community. The data consist of 721 adult Korean Americans residing in 63 communities in metropolitan Chicago who were sampled from 33 ethnic Korean churches. Findings from hierarchical linear models indicate that Korean Americans who confide in the local police in terms of their consideration for minority residents, evaluate the routine police protection more favorably, and reside in a community which has a bigger police force size are less likely to be afraid of crime at night in their neighborhood. Such fear-reducing effects of policing, however, were not significant for fear of crime during the day. This study also found that the black visibility in a community was a very critical factor for Korean Americans' fear, both at night and during the day.

Sociology of Crime, Law and Deviance, Volume 2, pages 63–87.

ISBN: 0-7623-0680-7

INTRODUCTION

Fear of crime is widespread in the United States, and has detrimental consequences for both individuals and communities (Hale, 1996). Further, fear of crime is conceptually and empirically distinct from actual rates and severity of crime (Lindquist & Duke, 1982; Smith, 1987).

Several scholars have noted the link between policing-related factors (i.e. rate of local police force, public confidence in the police or perceived adequacy of police protection), as an explanatory (or mediating) variable, and fear of crime. Some of them found evidences to show that the quality of policing in a local area might be negatively correlated with fear of crime (Baker et al., 1983; Baumer, 1985; Box et al., 1988).

The potential effect of policing on fear of crime is an important research agenda since it might suggest practicable (and effective) policy options to tackle fear. For example, individual and community factors such as age, gender, race, and crime rates, population composition, economic and housing conditions in a local area are difficult to manipulate as part of an intervention strategy to lower fear level, although a substantial proportion of the variation in fear can typically be explained by those factors. In contrast, however, it would be less difficult to alter public perception of the quality of policing (Bennett, 1994).

Although fear of crime has been a very important research area in criminology during the last three decades (see Ferraro & LaGrange, 1987; Hale, 1996), very little is known about the distribution and etiology of fear among just-immigrated and small minority groups such as Korean Americans. This constitutes a gap in the literature on fear of crime in particular and criminology in general since Korean Americans are a unique cultural group and constitute a rapidly growing minority in the United States. Poole and Pogrebin's (1990) ethnographic study of crime and law enforcement policy in the Korean community in Aurora, Colorado is especially deserving of note in these regards. These authors suggest that the quality of policing and cultural characteristics of an immigrant minority group might be very critical in explaining reactions to crime.

This study is to investigate the correlates of fear of crime among Korean Americans,[1] and especially examines the effects of policing-related factors on it to learn more about the nature of policing as a potential source of protection against public worries, drawing on Poole & Pogrebin's (1990) works. The current study uses survey data from a sample drawn from Korean American ethnic churches in the Chicagoland area.

THE POLICING AND FEAR OF CRIME

The literature review on the connection between policing and fear of crime reveals conflicting results, with some research showing a significant association between them while others show no significant correlation.

Several surveys in the United States and Great Britain reported significant negative effects of the policing on fear, based on the multivariate analyses. Baker et al. (1983) found, from survey data collected in Phoenix, Arizona, that confidence in the police, measured in terms of public perceptions of response times to assistance calls, was an integral component of fear, especially in the context of a crime wave. Baumer (1985), using data from a national sample in the United States, found that the perceived adequacy of police protection in a neighborhood was a critical factor of fear of crime. Baumer identified increased patrols or the shift to foot patrols as a means of increasing a sense of security and reducing fear. Box et al. (1988), examining data from the second British Crime Survey (BCS), found that confidence in the police and criminal justice systems significantly decreased fear of crime. The authors also concluded that this linkage might be used as a basis for an intervention strategy. They proposed that the police might become more visible on the streets and might form alliances with the public as a fear-reducing measure (also see Maxfield, 1987).

Other survey research has shown weaker associations or no significant correlation between policing and fear. In a national survey conducted in the United States, Block (1971) found that there was no significant correlation between support for the police and fear of crime. Garofalo (1979) analyzed aggregated data for eight North American cities (selected from National Crime Survey) and found that evaluation of police performance, measured in terms of whether people thought that the local police were doing a good job, was only weakly associated with fear of crime. Krahn & Kennedy (1985) also examined aggregated data from a sample of 23 Canadian cities and found that police force size (i.e. police staffing rates) did not covary systematically with fear of crime. Recently, Bennett (1994) analyzed aggregated data drawn from surveys conducted in four residential areas in England and found that while there appeared to be a negative correlation between confidence in the police and fear at the bivariate level this effect disappeared at the multivariate level when other relevant factors were taken into account. Similarly, Hale et al. (1994), using census and crime survey data from England and Wales, reported no significant correlation between public confidence in the police and fear of crime once structural and individual correlates had been taken into account.

Several major American police departments have implemented programs or experiments (i.e. as intervention strategies) that have been aimed directly or indirectly at dampening fear of crime. An evaluation of the Cincinnati Team Policing Experiment showed that following the implementation of decentralized policing the percentage of residents reporting that they felt 'very unsafe' walking in the neighborhood at night decreased (Schwartz & Clarren, 1977). An evaluation of the Citizen Oriented Police Enforcement (COPE) project which was implemented in Baltimore, Maryland in 1982 and consisted of three stages – directed patrol, crime prevention, and problem solving – found some evidence of success in reducing fear of crime. The research found that fewer respondents assessed their fear as 'very high' following the implementation of the project and that as the project evolved, the effects became more pronounced (Cordner, 1986). A study of the impact of increased foot patrols in Flint, Michigan found that a large proportion (about 70%) of neighborhood residents reported that they felt safer during the program than they did before, though more patrols might not reduce crime itself (Trojanowicz, 1986). Studies based on increased vehicle patrols, however, have been less successful in showing reductions in fear of crime. The Kansas City Preventive Patrol Experiment, which intensified the number of routine preventive patrols by two to three times, provided evidence that police patrolling in cars had little effect on crime or citizens' feelings of safety. In fact, people did not even notice the changes in patrol level maintained during the year-long study (Kelling et al., 1974).

The final kind of research on the association between policing and fear is on the quality of a police service for victims (Rosenbaum, 1987; Skogan, 1989; Skogan & Wycoff, 1987; Winkel, 1989). The findings from this kind of research, however, are doomed to be limited in that their subjects are confined to victims. Also, setting aside this limitation, such research mostly failed to find any significant association between the quality of victim services and fear of crime. For example, in the Detroit Victims Experiment, Rosenbaum (1987) examined the effect of improving the quality of police-victim interactions shortly after victimization by special victim-focused police training. While officers who received the training had more favorable attitudes, perceptions, and behavioral intentions toward the victims of crime than officers who did not receive such training, victims' fears were largely unaffected by this police intervention.

Thus, the available research evidence on the ability of policing to control fear of crime is not unanimous. What is significant at this stage, however, is that there are a number of studies which show that under some circumstances and in some conditions the quality of policing might be effective in reducing fear.

While the precedent studies are useful for assessing the connection between the policing and fear among the general population in a multicultural society, they seldom inform about ethnic minority groups such as Korean Americans who might suffer from higher levels of fear and need better protection. In the following section, we review Poole & Pogrebin's (1990) ethnographic study of crime and law enforcement policy in the Korean community in Aurora, Colorado. The negative effect of Houston callback program upon Asian (and Hispanic) victims may yield a useful lesson about the importance of sensitivity to cultural differences in the implementation of police programs (see Skogan & Wycoff, 1987).

Policing, Culture, and Fear of Crime in the Korean Community

The police in new immigrant communities purposively maintain a low visibility due to their lack of awareness or understanding of local conduct norms and community-specific problems. The role of police as enforcers of local community standards is thus undermined. To residents of these neglected communities, the police are perceived as insensitive to their needs, and their unresponsive policies may amplify tensions and conflicts in these communities. Also, police-initiated programs that focus on community involvement – such as neighborhood watch – are often viewed with suspicion and resisted. Moreover, since the police in fact establish community priorities in deciding how and where personnel will be utilized, the police in minority communities have frequently been criticized by the residents for their lack of commitment to developing and executing community-based programs.

Poole & Pogrebin (1990) present three problems in police-community relations that are particular to the Korean American community and require critical action and adjustment on the part of law enforcement agencies. First of all, most of the Korean American communities are composed of recent immigrants who have no knowledge of American criminal law or their legal rights and obligations. Since the legal process seems inconsistent with their expectations, Korean victims tend to be dissatisfied with the police. The second major problem is communication. Korean Americans say that they are intimidated about calling the police because of the language barrier. Furthermore, even if they call the police for assistance, the complications in communicating their problem over the phone present the major obstacle to receiving an appropriate response. Finally, there are distinct cultural barriers. Many Korean immigrants distrust and avoid the police because of their negative conceptions of and experiences with the police in Korea (e.g. law enforcement in Korea is largely seen as part of a corrupt political system). Also, there are perceived risks of retaliation if offenses are reported to the police. In sum, these barriers reinforce a sense of powerlessness among Korean Americans. Korean Americans are in effect inhibited from speaking up, demanding their rights, and believing that they have a role in determining the quality of police service provided to them.

Thus, Poole & Pogrebin's works suggest that the quality of policing and cultural traits, such as English language proficiency, knowledge of the American Justice System, and victimization experiences, might be very important in predicting reactions to crime among Korean Americans. Moreover, their research suggests that we consider aspects relevant for Koreans' immigrant situation when we measure the quality of policing (e.g. confidence in the police or evaluation of the police protection) as a predictor of fear of crime.

DATA AND METHODS

The research sites of this study are Chicago City and three suburban counties – Cook, DuPage, and Lake Counties – in the Chicago metropolitan area. According to the 1990 Census of Population and Housing, as of 1989 the total population of Korean Americans dwelling in the three counties, including Chicago City, is 35,197.[2] It is very difficult to design and practice a strict probability sampling in a study of recently immigrated, small-minority groups such as Korean Americans, since recent and accurate population data for them are not typically available and the members of such groups tend to migrate frequently (see Ackah, 1992).

The present study therefore utilizes a purposive sampling, based on a key Korean American ethnic association, Korean ethnic Christian churches. Korean Americans are well known for their active participation in ethnic Christian churches. Approximately 70% of Korean immigrants regularly attend ethnic Christian churches (Hurh & Kim, 1984, 1988; Kim & Hurh, 1993; Min, 1988, 1989, 1992). In Korea, Christians constitute only about 21% of the population (Korean National Bureau of Statistics, 1987), but the majority of Korean immigrants in the U.S. have had a Christian background in their home country (Park et al., 1989, cited in Min, 1992: 1376). A 1986 survey in Chicago (Hurh & Kim, 1990) also indicates that about 53% of Korean immigrants were Christians in Korea. By contrast, Buddhists constitute a very small fraction of Korean immigrants. The 1986 Chicago survey shows that only 4.2% of the subjects were affiliated with Buddhist temples. The growth in the proportion of Christians among Korean immigrants from approximately 50% in Korea to

70% in the U.S. indicates that roughly 40% of non-Christian Korean immigrants have become Christians since their immigration (Min, 1992).

According to the 1997 directory of Korean churches in Chicago area (from the Korean Churches Federation of Greater Chicago), there are 187 Korean churches in the three counties. From these, we selected 33 relatively large churches (18%). The number of churches selected by area roughly follow the proportion of Korean American population by area.[3] However, relatively more churches were selected for Chicago City, since Chicago City contains a large number of Korean Churches with smaller memberships than those in suburban areas.

The data were collected through a questionnaire survey, and respondents could select either an English or Korean version of the questionnaire. The survey was administered from August through October, 1997. The author visited each sampled church and distributed questionnaires to adults (over 18 years) attending church services. They were returned by postage-paid mail. 3,195 questionnaires were distributed and 780 responses were collected. While this response rate of roughly 25% is low, the actual response rate may have been somewhat higher to some unknown degree. 3,195 surveys were distributed at the churches, but this does not mean that there were actually 3,195 potential respondents. In any case, this response rate (along with the fact that this is a non-probability sample) should be a caution against generalization to the general Korean American population. 38 cases were excluded from the analyses due to missing data, or because respondents resided outside research sites. 21 cases were excluded since they were the only respondent from a community. This left 721 cases for analyses, and those subjects were nested within 63 out of 199 distinct community areas (as identified by city and county government maps) in Chicago City and Cook, DuPage, and Lake Counties. The distribution of the sample roughly follows the proportions of Korean-American population by area, but suburban areas were somewhat over-represented.[4]

Measurements

Fear of Crime. The dependent variable, fear of crime, was measured in a global measure as that in National Crime Survey (NCS): "How safe from crime do you feel being out alone in your neighborhood at night?" [Paralleled question for "during the day?"] This question is similar to that used in NCS, but it was modified in two ways. First, the question made explicit reference to crime, and second, it was also stated in a non-hypothetical format by dropping the word 'would' out as Ferraro & LaGrange (1987) suggested. The four response categories remained in the same format: very safe, somewhat safe, somewhat unsafe, and very unsafe.

Independent Variables. As we reviewed above, most previous studies used a single question item to measure confidence in the police, as a proxy measure of the quality of policing. For example, confidence in the police has been measured using responses to questions concerning evaluations of police performance (whether they are doing a good job: e.g. Garofalo, 1979), public satisfaction with the speed of police response to emergent (or non-emergent) calls for assistance (e.g. Baker et al., 1983), perceived adequacy of police protection (e.g. Baumer, 1978), and public support for the police (e.g. Block, 1971). The present study, however, constructed two separate variables measuring different aspects of the policing at the individual level: confidence in the police and evaluation of the police protection. Confidence in the police measures some aspects of the policing which are relevant for unique experiences that Korean immigrants, as a minority group, have in the multicultural society, as suggested by Poole & Pogrebin (1990). This variable is composed of five question items: (1) The local police treat people of all races the same; (2) The local police try to understand unique customs of small minority groups such as Koreans; (3) The local police try to hear with kindness and patience complaints of minority residents who are poor at English; (4) The local police try to develop policing programs relevant to unique characteristics of my community; and (5) The local police are relatively fair and incorrupt. Second, evaluation of the police protection measures respondents' assessments of police routine performances, and consists of three items: (1) The local police patrol more frequently in my community than other communities; (2) The local police respond to emergency calls for assistance speedily; and (3) The local police arrest suspects speedily and certainly. Factor analyses for these items support that the two variables measure quite different concepts (results available on request). For each question item, the response categories are 'strongly agree', 'agree', 'disagree', and 'strongly disagree'.

The current study includes two proxy measures of cultural traits among Korean immigrants that were highlighted by Poole & Pogrebin's (1990) works: English language proficiency and knowledge of the American Justice System. We predict that Korean Americans who speak English well and are familiar with the American justice system would have less fear of crime. At the individual level, we also control the influences of vulnerability or status characteristics (such as age, gender, health, and income), victimization (direct and indirect), and perceptions of incivility.

Many recent studies (Donnelly, 1988; Hale et al., 1994; Maxfield, 1984; Perkins & Taylor, 1996; Rountree & Land, 1996; Taylor & Covington, 1993; Ward et al., 1986) show that fear of crime can be better explained with community (or neighborhood) contexts rather than by simply concentrating on personal factors. It seems, therefore, appropriate that studies of fear examine the effects of such objective environments while examining people's perceptions of those ecological characteristics. In this study, we investigate the influences of four community level factors: crime rates, rate of police force size, and percents of black and Hispanic populations. Such ecological factors are often the indicators of social disorganization in a local community (see Bursik & Grasmick, 1993; Greenberg & Rohe, 1986; Lewis & Salem, 1986; Sampson & Grove, 1989; Skogan, 1986; Taub et al., 1984).

Although there is considerable evidence that fear of crime does not straightforwardly reflect crime rates in an area,[5] crime rates have still been treated as a critical ecological factor of fear. At least, in part, it seems reasonable to suppose that between-communities variations in fear are influenced by variations in crime rates between those communities. Ethnic heterogeneity or visibility of racial minorities (i.e. typically the ratios of black or Hispanic population) has been examined as critical ecological factors through the literature, and several studies (Covington & Taylor, 1991; Hale et al., 1994; Liska et al., 1982; Moeller, 1989; Taylor & Covington, 1993) generally supported their positive effects on fear, but the evidence of the associations was nearly unqualified (see Chiricos et al., 1997; Lavrakas, 1982; Thompson et al., 1992; Ward et al., 1986).

We would expect that people in a community with a large police force would be less afraid of crime. Police force size is a proxy measure of the quality of policing in the community level, since it can be regarded as a measure of official guardianship of residents (see Cohen & Felson, 1979; Krahn & Kennedy, 1985). The ecological data are from the 1990 Census of Population and Housing (issued by Bureau of the Census and City of Chicago[6]) and Uniform Crime Reports (1990–1996). Appendix 1 describes the measurements of the community- and individual-level variables.

Data Analysis

The data are analyzed by Hierarchical Linear Modeling. HLM has proved to be a very effective analytic technique for data having multilevel structures. Hierarchical linear modeling allows researchers to overcome a number of conceptual and technical difficulties (e.g. aggregation bias, mis-estimated standard errors, and heterogeneity of regression) that have plagued past analyses of multilevel data (for details, see Bryk & Raudenbush, 1992). Our data exhibit a two-level hierarchical structure. People are nested within a community whose characteristics might influence their fear. Therefore, both persons (level-1) and communities (level-2) are our units of analysis.[7]

In hierarchical linear models, each level is formally represented by its own sub-model. These sub-models express the relationships among variables and residual variability within a given level, and specify how predictors at one level influence relationships at another. Thus, we can examine relationships at each level and assess the amount of variation explained at each level.[8]

Hierarchical linear modeling encompasses a variety of sub-models that enable multilevel analyses of nested data (Bryk & Raudenbush, 1992). In this study, we use three kinds of HLM sub-models: one-way ANOVA with random effects, ANCOVA with random effects, and a random-intercept model with level-1 covariates. First, the ANOVA model with random effects (i.e. fully unconditional model) is the simplest hierarchical linear model, and shows what proportion of total variation in fear of crime exists between and within communities. The estimates of variance components at each level are used as criteria in evaluating the results of subsequent models. Second, the ANCOVA model with random effects shows the effects of person-level predictors on the dependent variables. We evaluate the effects by computing the proportion of reduction in variance of the outcome variable, from the previous ANOVA model. Third, and most importantly, the random-intercept model with level-1 covariates represents our full model, and shows the influences of both community-level and person-level predictors.

The data in this study have a small average number of level-1 units per level-2 unit, and thus an unbalanced structure. Due to this data limitation, we focus on the common effects that community characteristics (such as crime rates) exert on each person's fear within the community, rather than any differentiating effects (i.e. we cannot examine cross-level interaction effects). Such community effects modify only the mean level (i.e. the level-1 intercept, β_{0j}) of fear for each community. That is, we fix the slopes of level-1 predictors, and do not allow them to vary across level-2 units. We do this because of the small average number of level-1 units (i.e. about 11 persons) per level-2 unit, and also because we have no theoretical rationale for allowing specific slopes to vary.[9]

FINDINGS

Table 1 shows descriptive statistics on both individual and community factors. The ages of respondents are distributed between 18 and 82, and mean age is 43. The gender ratio of subjects is exactly 1:1. About 27% of respondents reported

Table 1. Descriptive Statistics of Variables

Variables	Mean	SD	Minimum	Maximum
Community-Level Variables (N = 63)				
Crime rate	7249.24	11191.22	1789.12	65834.09
Police force	354.47	384.50	174.88	2394.59
% Black	9.42	19.12	0.03	93.54
% Hispanic	8.29	10.02	0.83	61.98
Person-Level Variables (N = 721)				
Fear of crime – Night	1.98	0.82	1.00	4.00
Fear of crime – Day	1.46	0.63	1.00	4.00
Age	42.54	14.71	18.00	82.00
Female	0.50	0.50	0.00	1.00
Health	2.84	0.75	1.00	4.00
Income	8.28	4.20	1.00	15.00
Victimization	0.27	0.44	0.00	1.00
Indirect victimization	0.81	0.73	0.00	2.00
Perceived incivility	12.98	4.88	9.00	27.00
English proficiency	3.20	1.04	1.00	5.00
American justice system	2.75	0.90	1.00	5.00
Confidence in the police	16.03	4.06	5.00	25.00
Evaluation of police protection	10.56	2.35	3.00	15.00

being victimized during the past three years in the United States while 60% of respondents reported victimization of their family members (or relatives) or other Korean acquaintances. About 34% of subjects evaluated their English speaking ability as 'fluent' or 'good' while about 19% of them assessed their knowledge of the American justice system as 'pretty good' or 'good'.

Table 2 shows bivariate correlations between question items measuring confidence in the police and evaluation of the police protection and fear of crime. While all the associations, except for that between the first item – "The local police treat people of all races the same" – of confidence in the police and fear during the day, are negative and highly significant, the strength of the correlations is higher for fear at night than for fear during the day. The results also indicate that the image of the police as fair and incorrupt enforcers of social norms (as an item measuring confidence in the police) and satisfaction with the speed of police assistance to emergent calls (as an item measuring evaluation of police protection) are the most important factors influencing Korean immigrants' fear of crime both at night and during the day. Appendix

2 reports bivariate correlations between fear of crime and person-level predictors.

Since the present study adopted a purposive sampling based on Korean ethnic churches, we first examined the effect of religiosity on fear of crime. The results of these analyses, however, indicate that religiosity is not a significant predictor, and so we do not control for religiosity in the final analyses. The sample is fairly homogeneous in terms of religiosity – about 93% of the respondents reported that they attend church at least once a week.

Unconditional Model

The one-way ANOVA with random effects (i.e. unconditional model) provides useful preliminary information about how much variation in fear of crime lies within and between communities. The estimates of variance components in each level work as criteria in evaluating the results of subsequent models. These models have no predictors at either level. That is, the level-1 or person-level model characterizes fear (Y_{ij}) of individual *i* in community *j* with just an intercept (β_{0j}), which is the community mean of fear, and the random effect (r_{ij}).

Table 2. Correlations Between Confidence in the Police, Evaluation of the Police Protection, and Fear of Crime

	Fear – Night	Fear – Day
Confidence in the local police:		
The local police treat people of all races the same.	–0.106**	–0.072
The local police try to understand unique customs of small minority groups such as Koreans.	–0.118**	–0.096**
The local police try to hear with kindness and patience complaints of minority residents who are poor at English.	–0.139**	–0.111**
The local police try to develop policing programs relevant to unique characteristics of my community.	–0.159**	–0.120**
The local police are relatively fair and incorrupt.	–0.189**	–0.130**
Evaluation of the local police protection:		
The local police patrol more frequently in my community than other communities.	–0.151**	–0.095*
The local police respond to emergency calls for assistance speedily.	–0.238**	–0.172**
The local police arrest suspects speedily and certainly.	–0.153**	–0.136**

At the level-2 or community-level, each community's mean fear (β_{0j}) is represented as a function of the grand mean (τ_{00}) plus a random error (u_{0j}) associated with community *j*.

Table 3 reports the results from the ANOVA models for fear of crime at night and during the day. The estimates for the within-community or level-1 variance [i.e. Var (r_{ij}) = σ^2] in the outcomes are 0.497 and 0.318 respectively. The overall variability among the community means [i.e. Var (β_{0j}) = τ_{00}] is 0.183 and 0.074 respectively. These values, along with the intraclass correlation (IC $r = \tau_{00}$ / $\tau_{00} + \sigma^2$), indicate that about 27% of the variation in fear at night and 19% of the variation in fear during the day exist between communities. In other words, the community-level variation in each fear outcome is fairly large, though most of the variance is at the individual level.

Next, we test whether the estimated values of between-community variance, τ_{00}, are significantly greater than zero (H_0: $\tau_{00} = 0$), using a large sample χ^2 distribution with J − 1 degrees of freedom. In Table 3, the test statistic for fear at night is 314, with 62 degrees of freedom ($J = 63$ communities). We thus reject the null hypothesis ($p < 0.001$); significant variation to be explained exists among community means of fear at night. The null hypothesis for fear during the day is also rejected, but the between-community variance is somewhat smaller than for fear at night. The overall reliabilities of the sample means (β_{0j}) of the outcomes in any community *j* are 0.680 and 0.591

Table 3. Unconditional Model (One-Way ANOVA[a])

Fixed Effect[b]	Coefficient	se
Between Communities		
Grand mean, γ_{00}		
Fear of crime – Night	1.978	0.065
Fear of crime – Day	1.450	0.045
Random Effect	**Fear of crime – Night**	**Fear of crime – Day**
Between Communities, u_{0j}	0.183	0.074
df/χ^2/p	62/314/.000	62/234/.000
Within Communities, r_{ij}	0.497	0.318
Intraclass Correlation	0.269	0.189
Reliability, β_{0j}	0.680	0.591

[a] ANOVA = analysis of variance.

[b] All fixed effects are tested by the t-distribution.

respectively. The overall measure of reliability is the average of the community reliabilities.[10]

Models of Fear at Night

Table 4 reports the results from models for fear of crime at night. The random-effects-ANCOVA models (model 1 and 2) show the influences of person-level predictors on fear at night. Model 1 includes only vulnerability or status characteristics (i.e. age, female, health, and income), (direct and indirect)

Table 4. Results from Models for Fear of Crime – Night[a]

Fixed Effect	Model 1	Model 2	Model 3
Between Communities			
Grand mean, γ_{00}	1.979	1.987	2.025
Crime rate, γ_{01}			0.682*
Police force, γ_{02}			–0.641*
% Black, γ_{03}			0.266***
% Hispanic, γ_{04}			0.107*
Within Communities			
Age, γ_{10}	–0.001	–0.001	0.001
Female, γ_{20}	0.241***	0.235***	0.239***
Health, γ_{30}	–0.169***	–0.153***	–0.148***
Income, γ_{40}	–0.028***	–0.027***	–0.025***
Victimization, γ_{50}	0.034	–0.003	–0.007
Indirect victimization, γ_{60}	0.079*	0.077*	0.084*
Perceived incivility, γ_{70}	0.024***	0.022***	0.019***
English proficiency, γ_{80}		–0.036	0.036
American justice system, γ_{90}		0.012	0.007
Confidence in the police, γ_{100}		–0.018*	–0.013
Evaluation of police protection, γ_{110}		–0.031*	–0.034*
Random Effect	**Model 1**	**Model 2**	**Model 3**
Between Communities, u_{0j}	0.115	0.111	0.028
df/χ^2/p	62/231/.000	62/222/.000	58/100/.001
Within Communities, r_{ij}	0.448	0.437	0.434
Proportion of Variance Explained (R^2): $R_2^2/R_1^2/R_T^2$	0.372/.099/.172	0.393/.121/.194	0.847/.127/.321

* $p \leq 0.05$; ** $p \leq 0.01$; *** $p \leq 0.001$

[a] All of the level–2 predictors were standardized (mean = 0, standard deviation = 1).
All of the level–1 predictors were centered around their grand means.

victimization, and perceived incivility as the person-level predictors. In this model, the effects of female, health, income, indirect victimization, and perceived incivility are significant in explaining fear of crime at night among Korean Americans in the sample, and the directions of associations are the same as those in the previous studies for the general U.S. population. For example, Korean Americans who are female, experience poor health, have lower levels of household income, know victimization of family members (or relatives) or Korean acquaintances, and perceive higher incivilities in their neighborhood appear to be more afraid of crime at night. These results are constant through model 2 and 3. In the first model, the explained-variance statistics are 0.099 within communities and 0.372 between communities, and these are equivalent to a 17.2% reduction from the total residual variation of fear at night. The person-level predictors contributed to explain the between-communities variance fairly. In the hierarchical analysis, the level-1 predictors can explain both individual-level and community-level variances (Bryk and Raudenbush, 1992).

In model 2, we add cultural (or acculturation) factors (English speaking proficiency and knowledge of the American Criminal Justice System) and the policing variables (confidence in the police and evaluation of the police protection). The effects of two policing factors are negatively significant in explaining fear of crime at night among this Korean sample, while two cultural variables are not. These results support that Korean Americans who confide in the police and evaluate the police protection as favorable tend to be less afraid of crime at night in their neighborhood. The quality-of-policing variables explain the total variation of fear at night about 2% additionally.

We now examine the common contextual effects of community environments on fear at night with the random-intercept models with level-1 covariates as the full model (i.e. model 3). In the last model, the individual-level model is the same as that in model 2, but the level-2 model for the community mean (i.e. β_{0j}) of fear now includes a set of community characteristics as the level-2 predictors. All the ecological factors appear to be significant in explaining the community means of fear at night. In model 3, controlling for community characteristics did not produce any remarkable changes in the effects of the individual-level predictors, except for the minor reduction in the effect of confidence in the police (after controlling for the effects of community factors, the influence of confidence in the police is not significant any more). This is because, in the random-intercept models, the level-2 predictors explain only the level-2 variance of the outcome. Model 3 shows significant positive effects of crime rates and the ratios of black and Hispanic populations and a significant negative effect of the police-force rate as a proxy measure of the quality of policing in the community level on the community means of fear at night. These results inform that, as expected, Korean Americans who reside in a community which has high crime rates and high ethnic heterogeneity or high minority visibility (measured by the ratios of black and Hispanic populations) are more likely to be afraid of crime at night in their neighborhood. In contrast, Korean Americans who live in a community which has a higher rate of police force are less likely to be afraid of crime at night.

Controlling for some ecological factors greatly enhanced the goodness-of-fit of the model. The explained-variance statistics are now 0.847 between communities and 0.321 in total. The χ^2-test of significance of the residual between-community variance indicates that there remains still significant variance, to be explained, between community means of fear at night ($p < 0.05$) even after controlling for the four community factors. This may suggest that we need to search for other community factors that are potentially useful, but the actual amount of residual level-2 variance is now very small (0.028).

Models of Fear during the Day

Table 5 reports the results from models for fear of crime during the day. In the first model without cultural and policing factors, the effects of health, income, victimization, and perceived incivility are significant in explaining fear of crime during the day (again, these results are almost constant through model 2 and 3). In the comparison with the first model for fear at night (in Table 4), the influences of female and indirect victimization are not significant for fear during the day. Victimization experience is now a significant factor, instead of indirect victimization. It is interesting that the gender effect now becomes non-significant in the models for fear of crime during the day. This change, however, is not surprising in that it is not irrelevant to the fact that the subjects' fear levels during the day are generally lower than fear at night (see Table 1). In other words, during the day the subjects are likely to perceive lower risk than at night, and there is no big difference in fear by gender during the day. The explained-variance statistics in this model are 0.500 between communities, 0.057 within communities, and 0.140 in total respectively.

In the second model, including cultural and policing variables, only English proficiency is significant in explaining fear of crime during the day among these Korean samples. Korean Americans who speak English well are significantly less afraid of crime in their neighborhoods than their counterparts. As Poole & Pogrebin (1990) pointed out, language barriers can frustrate Korean Americans when they encounter crime victimization. Frustrated by language barriers, they often give up on calling police, and often believe they

Table 5. Results from Models for Fear of Crime – Day[a]

Fixed Effect	Model 1	Model 2	Model 3
Between Communities			
Grand mean, γ_{00}	1.450	1.455	1.485
Crime rate, γ_{01}			0.302
Police force, γ_{02}			–0.247
% Black, γ_{03}			0.126**
% Hispanic, γ_{04}			0.065
Within Communities			
Age, γ_{10}	–0.001	–0.003	–0.001
Female, γ_{20}	–0.020	–0.035	–0.031
Health, γ_{30}	–0.085**	–0.063*	–0.061*
Income, γ_{40}	–0.026***	–0.022***	–0.020***
Victimization, γ_{50}	0.126*	0.110*	0.109*
Indirect victimization, γ_{60}	0.016	0.012	0.014
Perceived incivility, γ_{70}	0.022***	0.020***	0.018***
English proficiency, γ_{80}		–0.081**	–0.080**
American justice system, γ_{90}		0.012	0.009
Confidence in the police, γ_{100}		–0.010	–0.007
Evaluation of police protection, γ_{110}		–0.017	–0.018
Random Effect	Model 1	Model 2	Model 3
Between Communities, u_{0j}	0.037	0.033	0.010
df/χ^2/p	62/151/.000	62/142/.000	58/87/.008
Within Communities, r_{ij}	0.300	0.295	0.295
Proportion of Variance Explained (R^2): $R_2^2/R_1^2/R_T^2$	0.500/.057/.140	0.554/.072/.163	0.865/.072/.222

* $p \leq 0.05$; ** $p \leq 0.01$; *** $p \leq 0.001$

[a] All of the level–2 predictors were standardized (mean = 0, standard deviation = 1). All of the level–1 predictors were centered around their grand means.

will not be understood and will not receive an appropriate response or assistance. These kinds of language-related situations may amplify fear of crime among Korean Americans who speak little English. The policing-quality variables are not significant for fear of crime during the day.

In the full model for fear during the day, the influences of community factors are not so much remarkable as those were in the model for fear at night. We may find the reasons for the weaker explanatory power of ecological factors from the level-2 variation in fear during the day smaller than that in fear at night (see the intraclass correlations in Table 3). However, the effect of the rate of black population (or the black visibility) in a community is still significant in this model. This may provide a very important implication for understanding fear of crime among Korean Americans. As well symbolized by the highly publicized events surrounding the 1992 L. A. Riots (or the 1991 Roseland picketing and 1992 Bulls riots in the Chicagoland area), relations between Korean Americans and African Americans have been strained in many large U.S. cities for at least two decades (Min & Kolodny, 1994). Korean immigrants in the U.S. have had a long history of victimization by the black host. Therefore, their stereotypes and hostilities toward African Americans might be another important factor in predicting their fear of crime. The effects of racial conflicts and prejudices on Korean Americans' fear would be an interesting research agenda for the future. As Jang et al. (1991) found that Korean Americans strongly tended to perceive their black neighbors as a potential threat, a high level of black visibility in a local community may be a critical factor for Koreans' fear in the community. Finally, model 3 shows that the quality of policing measured by confidence in the police, evaluation of the police protection, and the relative size of police force is not important factor for fear of crime during the day.

CONCLUSIONS

The current study has investigated the etiology and distribution of fear of crime among Korean Americans in the Chicagoland area by focusing on the influences of local policing. The quality of policing, measured by confidence in the police and evaluation of the police protection (at the individual-level) and by the rate of police force (at the community level), generally decreased fear levels among the subjects. The results indicate that Korean Americans who confide in the local police in terms of their considerations for minority residents, evaluate the routine police protection more favorably, and reside in a community which has a bigger police force size are less likely to be afraid of crime at night in their neighborhood. Such fear reducing effects of policing, however, were not significant for fear of crime during the day.

The effects of two acculturation variables (i.e. English language proficiency and knowledge of American justice), recommended as potentially important factors by Poole and Pogrebin's (1990) research, turned out to be non-significant for fear at night, but for fear during the day the influence of English proficiency was negatively significant. This result indicates that Korean Americans who speak English well are noticeably less afraid of crime during the day in their neighborhood.

All the contextual variables were significant in explaining fear of crime at night among these Korean samples. However, the effects of these ecological factors, except for the rate of black population, were not significant for fear during the day. We may attribute part of this change to the smaller community-level variance in fear during the day. The remarkable influences of the black visibility on fear, both at night and during the day might have some special implications for Korean immigrants' fear when we consider their long history of conflicts with and victimization from their black neighborhoods.

The present study has some limitations. Most obviously, the data represent a non-probability sample, and this limits the statistical generalizability of these findings to a general Korean American population. Second, small numbers of individuals per community prevented us from making optimal use of the merits of hierarchical linear models, such as testing for cross-level interaction effects. For example, we could not permit the relationships (i.e. level-1 slopes) between individual characteristics and the outcomes to vary across communities, and could not examine whether community characteristics affect the relationships between individual characteristics and the outcomes, either amplifying or attenuating them. Future studies that did so would obviously make significant contributions to our understanding of interrelationships between the social organization of communities and individual-level factors as they influence fear of crime.

NOTES

1. In this study, Korean Americans refer to Koreans who reside in the United States (i.e. Chicago area) at that time of the survey. They may or may not be U.S. citizens.
2. About 84% (29,687 persons) of this population reside in Cook County, 10.2% (3,587 persons) in DuPage County, and 5.5% (1,923 persons) in Lake County. Also, about 47% (13,863 persons) of Korean Americans in Cook County live in Chicago City.
3. 16 churches were selected for Chicago City, 15 for suburban Cook County, and 1 for each of DuPage and Lake Counties.
4. Among 721 respondents, 35% (254 persons) live in 23 communities of Chicago City and 65% (467 persons) reside in 40 communities of the three suburban counties.
5. Many studies show that individual fear levels are not strongly linked with local crime rates. People living in higher, as compared to lower, crime areas are not proportionally more fearful (See Donnelly, 1988; Gates & Rohe, 1987; Miethe & Lee, 1984; Rohe & Burby, 1988; Taylor & Hale, 1986; Thompson & Norris, 1992).
6. The census data from Chicago City is used for statistics of the 77 official community areas in Chicago City.
7. We recognize that neighborhoods or census tracts, as smaller units, are more reasonable contexts for shaping fear of crime. In this study, however, community areas would be preferable as the level-2 units because of the small number of Korean

Americans within each census tract. Moreover, many census tracts in the research sites have no Korean Americans.

8. These analyses utilize full-information maximum likelihood methods with a combination of Empirical Bayes (Dempster et al., 1977) and Fisher scoring algorithms (Longford, 1987) for fast and stable convergence to the solution optimum.

9. In addition, actually, this study tested the possibility that the slopes of level-1 predictors might vary across communities, but there was no evidence of such variations found with the present data.

10. The formula of the reliability of the sample mean in any community j is: $\lambda_j = \text{reliability}\ (Y_{.j}) = \tau_{00}/[\tau_{00} + (\sigma^2/n_j)]$.

REFERENCES

Ackah, S. (1992). *An Empirical Investigation of the Correlates of Fear of Crime Among Immigrants in the Washington D.C. Metropolitan Area: A Case Study of Ghanaians*. Ph.D Dissertation.

Baker, M. H., Nienstedt, B. C., Everett, R. S., & McCleary, R. (1983). The Impact of a Crime Wave: Perceptions, Fear, and Confidence in the Police. *Law and Society Review, 17*, 319–335.

Baumer, T. L. (1985). Testing a General Model of Fear of Crime: Data from a National Sample. *Journal of Research in Crime and Delinquency, 22*, 239–255.

Bennett, T. (1994). Confidence in the Police as a Mediating Factor in the Fear of Crime. *International Review of Victimology, 3*, 179–194.

Block, R. L. (1971). Fear of Crime and Fear of the Police. *Social Problems, 19*, 91–101.

Box, S., Hale, C., & Andrews, G. (1988). Explaining Fear of Crime. *British Journal of Criminology, 28*, 340–356.

Bryk, A. S., & Raudenbush, S. W. (1992). *Hierarchical Linear Models: Applications and Data Analysis Methods*. Newbury Park, CA: Sage Publications.

Bursik, R. J. Jr., & Grasmick, H. G. (1993). *Neighborhoods and Crime*. New York: Lexington Books.

Chiricos, T., Hogan, M., & Gertz, M. (1997). Racial Composition of Neighborhood and Fear of Crime. *Criminology, 35*, 107–131.

Cordner, G. W. (1986). Fear of Crime and the Police: An Evaluation of a Fear-Reduction Strategy. *Journal of Police Science and Administration, 14*, 223–233.

Covington, J., & Taylor, R. B. (1991). Fear of Crime in Urban Residential Neighborhoods: Implications of Between- and Within-Neighborhood Sources for Current Models. *Sociological Quarterly, 32*, 231–249.

Dempster, A. P., Laird, N. M., & Rubin, D. B. (1977). Maximum Likelihood from Incomplete Data via the EM Algorithm. *Journal of the Royal Statistical Theory, Series B, 39*, 1–8.

Donnelly, P. G. (1988). Individual and Neighborhood Influences on Fear of Crime. *Sociological Focus, 22*, 69–85.

Ferraro, K. F., & LaGrange, R. L. (1987). The Measurement of Fear of Crime. *Sociological Inquiry, 57*, 70–101.

Garofalo, J. (1979). Victimization and the Fear of Crime. *Journal of Research in Crime and Delinquency, 16*, 80–97.

Gates, L. B., & Rohe, W. M. (1987). Fear and Reactions to Crime: A Revised Model. *Urban Affairs Quarterly, 22*, 425–453.

Greenberg, S., & Rohe, W. (1986). Informal Social Control and Crime Prevention in Modern Urban Neighborhoods. In: R. Taylor (Ed.), *Urban Neighborhoods: Research and Policy*. New York: Praeger.

Hale, C. (1996). Fear of Crime: A Review of the Literature. *International Review of Victimology, 4*, 79–150.

Hale, C., Pack, P., & Salked, J. (1994). The Structural Determinants of Fear of Crime: An Analysis Using Census and Crime Survey Data from England and Wales. *International Review of Victimology, 3*, 211–233.

Hurh, W. M., & Kim, K. C. (1984). *Korean Immigrants in America: A Structural Analysis of Ethnic Confinement and Adhesive Adaptation*. London: Associated University Presses.

Hurh, W. M. (1988). *Uprooting and Adjustment: A Sociological Study of Korean Immigrants' Mental Health*. Final Report Submitted to National Institute of Mental Health.

Hurh, W. M. (1990). Religious Participation of Korean Immigrants in the United States. *Journal of the Scientific Study of Religion, 19*, 19–34.

Jang, S. J., Messner, S. F., & South, S. J. (1991). Predictors of Interracial Homicide Victimization for Asian Americans: A Macrostructural Opportunity Perspective. *Sociological Perspectives, 34*, 1–19.

Kelling, G. L., Pate, T., Dieckman, D., & Brown, C. E. (1974). *The Kansas City Preventive Patrol Experiment: A Summary Report*. Washington D.C.: Police Foundation.

Kim, K. C., & Hurh, W. M. (1993). *Beyond Assimilation and Pluralism: Syncretic Sociocultural Adaptation of Korean Immigrants in the U.S. Ethnic and Racial Studies, 16*, 696–713.

Krahn, H., & Kennedy, L. W. (1985). Producing Personal Safety: The Effects of Crime Rates, Police Force Size, and Fear of Crime. *Criminology, 23*, 697–710.

Lavrakas, P. J. (1982). Fear of Crime and Behaviorial Restrictions in Urban and Suburban Neighborhoods. *Population and Environment, 5*, 242–264.

Lewis, D. A., & Salem, G. (1986). *Fear of Crime: Incivility and the Production of a Social Problem*. New Brunswick, NJ: Transaction Books.

Lindquist, J. H., & Duke, J. M. (1982). The Elderly Victim at Risk: Explaining the Fear-Victimization Paradox. *Criminology, 20*, 115–126.

Liska, A. E., Lawrence, J. J., & Sanchirico, A. (1982). Fear of Crime as a Social Fact. *Social Forces, 60*, 760–770.

Longford, N. T. (1987). A Fast Scoring Algorithm for Maximum Likelihood Estimation in Unbalanced Mixed Models with Nested Random Effects. *Biometrika, 74*, 817–827.

Maxfield, M. G. (1984). The Limits of Vulnerability in Explaining Fear of Crime: A Comparative Neighborhood Analysis. *Journal of Research in Crime and Delinquency, 21*, 233–250.

Maxfield, M. G. (1987). *Explaining Fear of Crime: Evidence from the 1984 British Crime Survey*. London: Home Office.

Miethe, T. D., & Lee, G. R. (1984). Fear of Crime among Older People: A Reassessment of the Predictive Power of Crime-Related Factors. *Sociological Quarterly, 25*, 397–415.

Min, P. G. (1988). *Ethnic Business Enterprise: Korean Small Business in Atlanta*. Staten Island, NY: Center for Migration Studies.

Min, P. G. (1989). *Some Positive Effects of Ethnic Business for an Immigrant Community: Koreans in Los Angeles*. Final Report Submitted to the National Science Foundation.

Min, P. G. (1992). The Structure and Social Functions of Korean Immigrant Churches in the United States. *International Migration Review, 26*, 1370–1394.

Min, P. G., & Kolodny, A. (1994). The Middleman Minority Characteristics of Korean Immigrants in the United States. *Korea Journal of Population and Development, 23*, 179–202.

Moeller, G. L. (1989). Fear of Criminal Victimization: The Effect of Neighborhood Racial Composition. *Sociological Inquiry, 59*, 208–221.

Park, I. S., Fawcett, J., Arnold, F., & Gardner, R. (1989). *Koreans Immigrating to the United States: A Pre-Departure Analysis. Paper No. 114*. Hawaii: Population Institute, East-West Center.

Perkins, D. D., & Taylor, R. B. (1996). Ecological Assessments of Community Disorder: Their Relationship to Fear of Crime and Theoretical Implications. *American Journal of Community Psychology, 24*, 63–107.

Poole, E. D., & Pogrebin, M. R. (1990). Crime and Law Enforcement Policy in the Korean American Community. *Police Studies, 13*, 57–66.

Rohe, W. M., & Burby, R. J. (1988). Fear of Crime in Public Housing. *Environment and Behavior, 20*, 700–720.

Rosenbaum, D. P. (1987). Coping with Victimization: The Effects of Police Intervention on Victim's Psychological Readjustment. *Crime and Delinquency, 33*, 502–519.

Rountree, P. W., & Land, K. C. (1996). Burglary Victimization, Perceptions of Crime Risk, and Routine Activities: A Multilevel Analysis across Seattle Neighborhoods and Census Tracks. *Journal of Research in Crime and Delinquency, 33*, 147–180.

Sampson, R., & Groves, W. B. (1989). Community Structure and Crime: Testing Social Disorganization Theory. *American Journal of Sociology, 94*, 774–802.

Schwartz, A. L., & Clarren, S. N. (1977). *The Cincinnati Team Policing Experiment: A Summary Report*. Washington D.C.: Police Foundation.

Skogan, W. G. (1986). Fear of Crime and Neighborhood Change. *Crime and Justice: A Review of Research 8 (Communities and Crime)*, 203–229.

—— (1989). The Impact of Police on Victims. In: E. Viano (Ed.), *Crime and Its Victims*. New York: Hemisphere.

Skogan, W. G., & Wycoff, M. A. (1987). Some Unexpected Effects of a Police Service for Victims. *Crime and Delinquency, 33*, 490–501.

Smith, S. J. (1987). Fear of Crime: Beyond a Geography of Deviance. *Progress in Human Geography, 11*, 1–23.

Taub, R. P., Taylor, D. G., & Dunham, J. D. (1984). *Paths of Neighborhood Change: Race and Crime in Urban America*. Chicago: University of Chicago Press.

Taylor, R. B., & Covington, J. (1993). Community Structural Change and Fear of Crime. *Social Problems, 40*, 374–397.

Taylor, R. B., & Hale, M. (1986). Testing Alternative Models of Fear of Crime. *Journal of Criminal Law and Criminology, 77*, 151–189.

Thompson, C. Y., Bankston, W. B., & St. Pierre, R. L. (1992). Parity and Disparity among Three Measures of Fear of Crime: A Research Note. *Deviant Behavior, 13*, 373–389.

Thompson, M. P., & Norris, F. H. (1992). Crime, Social Status, and Alienation. *American J. of Community Psychology, 20*, 97–119.

Trojanowicz, R. (1986). Evaluating a Neighborhood Foot Patrol Program: The Flint, Michigan Project. In: D. P. Rosenbaum (Ed.), *Community Crime Prevention: Does It Work?* London: Sage Publications.

Ward, R. A., & LaGory, M., & Sherman, S. R. (1986). Fear of Crime Among the Elderly as Person/Environment Interaction. *Sociological Quarterly, 27*, 327–341.

Winkel, F. W. (1989). Responses to Criminal Victimization: Evaluating the Impact of a Police Assistance Program and Some Social Psychological Characteristics. *Police Studies, 12*, 59–72.

APPENDIX 1. Description of Variables

Variable	Description
Community-Level	
Crime rate[a]	[7 index crimes (murder, robbery, aggravated assault, burglary, larceny-theft, motor vehicle theft, arson)/population] * 100,000
Police force[b]	Rate of full-time law enforcement employees; (total police employees/population) * 100,000
% Black	Black visibility; percent of black population
% Hispanic	Hispanic visibility; percent of Hispanic population
Person-Level	
Age	Age in years.
Female	Dummy variable (1 = female, 0 = male).
Health	Self-evaluated health; Would you describe your present health as . . . ?; 4 = excellent, 3 = good, 2 = fair, 1 = poor.
Income	Total household income for the previous year; 1 = less than \$10,000, 2 = \$10,000 – \$14,999, 3 = \$15,000 – \$19,999, . . . , 14 = \$70,000 – \$74,999, 15 = \$75,000 and over.
Victimization	Experience of victimization during the past 3 years; dummy variable (1 = yes, 0 = no).
Indirect victimization	This variable is composed of two question items: one for victimization of family members or relatives, the other for victimization of Korean acquaintances. Each question item has two response categories (1 = yes, 0 = no). The range of the variable is between 0 and 2.
Perceived incivility	A composite variable consisted of the following 9 items that represent social and physical incivilities in a neighborhood: (1) Trash and litter lying around your neighborhood, (2) Neighborhood dogs running loose, (3) Inconsiderate or disruptive neighbors, (4) Graffiti on sidewalks and walls, (5) Vacant houses and unkempt lots, (6) Unsupervised youths hanging out on the street, (7) Too much noise, (8) People drunk or high on drugs in public,

	(9) Abandoned cars or car parts lying around. Each item has three response categories: 1 = not a problem, 2 = somewhat of a problem, 3 = very serious problem. The range of the variable is between 9 and 27. Alpha = .94. This variable is standardized when it is used as the outcome variable in a model.
English proficiency	Self-evaluated English speaking ability; 5 = fluent, 4 = good, 3 = fair, 2 = poor, 1 = not at all.
Amer. Justice system	Self-evaluated knowledge of the American Criminal Justice System, that is, its operations, people's rights and obligations, etc.; 5 = pretty good, 4 = good, 3 = fair, 2 = poor, 1 = not at all

[a] Forcible rape figures furnished by the state-level Uniform Crime Reporting Program administered by the Illinois Department of State Police are not in accordance with national UCR guidelines. Therefore, the figures are excluded from the total index crime rates. The data for suburban communities are collected from 1996 through 1990 Uniform Crime Reports, on the basis of latest-availability. The data for 77 community areas in Chicago City are acquired from 1996 index crime data by 25 police districts, from the Chicago Police Department. Although the borders of those districts do not coincide with the community areas, they can be matched with minimal discrepancies, by breaking down community areas to census tracts. All of the population data used in the computation of crime rates come from the 1990 Census.

[b] The data of full-time law enforcement employees are also collected in the community-level because the UCR data is not available in the neighborhood level. The data for suburban communities are from 1996 UCR while the data for 77 community areas in Chicago City are from a 1997 report by Chicago Police Department. Again, the population data come from the 1990 Census.

APPENDIX 2. Correlations Between Person-Level Predictors and Fear of Crime

	1	2	3	4	5	6	7	8	9	10	11	12	13
1. Fear of crime – Night	–												
2. Fear of crime – Day	0.653**	–											
3. Age	–0.013	0.036	–										
4 Female	0.171**	0.006	–0.130**	–									
5. Health	–0.203**	–0.146**	–0.186**	–0.136**	–								
6. Income	–0.302**	–0.296**	0.008	–0.023	0.162**	–							
7. Victimization	0.112**	0.161**	0.127**	–0.007	–0.080*	–0.076*	–						
8. Indirect victimization	0.141**	0.108**	–0.030	–0.047	–0.020	–0.045	0.318**	–					
9. Perceived incivility	0.265**	0.277**	0.042	–0.039	0.009	–0.195**	0.120**	0.127**	–				
10. English proficiency	–0.171**	–0.217**	–0.396**	–0.076*	0.297**	0.280**	–0.075*	–0.041	–0.113**	–			
11. American justice system	–0.095*	–0.097**	0.028	–0.156**	0.136**	0.191**	–0.038	0.049	–0.039	0.442**	–		
12. Confidence in the police	–0.182**	–0.136**	0.231**	–0.015	–0.023	0.060	–0.124**	–0.061	–0.118**	–0.113**	0.103**	–	
13. Evaluation of police protection	–0.231**	–0.171**	0.082*	–0.029	0.065	0.075*	–0.089*	–0.063	–0.111**	–0.002	0.079*	0.459**	–

** Correlation is significant at the 0.01 level (2-tailed).
* Correlation is significant at the 0.05 level (2-tailed).

ETHNOGRAPHIES OF LEGAL AND SOCIAL CONTROL ORGANISATIONS

MARKETS FOR LEGAL SERVICES AND THE RISE OF FRANCHISE LAW FIRMS[1]

Jerry Van Hoy

ABSTRACT

In recent years the literature on the legal profession has debated the consequences of a number of changes in the environment in which lawyers work. In this article I examine how franchise law firms innovated to take advantage of the glut of lawyers and U.S. Supreme Court rulings eliminating restrictions on advertising and eliminating bar association minimum fee schedules. My analysis supports the argument that legal work may be deskilled. However, I argue that specific conditions must exist for deskilling of the type found at franchise law firms to develop. These conditions include strong competition for clients, competition that is limited to the cost of services rather than the quality of services and the existence of a large potential market for services that may be easily mass produced.

INTRODUCTION

In recent years the literature on the legal profession has debated the causes and consequences of a number of changes that have transformed the environment in which lawyers work. These changes include the rapid growth of new lawyers since the 1970s, the reinstatement of advertising, and the elimination of bar association minimum fee schedules. Some view these changes as a sign that professional controls over the market for legal services are eroding and predict

Sociology of Crime, Law and Deviance, Volume 2, pages 91–105.

ISBN: 0-7623-0680-7

eventual deprofessionalization (Abel, 1989, 1988, 1981; Hartmann, 1993; Rothman, 1984; Spangler, 1986). For example, Abel (1989) envisions most individual practitioners being displaced by employment and law firms that become bureaucratic "capitalistic multiservice corporation[s]" (p. 244). Others disagree, noting that there is little evidence that lawyers are losing control over their professional jurisdiction or professional autonomy at the workplace (Freidson, 1994, 1986; Nelson, 1988; Nelson & Trubek, 1992; Patterson, 1988).

Yet while numerous studies debate why the legal profession is experiencing change, there have been few empirical attempts to examine how the glut of lawyers, advertising and elimination of minimum fee schedules may be affecting the practice of law. In this chapter I examine how changing market conditions helped to create the environment in which franchise law firms have developed and prospered by deskilling staff attorneys. Franchise law firms provide basic personal legal services at competitive, flat rate fees. Clients are largely gained through television advertising and legal services are standardized and mass-produced using boilerplate forms and proprietary computer programs. My analysis supports the argument that legal work may be deskilled. However, I argue that the conditions necessary for deskilling of the type found at franchise law firms does not characterize all types of legal work.

DATA AND METHODS

I combine a general discussion of the literature on the changing markets for legal services with the specific examples of firm organization and innovation at two of the largest franchise law firms in the US. Both Arthur & Nelson and Beck & Daniels (pseudonyms, used for confidentiality) were first organized as small, local legal clinics. However, both successfully adapted to the increasingly competitive market for legal services in the 1970s and 1980s by developing mass production systems for serving legal clients. Innovation in organizing the work of lawyers and secretaries allowed them to take advantage of the glut of lawyers and television advertising. In a relatively short period of time, these firms were able to successfully compete with other legal clinics and grow into national chains.

Data about firm organization, the organization of work, and the success of television advertising at Arthur & Nelson and Beck & Daniels are from observation at branch offices and interviews with 85 managers, lawyers and secretaries at both firms. Interviews were based on a semi-structured interview schedule. Interviews were audio tape recorded with the permission of the respondents (for a more detailed discussion of my methods see Van Hoy, 1997, 1995).

THE CHANGING MARKETS FOR PROFESSIONAL SERVICES

Evidence of changes in the markets for professional services and the employment patterns of professional workers have received extensive discussion and debate in recent years. Some have argued that there is an on-going shift from independent practice to employment in large organizations (Abel, 1989; Derber, 1982). Others have argued that new patterns of employment for professionals are more complex (Freidson, 1994, pp. 135–136, 1986). Nonetheless, most observers agree that legal markets are changing at a rapid pace and have not affected all lawyers equally (Abel, 1989, 1988; Galanter & Palay, 1991; Nelson & Trubek, 1992; Powell, 1985). Sander & Williams (1989, pp. 440–441, 474) show that from 1967 to 1982 business legal services grew at almost twice the rate of personal legal services. The demand for new lawyers at large law firms serving corporate clients sent associate salaries skyrocketing in the 1980s.

In contrast, from the early 1970s to 1985 the incomes of solo practicing attorneys declined by 30% while the number of sole practitioners has increased. The number of individually practicing lawyers increased 34% from 1980 to 1988 (Curran & Carson, 1991, p. 6). This sector of the legal profession appears to have absorbed the majority of new law school graduates as the number of lawyers has more than tripled in the last 30 years (from 218,000 in 1960 to 757,000 in 1988 [Sanders & Williams, 1989, pp. 432–433]).

In the 1970s successful court challenges to bar association minimum fee schedules[2] and ethical restrictions on advertising by lawyers[3] also increased competition for at least some practitioners. In both the *Goldfarb* and *Bates* decisions the U.S. Supreme Court tried to balance strongly opposed ideological views. Bar associations claimed that minimum fee schedules and restrictions on advertising and solicitation helped to maintain lawyer professionalism and protect consumers. Consumer and free market advocates, on the other hand, argued that a lack of information and at least a perception of high fees kept many consumers from seeking legal services when appropriate (American Bar Association, 1995; Macaulay, 1985).

The Court's decisions did not fully deny or satisfy either set of parties. The lawyers they were most likely to hurt (solo and small firm practitioners) often supported minimum fee schedules (Abel, 1989). With lawyer advertising the

Court used a 'balancing' test it had developed in previous commercial speech cases.[4] Blanket bans on advertising were deemed impermissible, but false, fraudulent or misleading advertisements may be banned. In addition, the Court has allowed states to regulate advertising as well as direct mail, telephone and in-person solicitations (American Bar Association, 1995; Andrews, 1980; Hill, 1993, pp. 58–72). It seems clear that the Supreme Court has attempted to satisfy consumer and free market advocates without completely denying bar associations and states the right to regulate professional activities.

Despite the Supreme Court's rulings, it is not clear that the elimination of minimum fee schedules has produced rabid price competition (Abel, 1989, p. 118). However, particular routine services have become increasingly competitive. Examples of legal services where consumers regularly shop for the lowest fees include uncontested divorce, simple wills and demand letters to landlords or tenants.

Similarly, relatively few lawyers take advantage of advertising (Abel, 1989, pp. 119–122; American Bar Association, 1995). The yellow pages of the telephone directory remains the dominant form of advertising, even for legal clinics (American Bar Association, 1995, 1990, p. 25). Franchise law firms, along with plaintiff's personal injury lawyers, seek a competitive advantage through television advertising. Thus, franchise law firms exploit a market niche characterized by standardized, routine personal legal services. Because services are basic, competition for clients is based on the price of the services offered and intensive television advertising.

Franchise law firms have not been the only response to the changing legal markets. Nelson (1988) argues that increased competition among large corporate law firms for business clients has led to increased specialization of both law firms and attorneys. In personal legal services an abundance of 'new providers' of legal services have arisen (Seron, 1992). New providers include legal clinics – both for-profit and not-for-profit, prepaid legal services plans marketed to individuals, prepaid legal services plans marketed as employee benefits and franchise law firms (Seron, 1992; American Bar Association, 1982, 1990; Singsen, 1984, 1985). The strategy of the new providers is to offer legal services to blue collar and middle income clients at reasonable fees. Franchise law firms are among the most innovative of the new providers. To a great extent, prepaid legal service plans and legal clinics offered new methods of selling personal legal services to clients. Franchise law firms go beyond such limitations by instituting mass production techniques to take advantage of the glut of lawyers in the personal legal services market as well as the benefits of television advertising and lower fees.

THE RISE OF FRANCHISE LAW FIRMS: MARKET ORIENTED INNOVATION

In an environment where both jobs and clients are scarce it is likely that professionals will seek to lower overhead costs, increase productivity, seek new clients and differentiate themselves from competitors. Indeed, the legal profession, including personal legal services, has experienced increased specialization (Heinz & Laumann, 1982, Jacob, 1990). Furthermore, the advent of inexpensive, personal computers allows for the streamlining of routine writing and drafting tasks for virtually all lawyers. Specialization and the use of computers undoubtedly makes it easier for professionals to delegate tasks to support staff who have little or no formal legal training (Harman, 1993; Haug, 1977). Franchise law firms use computers and specialized work roles to standardize the production and delivery of personal legal services to consumers.

Franchise law firms evolved as a competitive offshoot of legal clinics. While legal aid offices serve the poor and indigent, legal clinics were to increase access to legal services for middle income people. Taking their cue from legal aid, legal clinics were supposed to be 'high volume, streamlined, efficient, and impersonal' but profitable operations (American Bar Association, 1982, p. 6). Legal clinics defined a new market of legal service patrons to be served at retail style outlets.

Unfortunately, legal clinics never lived up to the expectations. American Bar Association (1982, 1990) studies conclude that legal clinics are not particularly high volume, efficient or profitable. These studies suggest there is little difference between legal clinics and other small law practices. Legal clinics simply appear to be small law offices owned by a single lawyer. In fact, by the 1980s solo and small firm practitioners were employing the few innovations legal clinics were known for (such as setting flat fees for routine services like personal bankruptcy, uncontested divorce and wills, and running advertisements in the yellow pages). Plagued by low profits, legal clinics may be disappearing from the personal legal services landscape (American Bar Association, 1990).

While many legal clinics have apparently failed to compete successfully, a small number of innovators have developed into franchise law firms. Franchise law firms are the high volume, streamlined, impersonal and efficient operations envisioned for legal clinics. One reason many legal clinics have not been able to expand and compete well in their market is that American Bar Association ethics rules forbid the lay public from investing in law firms. Franchise firms have had to use innovative methods to attract capital. For example, both Arthur & Nelson and Beck & Daniels had trouble growing beyond the cities where they were founded as legal clinics in the 1970s. By the 1980s Arthur & Nelson developed a management company that is independent of the law firm. The management company provides capital, technology and secretarial services to the law firm. More importantly, the management company could seek outside investors. By the late 1980s Arthur & Nelson – and Beck & Daniels, following a similar path – exploded into national firms with each operating about 300 branch offices.

In the 1990s Beck & Daniels has differentiated itself again by selling equity interests in its branch offices to the attorneys who manage the offices. Each franchised office now pays royalties to Beck & Daniels for use of the firm name and production system, as well as paying into a regional pool to buy television-advertising time.

BENEFITS OF THE FRANCHISE ORGANIZATION

The ability to attract capital from non-lawyer sources and expand a firm's presence to saturate a market area makes television advertising an economically viable strategy for attracting clients. However, such an expansion and the use of television advertising to help create a mass market for legal services also creates potential problems related to client service and quality control. These problems include staffing and organizing branch offices in a manner that provides service quickly to the large number of clients who respond to advertising, ensuring the quality of the work performed is at an adequate and predictable level throughout the firm's branch office network, and last, but not least, making sure that client service and quality control do not interfere with earning profits in a competitive market. Franchise law firms attempt to solve these problems by developing efficient production systems. Branch offices mass produce routine legal services which require minimal legal counseling and relatively inexperienced lawyers. Examples of services offered at Beck & Daniels and Arthur & Nelson include: divorce, simple wills, name changes, personal bankruptcy fillings, small business incorporations and letters to landlords or tenants regarding evictions or other disputes.

Franchise law firms have branch offices that are staffed by managing attorneys, secretaries and staff attorneys. A managing attorney is contracted to run each branch office and is responsible for hiring and evaluating the work of other branch office staff. Like most franchises, management above the branch office level provide computer systems and software, office furniture, other supplies, accounting and book keeping services, payroll services, insurance and advertising to the branch offices. At one time Beck & Daniels and Arthur & Nelson had three or four levels of management above the branch office managing attorney. However, since moving to the franchise structure both firms have eliminated all but one level of management between the firm owners and local offices. A small number of middle managers conduct yearly file audits and provide troubleshooting services to branch offices.

Instead of having a large cadre of managers who oversee the operation and quality of services provided, franchise law firms provide their branch office staffs with basic platforms for the production of every service offered by the firm. The basic platforms include computer programs with letters, forms and other boilerplate documents and worksheets for each legal case-type the firm offers services for. The worksheets provide lawyers with specific questions to ask clients during consultations. The information gained from a worksheet is then entered directly into the computer and the appropriate letters and documents print out automatically. Though this does not mean that all divorce agreements will look exactly alike, they all share a standardized base.

By standardizing each service the franchise firm gains a number of competitive advantages. First, standardization allows branch office staffs to service clients efficiently by assigning many legal decision-making and production tasks to secretaries. At franchise law firms secretaries take over responsibility for determining the type of legal problems clients have and produce the letters and documents necessary for dealing with the problems. In the following example from my field notes an Arthur & Nelson secretary dispenses legal advice to motivate a caller to make an appointment with the office:

> The telephone rings. Ginny answers, "Arthur & Nelson, how may I help you?" After a moment of listening into the phone Ginny's eyes widen and she says, "Ma'am, do you have collision insurance? Is your son covered for collision? Is your daughter covered for collision? . . . Yes? Then you have nothing to worry about, you're covered. You're covered. You're covered! What's the problem? . . . The initial consultation fee is 35 dollars, but what are you going to talk to the lawyer about? . . . You're covered! What's the problem?" Ginny holds the phone away from her ear for a few moments and then responds, "Is the nose broken? Is the nose broken? Ma'am, listen to me. *Are there any broken bones? If there are broken bones you have a personal injury case. If there are no broken bones you are covered by insurance and you shouldn't worry about it.* . . . That's right, if there are broken bones. But if there are broken bones you have a case, so why not make an appointment with the lawyer? It doesn't cost you anything to see a lawyer for personal injury cases. . . . Ma'am, listen to me! If the nose is broken you have a personal injury case. Is the nose broken? Is the nose broken? . . . You don't know yet? Well if it's broken you have a case and should come in and talk to a lawyer. . . . Yes, OK, you can call back." (Emphasis added; Van Hoy, 1995.)

Secretaries also interview clients when they arrive for appointments to determine what type of legal problem they have. Once the case type has been determined the secretary writes it in code on the initial consultation form. Lawyers then use the secretary's determination to prepare for the consultation by gathering the relevant worksheets or checking the fee schedule (Van Hoy, 1997, 1995).

Because all services are generally offered in only one format, attorneys are forced to offer clients relatively standardized solutions to their personal problems. For example, an Arthur & Nelson attorney who regularly informs his clients that the firm's simple will "has everything you'll ever need" has only that one will available within the firm's production system. The only choices are to sell the services that are available or sell no services at all. Problems that do not fit the standardized solutions, or clients who are hard to deal with, are usually turned away to avoid taxing efficiency and profits.

A second benefit of standardizing legal services is a corollary of the first benefit. The productivity gains brought about by delegating legal decision-making tasks to secretaries and limiting lawyer autonomy allows large numbers of clients to be processed through branch offices on a regular basis. Franchise law firms develop mass production systems which make it feasible to employ television advertising to attract new clients. Television advertising differentiates franchise firms in a competitive market and helps to consistently supply branch offices with clients. Managing attorneys at both firms believe television advertising accounts for the continuous flow of clients into their offices.

> . . . You see, everyone – every single practitioner, every lawyer – is out there scrambling for business. [But] I don't have to go out and scramble for clients. [Arthur & Nelson] advertises and that's how you get your clients.
>
> I would say there is . . . some hostility about [Beck & Daniels] taking the share of the market that they do among [other lawyers]. . . . I know we take clients who would go to them otherwise. Obviously, television advertising definitely works!

The success of television advertising allows attorneys to have little concern for repeat business and helps to insulate branch offices from direct price competition with other lawyers. Indeed, it is not altogether clear that television advertising mainly draws clients away from other attorneys who offer personal legal services. Most solo and small firm practitioners report that their clients find them through personal networks – referrals from friends, relatives or coworkers (Carlin, 1994; Seron, 1993, 1996). Consumers who are influenced by television advertising may not have access to information about lawyers from their personal networks.

Television advertising and standardized production systems help franchise law firms to develop economies of scale that make offering legal services at competitive fees (often a flat-rate for each service) profitable. However, the profit margin on each case is small. Thus, incentive systems at franchise law firms emphasize productivity by making each office or each lawyer a profit center with compensation tied to sales. Attorneys gain an interest in selling the services that can be processed quickly and efficiently because that is how profits – and higher compensation – are generated. A Beck & Daniels branch office managing attorney made this clear when he scolded a new staff attorney for taking on an estate case that was more work than the firm regularly handled: "I can't afford to have you waste two hours a day on that case.... We put in as much time as we can on those [cases] and close the file as quickly as possible." Another attorney explained

> You have to realize . . . that the profit margin on these small items is much higher than the profit margin on major litigated cases.... So for us to do a will, where we are going to get 100% of our fee, which essentially [my secretary] does – I check off the boxes . . . but she essentially does the whole thing – is a very high profit item for us. I have found by working here, if you do a lot of very high profit items where you get 100% of the fee, that makes a lot more sense for this type of practice than it does to get involved in some major type litigation.

On average, attorneys at franchise law firms spend only 15 minutes with each client. During this time they must sell a service and obtain the relevant information for secretaries to produce the documents and letters necessary for the particular case. Two hours a day, especially over an extended period of time, is a huge amount of unproductive labor in a high volume production system. An Arthur & Nelson attorney compares the flat fee system employed by franchise law firms to the hourly rate charged by other lawyers. "At 150 dollars an hour our consultation fee [of 35 dollars] doesn't even buy fifteen minutes and I sometimes give them twenty minutes." Similarly a Beck & Daniels attorney argues that "the initial consultation fee [of twenty dollars] buys about eleven minutes" of legal advice.

A third benefit of the franchise law firm organization is that standardized production allows branch offices to take advantage of the buyer's market for lawyers. Inexperienced lawyers who find they are unable to secure other employment are willing to work for relatively low pay to gain experience and steady work. In addition, a recent survey of New York lawyers (Seron, 1993) finds that younger attorneys are especially accepting of new business-getting techniques such as those used by franchise law firms. These are the lawyers who are subjected to the most competitive market conditions and who are most likely to be recruited by franchise law firms. A branch office managing attorney explains how the competitive market for legal jobs helps him to keep his overhead low:

> There's a great resource out there of young attorneys. Every year the law schools are turning them out.... This year we're going to have maybe 4000, 5000 passing the bar, being admitted and looking for jobs. They're willing [to start working] at the salary of $24,000. They're willing to work for 24 for the experience alone! ... I don't actually need an expert attorney. In one year he becomes pretty good.... So I could turn over an attorney every two years, and that's the way it works.

The mass production systems implemented by franchise law firms makes it possible for attorneys with little experience to immediately begin working with little supervision. Standardized production techniques make experts unnecessary because decision making discretion is limited, as a staff attorney explains, to "mak[ing] a sale or tell[ing] [clients] that we can't help them." When attorneys become too frustrated with their limited range of options and experiences and the low pay – which, as the above quotation by the managing attorney suggests, happens on a very regular basis – the production systems and market forces makes replacing them quick, easy and inexpensive.

ALIENATED LAWYERS

Just as McDonald's uses computers and other technology to reorganize restaurant kitchen work into a fast-paced production line (Garson, 1988, pp. 17–39), franchise law firms reorganize personal legal services work in a similar fashion. This production line approach provides predictability and control as well as efficiency in the work process (Garson, 1988; Ritzer, 1993, p. 25). Staff attorneys especially show the frustration and strain of their limited positions and incomes. Firm management and branch office managing attorneys at Beck & Daniels and Arthur & Nelson report that they almost completely turnover their staff attorneys every two years. During interviews staff attorneys spoke longingly of the day that they would quit and become 'real professionals' in solo practice. At the franchises they were 'doing their time' and 'paying their dues'. In reality, most of the staff attorneys in my sample did not move to independent practices, but to other firms with similar production techniques. This is no doubt due to a number of factors including the harsh competition among solo practicing lawyers for clients and the type of skills gained while working at franchise law firms (discussed below).

Managing attorneys are somewhat more insulated from this type of alienation. The franchise organization provides an ideological orientation that one can be an employee and enjoy the benefits of solo practice at the same time. Because incentive systems at franchise law firms tie managing attorney compensation to branch office profits and managing attorneys have the ability to hire and fire staff attorneys and secretaries, they develop an ideology of running their own business. Yet while many managing attorneys view operating a branch office as "the same thing as solo practice," upper management at franchise firms argue that "attorneys are bad businessmen, so we do everything for them." Indeed, many of the attorneys who accept the solo practice ideology also complained during interviews that they could not easily transfer their franchise skills to an independent practice. After a few years of focusing on a limited number of legal issues that are very routine, the attorneys believe they may not be competent to practice elsewhere.

> My concern is that even though I may want to be doing this [personal injury] work . . . its been a long time since I've really done that. And given what's at stake in personal injury cases you really wouldn't want somebody representing clients who was not fairly well trained and recently experienced in those areas.
>
> You know, it's funny. I'm a general practitioner by any definition of the law. But if you think of it, it's kind of a specialty. You're only taking a certain menu of cases, your clientele is only a certain type of people . . . and you're not representing businesses or banks or anything like that. So in a way it's kind of a specialty to acknowledge.

A number of branch office managing attorneys in my sample are very concerned about their ability to practice law outside of the franchise environment. These attorneys are attempting to 'break' the security measures built into their firm's proprietary software systems. If successful, they hope to set up independent practices using their employer's production systems. In addition to being unable to practice outside of the franchise production system, attorneys at franchise law firms seem to find the computer system a more interesting challenge than the legal work.

ARE PROFESSIONALS AT RISK?

The organization of legal work at franchise law firms shows that professional work may be subjected to rationalization and deskilling. The convergence of falling barriers to entry into legal practice, the abolition of minimum fee schedules and the advent of television advertising limits the degree to which some lawyers are protected from economic market forces. But not all professional workers are homogeneous and all are not equally subject to the same markets or competitive environments (Freidson, 1994, p. 35). Thus it is important to recognize that some professionals may be well protected while others have virtually no protection from competitive market environments (Abbott, 1988). The legal profession has long been divided into two major markets: personal services and corporate services (Carlin, 1994; Heinz &

Laumann, 1982; Reed, 1921).[6] As noted earlier, corporate lawyers have benefited from a market that is growing almost twice as fast as the market for personal services. This has led to higher incomes and a demand for new lawyers from well-regarded law schools rather than the degradation of work (Galanter & Palay, 1991; Nelson, 1988; Spangler, 1986). Nelson (1988) argues that corporate law firms have become more bureaucratic and specialized as they have competed for a piece of their fast growing market, but he finds no evidence of any loss of autonomy resulting from these changes.

The example of franchise law firms suggests that professional work is most subject to deskilling when three conditions are met. The first condition is strong competition among practitioners for clients. Competition often drives down the prices charged for services and leads practitioners to find ways to increase productivity. Not surprisingly, franchise law firms developed mass production techniques for the most competitive segment of the legal profession – personal services. The same competitive market conditions also provide a steady supply of lawyers who are willing to submit – at least for short periods of time – to a production system that alienates workers by limiting professional autonomy.

Second, in addition to competition among practitioners, the more basic or routine the services offered, the greater the opportunities for standardization of those services. The personal legal services market is acknowledged to provide services that require less legal skill than corporate legal practice (Carlin, 1994; Heinz & Laumann, 1982). The division of professional markets into more and less skilled segments is a common phenomenon (Abbott, 1988). In the less skilled segments of the professions, competition among practitioners is limited to the cost of providing services instead of being based on the quality of services. Franchise law firms limit their practices even more by offering only the most basic and routine legal services in a standardized format that limits lawyer autonomy and increases productivity. In contrast, corporate law firms have specialized to offer their clients more skill and expertise, not less (Nelson, 1988; Spangler, 1986).

Finally, deskilling such as that found at franchise law firms requires not only standardized products and standardized production techniques, but a mass market for services. The appeal of standardizing and devaluing a service is lost if there is no large potential market for that service. For this reason franchise law firms were pioneers of advertising on television by lawyers and remain one of the largest advertisers of legal services today. Where sufficiently large markets for services are not available, deskilling to achieve mass production is less likely to occur. In smaller markets professional services competition is likely to focus on providing tailored services to clients who are willing to pay a premium for such services. It is no mere coincidence that the most routine professional services are generally available to the largest segment of consumers – personal clients – while the most specialized services are available only to special or specific consumers – large businesses, the government, and other professionals.

CONCLUSION

Whenever government policies or court decisions threaten to alter the environment in which a profession exists there is concern and debate about the potential consequences. As the U.S. legal profession has lost some of its ability to control competition among lawyers there have been predictions of deprofessionalization (Abel, 1989). The analysis presented in this paper suggests that the negative consequences of removing minimum fee schedules and ending the ban on lawyer advertising may be more limited than previously assumed. The Supreme Court's attempts at balancing consumer/free market and bar association positions have limited the ability of bar associations to regulate the activities of their members. However, the markets for legal services have mediated the effects of this deregulation. The above framework suggests that a specific set of circumstances must exist before professional work becomes subject to processes of degradation similar to those experienced by other, less skilled workers.

Nonetheless, these transformations of professional work must be taken seriously if we are to understand how professional work is changing – and the roles of governments, courts and markets in such change. Many lawyers simply do not work in the high status positions that help to protect practitioners from competitive market forces. Much – though certainly not all – legal work is routine enough to eventually become subject to the kinds of market forces that encourage the reorganization of work processes for efficiency and profits.

NOTES

1. Some of the information in this chapter was originally published in Jerry Van Hoy, *Franchise Law Firms and the Transformation of Personal Legal Services*, (1997, Quorum Books).

2. In Goldfarb v. Virginia State Bar (1975) the US Supreme Court ruled that minimum fee schedules violate antitrust laws.

3. In Bates & O'Steen v. State Bar of Arizona (1977) the U.S. Supreme Court held that advertising by lawyers is constitutionally protected free speech.

4. Virginia State Board of Pharmacy v. Virginia Citizens Consumer Council (1976). For a discussion see Hill (1993), pp. 31–32.

5. Franchise law firm attorneys report that the most exciting part of their job is selling services to clients who may be a 'hard' sell. These attorneys refer to their work

as 'processing law' instead of practicing law because of the routine and competitive nature of the work (see Van Hoy, 1995).

6. In addition, Seron (1993) suggests that the bar may also be divided by urban and suburban practice locations.

REFERENCES

Abbott, A. (1988). *The System of Professions: An Essay on the Division of Expert Labor.* Chicago: University of Chicago Press.

Abel, R. (1989). *American Lawyers.* New York: Oxford University Press.

Abel, R. (1988). United States: The Contradictions of Professionalism. In: R. Abel & P. S. C. Lewis (Eds), *Lawyers in Society, Vol. 1: The Common Law World*, Berkeley: University of California Press.

Abel, R. (1981). Toward a Political Economy of Lawyers. *Wisconsin Law Review, 1981*, 1117–1187.

American Bar Association (1995). *Lawyer Advertising at the Cross Roads.* Chicago: American Bar Association.

American Bar Association (1990). *Report on the Survey of Legal Clinics and Advertising Law Firms.* Chicago: American Bar Association.

American Bar Association (1982). *Legal Clinics: Merely Advertising Law Firms?* Chicago: American Bar Association.

Andrews, L. B. (1980). *Birth of a Salesman: Lawyer Advertising and Solicitation.* Chicago: ABA Press.

Carlin, J. (1994). *Lawyers On Their Own: A Study of Individual Practitioners in Chicago.* San Francisco, CA: Austin and Winfield.

Curran, B. A., & Carson, C. N. (1991). *Supplement to the Lawyer Statistical Report: The U.S. Legal Profession in 1988.* Chicago: American Bar Foundation.

Derber, C. (1982). *Professionals at Work: Mental Labor in Advanced Capitalism.* Boston: G. K. Hall.

Freidson, E. (1994). *Professionalism Reborn.* Chicago:University of Chicago Press.

Freidson, E. (1986). *Professional Powers: A Study of the Institutionalization of Formal Knowledge.* Chicago: University of Chicago Press.

Galanter, M., & Palay, T. (1991). *Tournament of Lawyers: The Transformation of the Big Law Firm.* Chicago: University of Chicago Press.

Garson, B. (1988). *The Electronic Sweatshop.* New York: Penguin Books.

Hartmann, M. (1993). Legal Data Banks, the Glut of Lawyers and the German Legal Profession. *Law and Society Review, 27*, 421–441.

Haug, M. R. (1977). Computer Technology and the Obsolescence of the Concept of Profession. In: M. R. Haug & J. Dofny (Eds), *Work and Technology.* Beverly Hills, CA: Sage.

Heinz, J. P., & Laumann, E. O. (1982). *Chicago Lawyers: The Social Structure of the Bar.* New York and Chicago: Russell Sage Foundation & American Bar Foundation.

Hill, L. L. (1993). *Lawyer Advertising.* Westport, Conn: Quorum Books.

Jacob, H. (1990). The Contours of Divorce Lawyering. Working Paper on file at the Center for Urban Affairs and Policy Research, Northwestern University.

Macaulay, S. (1985). *Lawyer Advertising: 'Yes, but . . '.* Madison, WI: Institute for Legal Studies Working Papers Series (University of Wisconsin).

Nelson, R., & Trubek, D. (1992). Arenas of Professionalism The Professional Ideologies of Lawyers in Collective and Workplace Contexts. In: R. Nelson, R. Solomon, & D. Trubek (Eds), *Lawyers' Ideals and Lawyers' Practices*. Ithaca, NY: Cornell University Press.
Nelson, R. L. (1988). *Partners With Power: The Social Transformation of the Large Law Firm*. Berkeley: University of California Press.
Paterson, A. A. (1988). The Legal Profession in Scotland: An Endangered Species or a Problem Case for Market Theory? In: R. Abel & P. S. C. Lewis (Eds), *Lawyers in Society, Vol. 1: The Common Law World*, Berkeley: University of California Press.
Powell, M. (1985). Developments in the Regulation of Lawyers: Competing Segments and Market, Client and Government Controls. *Social Forces*, *64*, 281.
Reed, A. Z. (1921). *Training for the Public Profession of the Law*. New York: Carnegie Foundation for the Advancement of Teaching.
Ritzer, G. (1993). *The McDonaldization of Society*. Thousand Oaks, CA: Pine Forge Press.
Rothman, R, A. (1984). Deprofessionalization: The Case of Law in America. *Work and Occupations*, *11*, 183–206.
Sander, R. H., & Williams, D. (1989). Why Are There So Many Lawyers? Perspectives on a Turbulent Market. *Law and Social Inquiry*, *14*, 431–479.
Seron, C. (1996). *The Business of Practicing Law: The Work Lives of Solo and Small-Firm Attorneys*. Philadelphia: Temple University Press.
Seron, C. (1993). New Strategies for Getting Clients: Urban and Suburban Lawyers' Views. *Law and Society Review*, *27*, 399–419.
Seron, C. (1992). Managing Entrepreneurial Lawyers: A Variation on Traditional Practice. In: R. Nelson, R. Solomon, & D. Trubek (Eds), *Lawyers' Ideals and Lawyers' Practices*. Ithaca, NY: Cornell University Press.
Singsen, G. (1985). Personal Legal Services: Problems and Possibilities in the Marketplace. Unpublished Manuscript.
Singsen, G. (1984). New Providers in the Marketplace: Prepaid, Mass Market and Clinical Legal Services. Unpublished Manuscript.
Spangler, E. (1986). *Lawyers For Hire*. New Haven: Yale University Press.
Van Hoy, J. (1997). *Franchise Law Firms and the Transformation of Personal Legal Services*. Westport, CT: Quorum Books.
Van Hoy, J. (1995). Selling and Processing Law: Legal Work at Franchise Law Firms. *Law and Society Review*, *29*, 703–730.

CASES CITED

Bates and O'Steen v. Arizona State Bar. 1977. 433 US 350.
Goldfarb v. Virginia State Bar. 1975. 421 US 773.
Virginia State Board of Pharmacy v. Virginia Citizens Consumer Council, Inc. 1976 425 US 748.

TRANSFORMATION OF THE ORAL TRADITION OF THE POLICE SUBCULTURE THROUGH THE INTRODUCTION OF INFORMATION TECHNOLOGY

Albert J. Meehan

ABSTRACT

The oral tradition among the police is an important part of the police subculture and an agent of socialization for new recruits. While oral traditions generally have the character of remaining stable over time, information technologies (IT) are being incorporated into the oral tradition in policing in ways that change its character. Amusing or sensational audio and video tapes, often remarkable for the startling character of the images they portray than for their relationship to a tradition of shared values, are replacing the apocryphal stories that formerly characterized the oral tradition. Information technology is generally viewed as increasing the efficiency of the police as an organization. However, it may in fact be significantly altering and strengthening the police subculture at its core. This paper examines the officers' uses of audio and video materials through a detailed analysis of the materials themselves, extensive field work documenting the officer's uses of them and stories about them, and formal interviews.

Sociology of Crime, Law and Deviance, Volume 2, pages 107–132.

ISBN: 0-7623-0680-7

INTRODUCTION

The oral tradition among the police has always been an important medium of the police subculture (Shearing & Ericson, 1991; Waddington, 1999) and an important agent of socialization for new recruits (VanMaanen, 1978; Manning, 1977). Oral traditions generally have a resilient character which facilitates their capability to remain stable over time (Ong, 1982; Edwards & Sienkewicz, 1990) and promote group solidarity (Couch, 1989). Recently, information technologies (IT) are being incorporated into and supplementing the oral tradition in policing in ways that change their character. The apocryphal stories chosen by officers to highlight the shared values of the group, or instruct other officers about their work, are being replaced and/or supplemented by amusing or sensational audio and video tapes that are often remarkable more for the startling character of the images they portray than for their relationship to a tradition of shared values or ethos. Thus, the introduction of IT promises to alter the oral tradition of policing in important ways.

The concept of a distinctive occupational subculture has largely informed police research which uses this concept to explain the emergence of a distinctive and shared police world view and its consequences for understanding police behavior (Manning, 1989; see Herbert, 1998 and Waddington, 1999 for a review). Further, the police subculture is often considered a formidable obstacle to organizational change (Goldstein, 1990) and a cause of police deviance (Kappeler et al., 1998). In recent years, the concept of 'police subculture' as traditionally conceived has been recast by researchers (e.g. Fielding, 1988; Shearing & Ericson, 1991; Waddington, 1999) in interactionist frameworks of varying types. These approaches consider the study of the police subculture *to be* the study of the talk or other linguistic forms (e.g. stories, narratives, myths), occurring within this context that demonstrate the rich and complex meanings that permeate the texture of the police subculture.

These arguments propose that the study of the police subculture focus upon interactional occasions where naturally occurring talk displays (sub)cultural competencies (Fielding, 1988); that the sharing of police stories and myths are figurative guides for action in the police subculture (Shearing & Ericson, (1991); or that police talk is a form of rhetoric that is palliative by providing meaning to experience and preserving occupational self-esteem (Waddington, 1999).

The present study considers how various forms of information technologies, or what Ong has characterized as 'secondary orality' (1982, p. 11), have emerged alongside the oral tradition in policing. The question of what effect this will have on policing itself and its oral traditions is an important one.

Information technology is generally viewed as changing the organization of police services by creating more efficiency and greater accountability (Manning, 1992; Meehan, 1998). However, technology affects the police subculture in unintended ways. Officers use these technologies, like they do other police practices, for their own purposes such as resistance to authority and entertainment (Fielding, 1988; Manning, 1992). While some technologies alter the police subculture, officers also mold various information technologies to fit and reinforce the existing police subculture which is often at odds with administration policy.

Officers have been collecting and circulating audio materials for at least thirty years since audio technologies were first introduced into policing in the 1960's and, in recent years, have begun compiling collections of video materials.[1] This paper demonstrates how audio and video records are becoming an important part of the oral culture or tradition in policing. This is reflected in the following ways: (1) law enforcement personnel are making their own personal copies of tapes consisting of encounters with obvious entertainment value, illustrative of 'real' police work, and in some cases, for their own personal protection from citizen's complaints and/or administrative review; (2) they are showing these tapes to each other, sharing copies of them and referencing them in their stories about police work to one another as well as with researchers/outsiders; (3) sergeants and other supervisory personnel are using tapes to 'informally' train officers and in some cases, to inform command decisions; and, (4) tapes that are formally incorporated into training, or shown by the mass media, are being bootlegged and distributed, and then discussions of those tapes make their way back into the oral culture.

The oral tradition or culture of the police constitutes an important phenomena relevant to understanding continuity and change within the police occupation. In spite of the fact that we are a society whose culture appears to be primarily written, oral cultures within various institutional organizations and groups maintain their strength and importance. The oral culture enhances the meaning and significance of social life for it's members (Hannerz, 1992) and, contrasted with literary forms, it involves the sharing of emotion which promotes group solidarity (Sehested, 1975; Couch & Chen, 1988). The oral tradition has gained importance in the face of increased bureaucratization, legal-rational organization, and professionalization. Elsewhere, I have shown how the demands of recordwork and recordkeeping have increased the importance of the interactional or running record among officers (Meehan, 1986). In this chapter, I am interested in how the oral tradition has adapted to and adopted various information technologies.

By oral culture or tradition, I am referring to occasions where the use of stories, narratives, or instructions (both formal and informal) predominate and are focused upon the work-relevant, shared values and normative orientations of the police occupation (Shearing & Ericson, 1991). It encompasses situations where expertise or the 'time tested' stock of occupational knowledge relevant to surviving on the streets against the 'bad guys' is shared. It often takes the form of humor (Pogrebin & Poole, 1988). It also includes shared knowledge about the organizational and political environment of the department, courts and community at large.

The oral tradition is not encumbered by department rules and regulations, paperwork or the threat of organizational censure: indeed the oral tradition is the resource for managing these constraints (Meehan, 1986; Shearing & Ericson, 1991). It occurs in the locker room, at roll call or in the briefing room (Pogrebin & Poole, 1988), in the car, and at calls for service or stops (particularly where officers huddle up to "talk the situation over"). It also occurs at 'coffee', dinner, or in the United Kingdom, in the 'canteen' (Waddington, 1999) and after work at the bar. The resiliency of the police subculture and its resistance to organizational change and formal mandates is in large part due to a strong oral tradition that is enshrouded with an element of secrecy (Manning, 1977).

Among the police, the oral tradition is recognized as an important aspect of policing. VanMaanen (1978) observes that the feature of 'officer training officer' insures the continuity of the police subculture from one generation to the next regardless of the specific content of the academy and other formally sponsored training. The 'real' socialization of police officers, which occurs after 'formal' academy training, stresses the importance, if not predominance, of the oral tradition (VanMaanen, 1978). The academy-trained recruit first encounters the importance and strength of the oral tradition when they are told by their field training officer or first partner: "forget everything you learned at the academy, you're on the streets now, we'll show you how it really works" (VanMaanen, 1978; Bayley & Bittner, 1984). Thus, the actual work practices of police, through which officer competency is achieved and assessed by their peers (Fielding, 1988), are transmitted from generation to generation through this oral tradition.

The question addressed in this chapter then is how audio and video records are affecting the oral tradition of policing. As collections produced **by** the police, and ostensibly "for each other," the materials have an obvious self-serving character. However self-serving they are, because they are audio and video records, they are a valuable resource with which to examine the police world view. I will focus on both the 'work' related and the 'playful' uses of audio and video records and:

(1) how they reflect, and on particular occasions of their use, reinforce, what has been termed the values or ethos of the police occupational subculture: notably solidarity, bravery, secrecy and loyalty, cynicism;
(2) how the materials are used to lend an authenticity and credibility to the police world view on the occasions of their playing as well as when they are referenced in the telling of stories; and
(3) how they can also enable research into the work site practices that are typically missed or glossed in stories told by the police.

DESCRIPTION OF THE DATA AND TECHNOLOGIES

For the past five years, I have been conducting field research in two police departments: one a suburban community of approximately 25,000 residents (55 police officers), and the other a large city of 180,000 residents (250 police officers) which also shares a border with a major Midwestern city. I have conducted approximately 400 hours of ridealongs with patrol officers in these two departments, taped interviews with various segments of upper and lower management, as well as specialized units within these departments (e.g. detectives, traffic, records, training, etc.), and have had various sorts of records produced by information technologies made available to me (e.g. print out of MDT communications, videotapes from the camera cars, audio tapes of calls and dispatches).

During the course of this research, I was made aware that officers in these departments collect audio and video episodes of different incidents that occur within the department as well as in other departments. The existence of tapes in 'collections' (i.e. culled from departmental sources for personal use) was initially revealed through stories told of one officer's bizarre traffic stop that would include the comment: "you oughta see the tape on it." Recurring stories about this and other tapes led me to ask officers in other departments about this phenomenon. Those officers in turn gave me copies of tapes they had or told me about tapes they knew about. A total of twelve different audio and video tapes from officers' collections were obtained through this informal method of collecting collections (several of these were collections of multiple encounters edited onto one tape). Several tapes were in the collections of a number of officers from different jurisdictions. In addition, stories were told to me by officers in the course of field work about camera car tapes that officers have seen but did not possess.

In these collections, there are no episodes utilizing visual mass media sources, although one tape does contains radio news broadcasts of riots during the 1960s interspersed with dispatch audio tapes. The media actively monitor police frequencies and mobilize action cameras, both on the ground and increasingly in the air, in order to visually capture the 'action'. In high profile or celebrity cases, audio and video tapes are sometimes obtained by the media either through filing a Freedom of Information Act claim, or because it is in a department's best interest to release them to the media for public consumption. In the latter case, the tapes typically lend credibility to the police position. There are also cases where the managerial position may not be in support of the line officer, as in the case of camera car footage of a South Carolina trooper who stopped and physically abused, without provocation or resistance, a black female motorist. By releasing this tape, the officer is sacrificed to the public for the purpose of preserving the organization's legal liability as well as promoting the appearance of professionalism.

The mass media airing of actual tapes or camera car footage to show "how it really happened" for public consumption is becoming quite common in news coverage of crime. For example, there are very few people in the U.S. who haven't seen the Rodney King beating tape; heard Nicole Brown Simpson's 911 call, or seen the Riverside County Sheriffs's pursuit of a truckload of illegal immigrants and the subsequent beating of the driver and passenger. It is not that these tapes are uninteresting to the police: one officer gave me a bootlegged copy of Court T.V'.s analysis of the Rodney King beating that contained as he said "the whole beating, not just the parts that showed up on TV that make the cops look real bad."

In general, however, audio/video tapes originating from the mass media have a different character than the ones in the collections I have received. The publicly available police videos and phone calls are 'mediated' products, and the police and media have a rather hostile and suspicious relationship, beneath a veneer of apparent cooperation. For example, after the public release of Nicole Brown Simpson's 911 domestic violence call to the police, which had occurred a year before her murder, an officer who had responded to the call stepped forward with a tape recording of his interview with Nicole that he made on his personal tape recorder. This suddenly became 'news' – even though undoubtedly the 911 tape's existence, as well as the practice of officers making their own 'personal' audio records, is common among the police.[2]

"Real life portrayals" of policing like the show 'Real Stories of the Highway Patrol' and 'COPS' are also a potential source for collections as well as stories. The COPS producers have produced two tapes available for purchase by the public: 'Too Hot for TV' which contains naked women, and other bizarre police encounters with drunks, etc., and 'In Hot Pursuit' which is a compilation of the 'best chase scenes' filmed by the show's camera crews. However, these are not camera car tapes, they are tapes recorded by the show's cameras, in the presence of a television crew accompanying selected officers in accommodating cities (Hallett & Powell, 1995). In my experience, these tapes are discussed by officers in the field and, in one department, reviewed by sergeants who scrutinize the tapes for violations of general orders as a 'game' played during down times on the shift.

However, tapes from the media were not in the collections officers' I studied. Rather, the collections compiled by officers that have been shared with me primarily utilize two technologies: (1) the 911 Computer Aided Dispatch System which records all phone conversations and radio dispatches to cars; and (2) audio-video equipment that is in police cars, typically called 'camera cars'.

The systems for recording police phone calls and radio dispatches are approximately thirty-five years old. Over the past ten years, however, storage and retrieval of these have been enhanced by the invention of small tapes with enormous memory and computer technologies that now index all citizen's calls and dispatches. This has enabled departments to store more calls/dispatches for longer periods of time and to access them more readily. Thus, audio collections are far more common (i.e. I have approximately 125 audio recordings of phone calls and dispatch sequences in these collections as compared to only six camera car episodes). Audio communication tapes are more easily accessible and copying equipment is built right into the 911 system. By contrast, camera car tapes have tighter supervision rules and thus are more difficult to copy. However, they are appearing on news and television police shows with greater frequency. One could expect that the proliferation of camera cars will also eventually change the proportion of audio to video segments in officers' collections.[3]

Camera cars were introduced into law enforcement approximately nine years ago. In the two departments I studied, only a selected number of patrol cars are equipped with audio-video equipment primarily due to its cost of approximately $6,000 per car. Officers wear a wireless microphone, activated at their discretion, which transmits a signal to the trunk of the car where the tape deck is stored. Although the general orders in both departments state that the microphone must be activated at the beginning of all citizen encounters, this procedure is *not* routinely followed. In fact, in the departments I observed, microphones are off most of the time and officers who are assigned to camera cars will typically inform other officers at the scene that "we are live," if they have in fact turned their mike on. These microphones have a transmission range of up to 300 feet and can also transmit from inside buildings. The camera is mounted on a swivel base, above the dashboard and typically points forward. There are General Orders in both departments that require officers to turn the camera toward the back seat whenever someone is taken into custody. The VCR unit is activated as soon as the officer starts the car, although officers do have the capability to turn it off, rewind, and fast forward tapes from the front of the car. There is an override function however: whenever the officer switches on their bar lights/siren, the camera, but not the microphone, is activated and stays on until the switch is turned off. Of course, if they 'forget' to put in a tape (which happened frequently in one department I studied), the system will not record anything.

Humor

The recordings in these collections are varied in both size and content. However, they have been broadly characterized by their owners, and by other officers who have viewed or heard them, as demonstrating either 'real' police work, or 'playful' or 'humorous' aspects of policing. I utilize this dichotomy recognizing that the line between what is considered 'real' and what is considered 'play' is often blurred (Huizinga, 1955). 'Real' police work may have its humorous and playful moments. Humorous situations also have their all too 'real' and serious consequences. Pogrebin & Poole further blur the distinction noting that police humor "reflects rather serious issues in policing" (1988, p. 205). They identify four strategic uses of humor: jocular aggression toward superiors, degradation of the citizen audience, diffusion of danger/tragedy, and normative neutralization of the law and police bureaucracy. All serve to dramatize the serious aspects of police work and participation in humor is a measure of group solidarity, as reflected in the extent to which the group responds with laughter rather than silence. Further, they argue that humor preserves the status quo in that it:

> . . . provides a forum for presentation of concerns without directly threatening the system that fosters them; that is, unlike the formal repercussions that could result from taking a serious position on an issue, humor affords expression (or even diffusion and defusion of concerns) without changing the terms of the organizational relationships. (Pogrebin & Poole, 1988, p. 206–7)

Examples from officers' collections contain humourous or playful police encounters with citizens (e.g. phone calls from mentally ill persons reporting bizarre happenings, strange traffic stops), reflecting what Pogrebin & Poole, following Goffman, call audience degradation: "Humorous putdowns of complainants served to promote the police sense of moral superiority and to maintain the dichotomy between police and policed" (1988, p. 206–7). In this respect, the collections mirror the oral tradition. However, the technological capacity to record, copy and circulate these encounters enables the police to archive *actual* experiences, to be re-played on various occasions, and elaborated upon through more stories. Thus the technology not only supplements the oral tradition, it also contributes to it.

For example, a collection of thirty calls to the police, put together by dispatchers in a large Midwestern city, consists of calls about or by chronically mentally ill persons, drunks and 'strange' citizens. Excerpts from two calls provide a sense of the collection:

Example 1

(A man calls reporting that he was attacked by juveniles several days earlier. He states that he was talking to these gang members "about he possibility of my running for office in Midcity" The caller (C) mentions that he has his own business to which the dispatcher (D) responds)

D: What kind of business do you have?
C: I am setting up a network hotline
D: What kind of network?
C: Its Apollo Nine Associates
D: Who are your Associates?
C: It has to do with the escaped aliens from outer space . . . and the refugees from Guatemala but we haven't gotten that set up yet

Example 2

C: There is a woman I saw in Redwood City and she keeps on making me move with my schizophrenia
D: How's she making you move?
C: she does it with her forehead . . . she just sticks some mover on her head and I have to move

This collection would be played for officers who visited the police dispatch center during the shift, as well as at police parties for entertainment value.[4] Playing the tape would provide an occasion for telling more stories about the city's large chronically mentally ill population, which was at times a source of extreme irritation for the police (see Meehan, 1995).

The tapes not only begin to circulate among officers, but also the narratives/stories that are created from the tapes are altered and imported into 'local' lore (see Maines & Bridger, 1992, on the relevance of narrative to locality). For example, I had received a 'classic' audio tape, called the 'Deer Tape' which is a call to the New York State police. On this tape, the caller, a black male, picked up a deer that he struck with his automobile and put it in the back of his Cadillac. The deer however, regains consciousness, kicks out the windows of the car and injures the driver who is soliciting police assistance. To add insult to injury, a dog confronts the driver as he is trying to get out of the car at the phone both near a gas station. The officer who gave me this tape remarked how funny it is because the citizen was not only 'stupid' for trying to put the deer in his car, but also has 'no idea' where he is calling from:

Example 3

C = Caller (A black male)
D = Dispatcher

C: This mutha fuck'n. This mutha fuck'n dog chased this deer, understand, and this mutha fucka chased this deer into my car. Now I picked the deer up and put him in the back seat. So I'm gonna take him to get him fixed, you know what I'm saying. That mutha fucker tear up my god damn car apart.
D: Where's all this takin place?
C: You know where the phone booth . . . I got me a mutha fuck'n Cadillac out here. This mutha fucker is tearin the shit out of the back seat. This mutha fucker bit me in the neck, shit
D: Where are you at?
C: At the damn telephone booth. The mutha fuck'n dog chased me in a telephone booth
D: Now where are you at?
C: In the damn telephone booth, god damn it. I just told you, shit.
D: Where abouts? Are you in the town of Poughkipsie?
C: No. huh?
D: Are you in the town of Poughkipsie?
C: By the Mobil station
((after a long sequence of establishing the exact location))
D: Hang, hang in there, and I'll send a car over
C: This mutha fucker bit me too. Send somebody with a gun to shoot the thing
D: O.K.
C: I hit him with a tire iron, and he bit me good.
D: O. K. we'll send, we'll send a car over

In the metro Detroit area, I have heard this same story told by different police officers with identical details except a different rural location in Michigan is substituted for the original one and they describe the motorist as a black male from Detroit who gets lost 'out in the sticks'. The motorist's predicament is humorous. But, the fact that the motorist is African American, driving a red cadillac in a rural setting, and confronting a hostile deer and dog is also relevant to its humourous import in a degrading sense.

Humor that plays off of race stereotypes is also illustrated by another call on a tape provided to me by a suburban police officer who worked for a university located in a Midwest inner city. The tape contains a series of calls by a distraught white suburban mother calling the university police department. The woman is afraid her son (who attends the university) has met with foul play because he should have been home two hours ago:

Example 4

D: Maam calm down
C: I'm getting hysterical
D: Maam
C: I am hysterical because I hate that inner city!

A short while later, when asked the age of her son, the mother replies "twenty two, please don't say he knows how to take care of himself, in the inner city age doesn't make any difference." The woman wants an "all out search" conducted by the police between the university and their suburban home, and has even sent her other two sons to look for their brother on a nearby freeway. It turns out that the son's car was locked in the university parking structure and he is safe. The overprotective, distraught mother type of call was apparently 'commonplace' according to the officer who gave me the tape finding it funny that "these parents send their kids to the university in a high crime city, then they expect us watch over them when they don't check in."

One tape which I haven't seen, but that officers from different departments have, and have told stories about, was produced with a camcorder by officers from a state police barracks that has responsibility for inner city freeway patrol. The tape contains interviews with prostitutes, races between drunks and winos that were sponsored by the officers, and a sequence that, using careful editing, shows rockets mounted to a state police car exploding a car under a viaduct. Each officer reportedly receives a copy of this tape upon leaving the assignment at the barracks as a memento. The title of the tape, 'Doing time in 79', refers to the barracks post number (a pseudonym in this reference) and importantly, the idea among officers that they, like prisoners in a jail/prison, are sentenced to 'doing time' in this city. It is common practice that all State Police officers will spend their first years on the force at this barracks. From the officers' perspective, it is the least desirable assignment, but most desirable in terms of socializing officers because it has more 'action' and opportunities for police work.

Finally, tapes that are shown for professional training purposes are coopted into the oral tradition, as well as copied and distributed among officers for "their own use." A camera car tape titled 'The Maryland State Trooper' (but which really occurs in Maine) contains a traffic stop of an unemployed person. The officer on the tape, in the words of one officer, "maintains his cool in the face of this guy going ballistic" (e.g. screaming and swearing at the officer). The officer who gave me my copy had bootlegged it from his department's training officer. I had just assigned VanMaanen's classic article 'The Asshole' in class and my student/officer handed the tape to me and said "watch this, here is a real asshole." In my research, fifteen law enforcement officers from five different departments, who had seen this tape in training sessions, reported that it was typically used as 'an ice breaker' 'to loosen everyone up', 'get them laughing' and move on to some other topic.

When it is used for training officers about how to professionally conduct a traffic stop, its message is often rejected or is the subject of derogation. A sergeant who was a training officer for a Sheriff's department used it to train approximately 600 officers. He reported that most officers' reaction to the tape is that the officer should have "reached into the car and jacked that asshole right out the window" or "he definitely should have been arrested." They do not agree that the reaction he received from the officer, who is portrayed as remaining calm and wrote a ticket, was appropriate. For these officers, as well as for the majority of officers that I have heard talk about this tape, the motorist is considered a classic illustration of the 'asshole' (VanMaanen, 1978), and deserved more punishment for his demeanor which displayed an overt lack of respect for police authority.[5] Pogrebin & Poole (1988: 190–1) similarly observed officers receiving minority relations training via a videotape of the policy featuring the Chief was the subject of wisecracks, hisses and boos demonstrating jocular aggression towards their superiors.

It is clear that the humor in many of these tapes is derived from a negative and cynical police view of the predicaments of the citizens who are the subjects of these tapes. In these cases, the citizens are typically minorities, people with little power or resources (e.g. the mentally ill, drunks, the unemployed), or 'know nothing' citizens (VanMaanen, 1978) who do not understand the police.

Officers themselves are also the subject, as well as object, of humor using camera car tapes. One camera car tape I was given is referred to as 'Bambi' by officers in the department. In it, the officer runs over a deer and then has to shoot it in the head. However, this fails to kill the deer, so the officer has to leave his car and shoot the deer again. In another department, officers returning from dropping off a prisoner at the county jail fifteen miles away neglected to re-position the camera forward (department regulations require that the camera point to the back seat whenever there is a prisoner in custody). The tape showed the one officer doing paperwork and smoking cigarettes throughout the return trip. A sergeant reviewing the tape showed it to roll call the next day with the quip: "now *this* is real police work!" The comment reflects officer's routine complaints that the job is more paperwork than police work. In another case, a prisoner in the back seat of a camera car "slipped the cuffs" while officers were investigating a scene further, climbed through the window opening of the front-back seat partition, and stole the scout car. All of this was on tape and was shown at the next roll call. While viewed 'seriously' at roll call, the humorous character of it as poking fun at the officer, was obvious in discussions of the tape both on the streets, and in the sergeant's office.

Whether 'real' or 'playful', on the occasions of their playing, as well as in the telling of stories, there is an important sense in which the materials are used to lend authenticity and credibility to the police experience and world view. The tapes, as personal collections by the police, stand as documentation of "life as it really is for us" on the streets. As collections created by "one of us," they demonstrate in sometimes very explicit, and other times more subtle ways, what has been termed the values or ethos of the police occupational subculture: notably solidarity, bravery, secrecy, loyalty, and cynicism. In the balance of this paper, I discuss aspects of tapes considered 'real' police work by officers. In doing so, I demonstrate how these tapes can also reveal specific work site practices of the police which would otherwise be glossed in telling a story orally about some incident or phenomena.

'Real' police work

It is generally acknowledged by police researchers and patrol officers alike that there are few, indeed, rare occasions where officers are engaged in the high profile 'crime fighting' role (e.g. stolen car chases), or heroic protection of the public that they were trained for (e.g. rescuing citizens during an explosion). In fact, when I asked one officer how particular excerpts were chosen for her tape collection, she replied: "because these are what would be considered 'real' police work." Examples of this sort from these tapes include communications sequences of:

(1) running a car plate that turns out to be stolen and the ensuing radio traffic chase;
(2) surveillance of suspicious persons who are breaking into cars;
(3) a police foot chase of purse snatchers;
(4) a dispatcher staying on the line with a caller who is being confronted by a mentally ill person who has broken into her house;
(5) calls and dispatch radio traffic that occurred when a gas explosion leveled a whole city block;
(6) a series of calls and dispatches involving an incident where two girls are apparently being held hostage by a boyfriend with a gun;
(7) calls and dispatch involving an officer's rescue of a driver of a gas tanker that overturned on a major interstate and exploded;
(8) police calls and dispatches during a riot and other civil disorders that occurred subsequently, juxtaposed with excerpts from radio broadcasts about those incidents.

'Real' police work, however, is not solely determined by the type of incident or by the fact that it does not occur with great frequency. 'Real' police work implies an assessment that "a good job was done," that what we are trained for, "really happened," and what we did and how we did it worked! The following excerpt is taken from a dispatch tape in which several cars are involved in a stolen car chase. A "matter of fact" tone of voice is used by both the officer in pursuit and by the dispatcher throughout the chase. They calmly call out the direction of the pursuit, description of the car and the occupants, and no one is "stepping on each other's transmissions." Officers involved in police chases, or other officers interested in joining in the action, tell stories about officers who scream out locations and descriptions unintelligibly into their mikes, or forget to roll up their windows (causing more siren than voice to be heard) because "they are really pumped up and are trying to drive the car and figure out what's going to happen next at the same time."

The calmness of the voices on this tape can thus be heard as illustrative of 'real policing' that is competently done. The practice is produced in a concretely recognizable way. In this example, the officer is joined by two other units who help capture the car and its occupants. The sequence ends with the following:

Example 5

73: We have car and man in custody
D: Any unit making Philadelphia between 14th and LaSalle Gardens go easy
73: Thirteen-six and I have thirteen-seven no other units are necessary
D: Okay sir, any cars making Philadelphia 14th and La Salle Gardens thirteen seven-three, six and seven have everything under control. They have the man and the car
(?1): excellent broadcasting
D: thank you
(?2): nice job seventy three
(?): (two mike clicks)

After the chase and capture has concluded, an officer (unidentifiable from the tape but likely known to those officers working that evening) calls out "excellent broadcasting." A compliment to the dispatcher who accepts with the same matter of fact tone. Another officer calls out "nice job seventy three" and the response is two sound clicks of the microphone.

The use of mike clicks is a separate paper, as the practice requires a fuller, more detailed elaboration. Essentially though, microphone clicks are a technological indexical: that is, their meaning is tied to the immediate sequentially relevant conversational environment. In this context, they are hearable as an acknowledgment of the compliment. In the context of receiving a dispreferred assignment from dispatch such as delivering City Council meeting minutes/agenda to members of city council, they are hearable and heard as sarcasm. Mike clicks that suddenly occur "out of the blue," however, for which sequential relevance cannot be established, are heard for the possibility that an officer is in trouble such as rolling around on the ground in a fight unable to hit the panic button on the radio.[6] In short, the presence of the tape in the collection, in addition to the explicit compliment and mike clicks, indicates that this is an example of the recognizably adequate dispatch and mike work during the course of a car chase. As the sergeant who provided this tape to me said: "I play this for the new guys here to show them this is how it is done."

Dangers of Police Work: Officer Shot

The danger of police work is a theme that permeates discussions of the police subculture. The patrol officers' view of their work as dangerous is often contrasted with the reality of police patrol work which is characteristically mundane, repetitive, and boring. These excerpts, because they reflect the officers' view, dramatize the dangers of police work even though statistically policing is not the most dangerous occupation (Manning, 1997: 285). The rate of police officers killed in the line of duty has decreased over the past thirty years and is comparable to the homicide rate of the general population (Bayley, 1994: 71; U.S. Department of Justice, 1997: 3) and eleven officers per hundred are assaulted in the line of duty (U.S. Department of Justice 1997: 69).

One tape contains a sequence of recordings that document the original citizen's calls and radio communications from three incidents where an officer was shot and, in one incident, killed. These segments are quite dramatic, each records the officer's initial broadcast that they have been shot. They also display the rapid response of those units formally dispatched and of others who are responding to the scene. They demonstrate the show of solidarity which typifies the police. At this moment political and racial divisions, rank and brass, shift disputes and personal problems dissolve into an energy that air traffic over the radio can't begin to capture. "Who shows up" is noticed, and accountings are made for those who don't get involved in the action.

During my field research, an officer was shot in the head by a motorist during a routine traffic stop. Numerous stories were recounted to me about that evening, including several by the homicide detective who accompanied the shot officer on the ambulance ride to the hospital and who questioned him about his assailants receiving squeezes of the hand as replies to the detective's yes-no formatted queries. One squeeze meant yes, two squeezes meant no. Over the years of my research in seven different police departments, I have heard stories about other officers who have been shot or slain. These stories have typically included the circumstances surrounding the shooting, the outcome, the fact that officers rushed to the scene to aid a fellow officer; a show of the expected solidarity that characterizes the police subculture.

What the audiotapes of these shootings in these collections contain, however, is another aspect of the police response, one which over the various tellings of stories is "left out," "perhaps not deemed important" for the story's recipient, or glossed by "so we got him to the hospital." The left out part was the blocking of intersections for the route to the hospital by officers and the coordination and management of that task by the dispatcher. When a shooting was in these collections, this task, this work was 'on the tape': the route was announced, units would called out claiming intersections, noting the passage of the ambulance as illustrated in the following example:

Example 6

1372 = officer with EMS unit accompanying shot officer to hospital
S = supervisor at dispatch
D = dispatcher Other numbered Units = officers in various patrol cars

1372: EMS is gonna go through to the Boulevard. Stevens to the Boulevard.
S: Okay, EMS is taking the Stevens to the Boulevard.
D All units please come in too.
S: Any units available to uhh block EMS going Stevens to the Boulevard.
1372: One three seven two at Stevens and the Boulevard now
S: Okay, One three seven two at Stevens and the Boulevard.
132 One three two is at the Stevens and Forest.
S The Stevens and Forest One three two.
132 and across from third now and Alex
S: Okay, yer uhh on Alex now across from third.
6885: Six eight eight five
S: Six eight eight five
6885: They're North on the Stevens Service drive from Stevens
S: Ok, they're now north bound. Park City ((hospital)) has been notified. Park City has been notified.
135: radio–135's aware of that EMS wagon.
S: Ok, 135's aware of EMS
135: 135 – EMS has now passed the Forbes Freeway
S: Ok. EMS now crossing the Forbes north on the Stevens service drive. EMS now crossing the Forbes north on the Stevens service drive.
685: Six eight five
S: six eight five
685: (siren in background) I'm on north Stevens at the Boulevard.
S: Ok, north Stevens at the Boulevard.
685: At the Boulevard. at the Stevens
S: Ok, at the Boulevard. at the Stevens.
Air1. From air one all the routes seem to be blocked off.
S: Ok, information from air one, all routes seem to be blocked.

All routes appear to be blocked. All intersections appear to be blocked at this time.

S: Ok, information

1372 EMS has arrived at Park City, EMS has arrived at Park City. Thank all units for blocking intersections.

At least five units and the police helicopter became involved in this conveyance of the EMS unit to the hospital. Undoubtedly, there are others involved in this situation who are not mentioned on the air. The department helicopter, surveying the route from above, announces to dispatch that all intersections to the hospital are blocked by police cars: informing all listeners and potential participants that there was no further need to scramble to some near or distant intersection. Thus, those officers who could not make it to the scene, could participate in the event by assuring safe and speedy passage to the hospital for the wounded officer. The blocking of intersections, reflected in the above dispatch tape, not only demonstrates the orderliness of a police practice, but also provides an illustration of the sort of competencies required to make it happen.

One axiom in sociological as well as anthropological field research is that normative structures are often made visible when violations of order occur and the breach reveals the taken for granted character, and underlying stability of concerted social action (see Garfinkel, 1967). The following example, taken from a dispatch tape in the same city as the previous example, only occurring some fifteen years later, was circulated around the police department after the shooting deaths of two police officers, and the wounding of a third. These officers, in two separate cars, were following a van suspected of being involved in a kidnaping when the occupants opened fire on them at a traffic light. The tape illustrates what happens when the expected action (the dispatcher coordinating the blocking of intersections) does not occur.

Prior to the excerpt below, you hear on the tape the radio traffic of two of the three officers in the car calling out that they have been shot and requesting assistance. You can also hear the response of various units moving to the scene of the shooting. Two of the first units on the scene (units 91 and 913) immediately decide not to wait for the EMS ambulance given the seriousness of the officers' wounds. They each begin conveying the wounded officers to the hospital. The officer in Unit 91 announces on the air "Nine one's going to St.Mark's." After this, the dispatcher communicates with other units who are responding to the scene, as well as with the third officer who is still at the scene and awaiting more units and the EMS. His wounds were not as serious as the other officers. Several minutes later, one hears the following:

Example 7

(91) and (913) = Units conveying wounded officers
(D) = Dispatcher
( ) = various units who do not call out their designation
// = indicates that the person's talk is being overlapped by the next speaker
Bold parts of transcript are highlighted for discussion in text

(91): Nine one is at Six and Shoemaker
(D): Nine one is at Six and Shoemaker – they're conveying – nine one at Six and Shoemaker//we're goin to CRH (initials for City Receiving Hospital)
( ): what hospital?
(91): Radio Nine-one, Six and Grable, **still need blocks though**
(D): They're going to CRH they're going to CRH we've got 913 and 91// conveying
(91): Radio nine one's going to St. Marks
(D): Okay, nine one's going to St. Mark's units//we
(913): Nine thirteen's going to St Mark's with (gives officer's name)
(D): Okay Nine thirteens' also going to St Mark's. **We don't know where the shots came from we do//not**
( ): **Call them out!**
(D): we do-//
( ): **get those intersections!**
(D): okay ((there is a 2 second pause here in which you can hear talk in the background at the dispatch center – this is not audible to officers)
( ): **Tell us! Your're suppose to give us – give us the route – give us the route!**
(D): Okay, we need some units to block traffic wherever you are. Block the traffic units, block traffic. They're going to St Mark's

After this, the officer in one of the cars conveying the officer begins calling out where he needs cars to block intersections. In effect, taking over the dispatch function. Soon, a patrol supervisor announces on the air that he is with the conveying units and specifically begins calling out the route to the hospital and where they will need cars to block intersections. Units begin responding to these specific requests, the intersections are being covered.

In this excerpt, the dispatcher, who is a female civilian, is not directing the officers in the field by providing the necessary information and direction officers need to participate in the practice of intersection blocking. At the point where Unit 91 announces their location and states "still need blocks though," this is the first request to coordinate the units. However, she provides the wrong hospital (CRH), which is in the opposite direction of St. Mark's. Further, after both units have made it clear where they are going, the dispatcher still does not begin coordinating the intersection blocking but rather states, "we don't know where the shots came from." This is then followed by three very explicit utterances by different unidentified officers in the field, who make visible the breach by requesting the expected action: "Call them out!," "get those intersections!" and "Tell us! Your're suppose to give us – give us the route – give us the route!" It is after this that the dispatcher begins to request the intersection blocking, but does not direct it. This task is eventually taken over by a supervisory officer in the field.

After listening to the tapes, I revisited the stories I had been told previously by officers in which the topic of intersection blocking was "noticeably absent." in its original telling. When I asked: "did the officers on duty block the intersections for the ambulance?" One officer responded "of course" and the other replied with an emphatic and enthusiastic "yes – that's what we do for each other." The question, opened up the practice to increasing detail. One of my officer interviewees had been shot and he replied the intersection blocking was coordinated between three different departments. He commented "I was a little worried at first because we were zooming through the streets of the next town. I asked the medic about it. He said don't worry. They've got you covered all the way to the hospital," which was yet another city away. The 'of course' and 'yes' responses in the interview serve as a gloss or summary of the concerted work-site practices, the details of which are invisible in the story, but are available through an examination of such tapes which preserves the practice for interactional analysis.

I should note here that there is no formal rule or general order about this type of activity. Officers do not receive formal academy or in-service training about it. In fact, one supervisor told me that this sort of practice opens departments up to lawsuits, especially if the vehicles become involved in an accident. Therefore, it is entirely unofficial, and off the record. There is no paperwork written by supporting units. Also, this supervisor noted: "just imagine someone from the public asking 'why don't you do that for everyone?' "[7]

CONCLUSION

Collections such as the ones I have discussed here are the product of the increased presence of information technologies both inside and outside policing. They provide a valuable window into the police subculture, as these technologies have been adapted by the officers for their own purposes. As data, the details of police work available in these tapes are typically missed in stories about incidents, and in the field notes collected by researchers.

It is often observed that the occupational subculture of the police is a powerful and enduring force. The resistance of the police subculture to change and the ability of the police subculture to maintain its strength in the face of numerous reforms, commissions, professionalization, regulations, management theories and models of organizational control, point to a phenomenon worthy of sociological investigation. Manning (1992) suggests that when officers use technology in ways other than those the organization intends, this constitutes a form of resistance to compliance with organizational mandates. While it is clear in some instances that line officers are resisting the formal, mandated uses of information technologies, they are also discovering their own work-related and playful uses for the information technology, particularly as a means for supplementing their strong oral tradition. That is, where officers used to rely upon stories about their work, and encounters with citizens (and of course still do), and researchers collected and wrote about those stories, now officers are using the technologies to document or present these experiences to each other and to new recruits (and also potentially to researchers who gain access to such data).

Discussions of the police subculture (e.g. Kappeller, 1998) typically portray it as a negative force and focus on its 'condemnatory potential' (Waddington, 1999, p. 293). In particular, its values of solidarity and secrecy are thought to contribute to corruption, abuses of authority and resistance to organizational change. While indeed this characterization may sometimes be warranted, it overlooks important reasons why the police subculture has developed as it has over the years and how it has maintained and perhaps even strengthened itself over time.

A strong oral tradition facilitates the ethos of secrecy, autonomy and solidarity; values recognized as important both to the police subculture and police as an organization. Instances where the oral tradition is breached are perceived as threats to solidarity. Because the police place such a strong emphasis on the oral tradition and treat written records and documentation as involving them in a larger network of bureaucracy, the threat of "writing one up," or "taking memos," can be used as a form of discipline. Rubenstein (1973) reports that when a sergeant was unhappy with the productivity of his platoon because they were becoming lax, he threatened to 'take memos' from every man who was late even though he had no real intention of placing them in their personnel files. As Rubenstein (1973:40) notes "policemen regard anything that requires them to commit statements to paper as a threat."

This is not to suggest that the written tradition within policing is considered unimportant. Indeed, from the police perspective, the written tradition, as reflected in the vast array of documents which constitute their internal as well as external accountability framework, plays a significant role in the occupation, requiring members of the subculture to be sensitive to their construction, meanings and use (Meehan, 1986). Indeed, I would argue that the very constraints posed by the written tradition have sustained the importance of the oral tradition. That is, there is a sense that as the formal written elements of the police organization have assumed an importance over the years, corresponding to the growth of the police as a bureaucracy, the oral tradition has become even more important. This, I would argue, is due to the fact that the written tradition is perceived as serving legal and/or bureaucratic purposes that conflict with police work on the streets.

Over the years, the increasing emphasis on formal training for the police as a part of their professionalization, has led to the development of such entities as state training councils, police academies, the FBI academy in Quantico, Virginia, professional conferences (e.g. Police Chiefs), and most recently the national accreditation of law enforcement agencies. Such efforts toward professionalization, while having some desirable effects and impact, may have ironically strengthened the informal occupational subculture of the police by providing occasions whereby the police from many different jurisdictions can meet and bond, not to some 'professional' model but rather to the street officer's model of policing. For example, the Maryland State Trooper mentioned earlier was bootlegged by the Chief from a tape shown at a national police chief's convention and given to the training director for his police department. It was then passed on to officers, who have developed their 'own view' of it.

Sociologists and Criminologists have yet to fully appreciate the degree to which the police subculture, like other subcultures, resists attempts to control it, 'stretches' the values of the mainstream to accommodate its own shared values, and otherwise manages to retain whatever control it can over their core shared values and occupational norms. This resistance is inevitable and perhaps necessary and not merely a deviant phenomenon. It's necessity derives from the dialectical tension which exists between the public and the police, who among themselves, embrace more feudal, or traditional mechanical solidarity, in opposition to the organic solidarity which characterizes the society at large. Durkheim [1893] (1933) proposed that occupational groups in modern society would provide the very sort of moral solidarity and social cohesion among its members that is displayed by the police subculture. Durkheim argued that this type of occupational solidarity was a necessary counterpart to the contractual-legal organization which he argued produces social cohesion in the modern world. Within policing, we see the managerial segments (middle and upper management) of the police representing the contractual-legal perspective much to the consternation of the line officers. Management reflects the professionalization movement and the implementation of rational-legal organization in policing. Management's embrace of technology to control officers, however, has been incorporated into the police subculture in ways that have reinforced the very values it was designed to control.

ACKNOWLEDGMENT

This is a revised version of a paper presented at 1996 SSSI Stone-Couch Symposium at the University of Nottingham. I thank David Maines and Anne Warfield Rawls for their helpful comments and editorial suggestions.

NOTES

1. The emergence of the practice of making collections of audio and video records is undoubtedly related to the widespread availability of audio and video technologies that are more efficient, smaller (hence enabling an easier bootlegging from the organization) and economically affordable (see Felson, 1998, pp. 184–187, for how the role of technology similarly facilitates theft by criminals). Also, making and circulating tapes is not solely a police practice. After hearing my lecture on this topic, a student who works as a retail fraud investigator for a large national department store chain said that at work they have similar collections. The next day he brought me a copy of a collection of 'busts' made at their store that was edited from surveillance camera tapes, and overlaid with graphics and music. Copies of this tape have been given to other store investigators and is played during the coffee breaks for store personnel in the staff lounge. Its purpose, he notes is "educational, but it obviously has real entertainment value! Everybody loves watching it and gets a kick out of how stupid people who get caught are!"

2. A similar example occurred in the Detroit area when the University of Michigan's head football coach was surreptitiously taped by police during a drunken, disorderly encounter in a local restaurant. The tape's release to the media and public airing left little doubt about the coach's behavior.

3. Perhaps in a tacit acknowledgment of this limitation, and in an effort to tap into the phenomenon I am describing here, the producers of the show COPS announced in 1996 a contest in which officers are being asked to send in their "amazing, unusual, exciting, or weird video tape of crazy arrests, angry suspects, hot pursuits and bloopers from 'in car' cameras, camcorders, surveillance cameras, actuals, for a new home video." First prize: a trip to Hawaii, second prize, a trip to Hollywood. Now, television shows that use actual camera car footage are becoming more common in the mass media, thus enabling a selective diffusion of the police subculture.

4. This tape was the first tape 'collection' I had received from the police. This was in 1986. The idea that there were collections of tapes made and passed around did not attract my interest at the time. In fact, the tape was given to me because I was researching the police handling of the chronically mentally ill (CMI), and one officer during a ridealong commented: "oh we have a lot of CMI calls on a tape." In 1994, when I heard about the existence of other tape collections, I 'realized' I already had one. In this tape collection, in one of the calls to dispatch, an officer plays a joke on the dispatcher by pretending to be a citizen complaining about how long it took the dispatcher to answer the phone. The officer provokes the dispatcher to the point where she becomes rude and then he identifies himself. The dispatcher, laughing and appreciating the practical joke says "Oh I thought you were some asshole calling in a load party!" The officer later says: 'now you can put that on your tape collection'.

5. This disjunct between what the police 'say' they would do, and what they 'actually' would do is at the heart of Waddington's (1999) argument about the police subculture as rhetoric. In a training session, how many officers, in the presence of their peers, will mitigate the behavior of an 'asshole' for the sake of professionalism? The officer's general reaction, from Waddington's perspective, may be more a means of preserving self-esteem (i.e. I'd never let anyone treat me like that without them paying for it with more than a ticket) than an actual guide for action. This particular tape is somewhat legendary in police circles. In the past two years, I have seen it aired nationally on police television shows. When I received this tape in 1996, it was being circulated by the police only and had not made it's media debut.

6. One evening, an officer who was using a public bathroom inadvertently hit the panic button on his radio. This button sends a silent signal to dispatch the officer is in trouble. Since the officer had not requested a break over the air, there was no 'last known location' for the officer. Further, because the officer had turned his radio off in the bathroom so that the radio transmissions would not be broadcast while he was on the toilet, he was completely unaware of the radio traffic occurring. Officers throughout the city were converging upon his sector, checking out the favorite coffee stops, hiding places, and other locations where they thought he could be. Thus, careful attention to radio traffic, and the subtle meaning of clicks, or the absence of response, is an important police practice and measure of competence.

7. One officer, after hearing me discuss intersection blocking, observed that officers also block intersections in other emergency situations as well. He told me the following story: "The night my daughter was born, I was on midnights – we were a real close group on that shift. I got the call from dispatch that my wife was in labor and on her way to the hospital. So I had to rush back to the station to get my car and as I started driving to the hospital (which was two cities away) I got an escort from the guys working – they were blocking the intersections for me along my route from the station to the next city because they wanted to make sure I at least got out of the city in one piece. It made me feel real good to see them do that."

REFERENCES

Bayley, D. (1994). *Police for the Future*. New York: Oxford University Press.

Bayley, D., & Bittner, E. (1984). Learning the Skills of Policing. *Law and Contemporary Problems*, *47*, 35–59.

Couch, C., & Chen, S. (1988). Orality, Literacy and Social Structure. In: D. R. Maines & C. Couch (Eds), *Communication and Social Structure*. Springfield IL: Charles C. Thomas, pp. 155–171.

Couch, C. (1989). Oral Technologies: A Cornerstone of Ancient Civilizations? *The Sociological Quarterly, 30*(4), 587–602.

Durkheim, E. [1893] (1933). *The Division of Labor in Society.* New York: Free Press.

Edwards, V., & Sienkewicz, T. (1990). *Oral Cultures Past and Present: Rappin and Homer.* Oxford, UK: Basil Blackwell.

Fielding, N. (1988). Competence and Culture in the Police. *Sociology, 22*(1), 45–64.

Felson, M. (1998). *Crime and Everyday Life.* (2nd ed.). Thousand Oaks CA: Pine Forge Press.

Garfinkel, H. (1967). *Studies in Ethnomethodology.* Englewood Cliffs, NJ: Prentice Hall.

Goldstein, H. (1990). *Problem Oriented Policing.* New York: McGraw-Hill.

Hallett, M., & Powell, D. (1995). Backstage with 'COPS': The Dramaturgical Reification of Police Subculture in American Crime 'Info-Tainment'. *American Journal of Police, 14*(1).

Hannerz, U. (1992). *Cultural Complexity: Studies in the Social Organization of Meaning.* New York: Columbia University Press.

Herbert, S. (1998). Police Subculture Revisited. *Criminology, 36*(2), 343–369.

Huizinga, J. (1955). *Homo Ludens: A Study of the Play Element in Contemporary Civilization.* Boston, MA: Beacon Press.

Kappeler, V., Sluder, R., & Alpert, G. (1998). *Forces of Deviance: Understanding the Dark Side of Policing.* Prospect Heights IL: Westview Press.

Maines, D. R., & Bridger, J. (1992). Narrative, Community and Land Use Decisions. *Social Science Journal, 29*, 363–380.

Manning, P. K. (1989). Occupational Culture. In: W. Bailey (Ed), *Encyclopedia of Police Science.* Dallas: Garland, pp. 360–365.

Manning, P. K. (1992a). Technological Drama and the Police. *Criminology, 30*, 327–346.

Manning, P. K. (1992b). The Police and Technology. In: M. Tonry and N. Morris (Eds), *Modern Policing* (pp. 348–398). Chicago: University of Chicago Press.

Manning P. K. (1997). Police Work.(2nd Edn) Prospect Heights IL: Waveland Press.

Meehan, A. J. (1986). Record-Keeping Practices in the Policing of Juveniles. *Urban Life, 15*(1), 70–102.

Meehan, A. J. (1995). From Conversion to Coercion: The Police Role in Medication Compliance. *Psychiatric Quarterly, 66*(2), 163–184.

Meehan, A. J. (1998). The Impact of Mobile Data Terminal (MDT) Information Technology on Communication and Recordkeeping in Patrol Work. *Qualitative Sociology, 21*(3), 225–254.

Ong, W. (1982). *Orality and Literacy: The Technologizing of the Word.* London: Routledge.

Pogrebin, M., & Poole, E. (1988). Humor in the Briefing Room: A Study of the Strategic Uses of Humor Among Police. *Journal of Contemporary Ethnography, 17*(2), 183–210.

Rubenstein, J. (1973). *City Police.* New York: Farrar, Strauss and Giroux.

Sehested, G. (1975). The Evolution of Solidarity. In: C. Couch & R. Hintz (Eds), *Constructing Social Life* (pp. 65–79). Champaign IL: Stipes.

Shearing, C., & Ericson, R. (1991). Culture as Figurative Action. *British Journal of Sociology, 42*(4), 481–506.

VanMaanen, J. (1978a). Kinsmen in Repose: Occupational Perspectives of Patrolmen. In: P. K. Manning & J. VanMaanen (Eds), *Policing: A View from the Street.* Santa Monica, CA: Goodyear Press, pp. 115–128.

VanMaanen, J. (1978b). The Asshole. In: P. K. Manning & J. VanMaanen (Eds), *Policing: A View from the Street* (pp. 221–238). Santa Monica, CA: Goodyear Press.

Waddington, P. A. J. (1999). Police (Canteen) Sub-Culture: An Appreciation. *British Journal of Criminology, 39*(2), 287–309.

A DUTY TO KILL: AN OCCUPATIONAL PERSPECTIVE FROM THE FRONT LINES OF A KILLING INSTITUTION

Elizabeth McLin

The American's Creed

I believe in the United States of America as a Government of the people, by the people, for the people; whose just powers are derived from the consent of the governed, a democracy in a republic; a sovereign Nation of many sovereign States; a perfect union, one and inseparable; established upon those principles of freedom, equality, justice, and humanity for which American patriots sacrificed their lives and fortunes.

William Tyler Page

The world wants to be deceived. The truth is too complex and frightening; the taste for the truth is an acquired taste that few acquire.

Walter Kaufman, I And You: A Prologue, 1970

INTRODUCTION

Every society faces the perplexing dilemma of dealing with its citizens who step outside the boundaries defined by law. A quagmire of conflicting values and emotions surrounds societal dealings with violent offenders and decisions about when and if the death penalty will be imposed. In the United States, survey after survey indicate strong contemporary support for the use of capital punishment (Dobrin et al., 1996). On the surface it seems simple that, in a democracy, when the majority support a concept it should be maintained.

Sociology of Crime, Law and Deviance, Volume 2, pages 133–166.

ISBN: 0-7623-0680-7

However, state executions do not occur in public, but behind closed doors, where only a few are privy to the actual process. In this case, the dispensation of justice operates behind a veil, thus discourse and public opinion emerge in a vacuum, with little known fact or thought given to the actual practice as experienced by the condemned or those whose duty it is to personally carry out executions. Citizenry are able to hold at arms length the concept of capital punishment and find comfort in the knowledge that their fears of crime and victimization can be dispensed with through the State's willingness to act on their behalf by eradicating offenders from the face of the earth.

The majority of Americans support capital punishment; yet the institution is semi-covert. Therefore, conversations occur and support is given without complete awareness of the actual practice despite common acknowledgement of the capricious nature of the system. National studies indicate that half of those opposing capital punishment would support it if changes were made (Bohm, 1989). Hence, while executions occur where we cannot see or experience them, as a nation, we support the act. Ideally, the decision to support or oppose a proposition should be given only after different perspectives are incorporated into the public's conversation. It is not enough to be frightened and ask the State to assuage our fear by killing. Killing those convicted of capital crimes one-by-one will not appease a nation's fears if new criminals continue to emerge. Public debate in support of capital punishment revolves around the need for punishment, the possibility of deterrence, retribution for victims, and the economic burden of maintaining the lives of offenders. It is with sadness that some supporters ask, "What else can we do? Something has to be done and we can't just keep them in prison for the rest of their lives." On the other side, opponents argue against the death penalty, citing the capricious and arbitrary way it is dispensed and fearing for the lives of those falsely accused. Regardless of these two sides, a more fundamental issue must be addressed because the real question is whether capital punishment is the way we want society to address problems associated with crime. And if so, we must understand the institutions that are requisite for support and the lives of the human beings who are impacted.

Only those unequivocally opposed to any taking of life will find a clear-cut answer. Otherwise, how can we support or oppose an institution we know so little about? How can an opinion be rendered in the absence of information on the practices, the handlers, and the condemned? To sanction State killing, as we have, we must not turn away from knowing the details. As a nation, we prefer to be distanced from unpleasant realities. We purchase things one might not otherwise buy because we have access to virtual money. We can take it home tonight by simply 'swiping' a card and never feeling any cash exchange hands.

For many of us, we would never prepare meat or foul if we had to slaughter and prepare it ourselves. Instead, we can purchase neatly packaged trays of meat at the nearby grocer with the butcher handling all the details for us. Similarly, public debate on capital punishment occurs with both sides avoiding the details and neither side wanting a comprehensive view. Proponents and opponents argue, giving theoretical reasons for their views, while executions are carried out with both sides preferring an abstract conception rather than detailed account. The discourse is too theoretical arguing whether executions deter and if the system is fair. Such approaches are devoid of the human dimensions that deserve attention. State-sanctioned killing is a serious business and worthy of careful consideration of the complexity of issues, including all humans affected by our decision to authorize the killing.

Part of the discourse should include the lives of human beings touched by the institutions and practices necessary to carry out a death sentence. Recent writings have begun to offer the general public thought-provoking exposure to death row and the process of state executions. Works such as John Grisham's *The Chamber* and Sister Helen Prejean's *Dead Man Walking: An eyewitness account of the death penalty in the United States* provide exposure to issues relevant to the death penalty debate in a format accessible for general consumption. Grisham captures a fictitious convict's battle to avoid death by lethal gas, while Sister Prejean takes a poignant look at the emotions of the real-life families of both the victims and the condemned in a capital crime. Such writings offer a glimpse behind the veil at the institution, by touching upon the very real human agony as experienced by a doomed man and those around him. These contributions inform the public by adding a human dimension to the debate.

Still, there are two other considerations related to humanity that seem absent from contemporary discourse. First, those most likely to land a seat on death row do not just wake up one day and start killing. Most often, public discourse refers to death row inmates as ruthless individuals, rotten to the core, with no redeeming qualities. No consideration is given to the process that 'created' that kind of human being. In the works of Lonnie Athens and Joel Norris, readers are given insight into how the environment may interact with genetics or social norms to create a person who is not humane. If violent humans are created, not born, discourse on the death penalty should, at the very least, include an acknowledgement of the process. By ignoring this process, it becomes possible to view death row inmates as less than human, and once dehumanized, our decision to kill becomes simpler and our discomfort eased. Even with this added dimension, the picture is not complete, for an important component is missing: Who are the caretakers of our wishes; those asked to handle the details so we can be spared?

When it comes to the intentional taking of human life, the rightness or wrongness of the killing becomes relative to the context in which it occurs. State executions, war, and certain use of force are largely considered 'legitimate' violence where representatives of the 'State' engage in the duties of their jobs. In fact, most of us would prefer to leave the handling of our 'enemies' to others. With criminals, society demands justice, and for those unfortunate enough, too unsavvy, or too poor to hire costly attorneys, society sentences them to death, at the hands of the State. Somebody must represent the State and assume the unsavory duties required to carry out our wishes. To some, these handlers become heroes who are willing to do 'courageously' and 'selflessly' the dirty work for us. Most of us never interact with, or even think about, the humans tasked with this assignment. Probably we think of these duty-bound officials even less than we think of the men and women on death row, and far less than we think of their victims. This project initiates the inclusion of death workers into the conversation by seeking the perspective through the eyes of a key player. By capturing the world view from one so close to the process the hope is to start filling the void that surrounds and persists in common discourse about capital punishment and remind us of our responsibility as participants in the sanctioning of this social institution.

METHOD

In an effort to capture thoughts on the death penalty from a strong proponent, an interview was scheduled with a high-level state official charged with active participation in executions. Bob (not his real name) agreed to an interview in his home on November 11, 1996. The interview was semi-structured, using a protocol (Appendix A) covering six areas pertinent to the death penalty: (1) general stance; (2) the present justice system; (3) publicity of executions; (4) duty to witness; (5) logistical issues; and (6) feedback on pro and con statements from students. The interview was recorded and the tapes transcribed. To ensure that Bob's views are accurately captured, without researcher bias, his own words will be used wherever possible. It should be noted that Bob's position is a political one, consequently, his views may most often reflect the predominant political perspective of his state.

At the time of the interview, Bob's state had 122 inmates on death row, representing 0.6% of the total prison population. In response to *Furman v. Georgia* changes were made during the 1970s to ensure that death penalty statutes would be deemed constitutional. Over the period of a few years, the classification of crimes considered for capital punishment in Bob's state dwindled from seven to one, when train robbery, kidnapping, assault, and bombing a building were deleted in 1973. Perjury and treason were subsequently removed in 1978, leaving first degree murder, with consideration of aggravating and mitigating circumstances, as the only capital offense.[1] In 1993, lethal injection became the method of execution, whereas lethal gas had been the method of choice for 60 years, and prior to that, death by hanging.

ONE VIEW FROM THE FRONT-LINES

The interview began with an inquiry into Bob's opinion of the death penalty and under what circumstances he felt it was justified. Bob strongly supports the death penalty, and as a governor appointee, agrees with the present law defining only pre-meditated or first degree murder as a capital offense. In Bob's position, it would be virtually impossible to carry out his duties were he unable to publicly, and at some personal level, support capital punishment, therefore, this section of the interview contains no surprises.

When asked about the benefits derived from its use, Bob cited three: punishment, deterrence, and the safety of other prisoners and prison staff combined with the cost. All are frequently cited in the literature, with punishment and deterrence the most common. Towards the end of the interview, Bob reiterated his support by summarizing:

> BOB: "The reason I believe in the death penalty is first, that it is an appropriate punishment, and I think that's based on Christian Judeo ethics. Secondly, I think it is a deterrent, although the deterrent impact is mitigated, and third, because that human being has to be maintained with other human beings and by human beings at great expense, I think there is a reason for executing them."

In explaining his reasons for support, Bob moved back and forth between punishment and deterrence. He made a distinction between going to prison *as* punishment versus *for* punishment. It is unclear, but appears that Bob believes that incarceration by itself, is not the punishment. Instead, those convicted will receive punishment upon incarceration. At some point, Bob incorporated deterrence into the mix, citing the legal system as a barrier to the positive effects of capital punishment for both deterrence and punishment.

> BOB: "I think it is kind of unfortunate, because for a number of years, people viewed corrections and incarceration and the implementation of the death penalty a lot differently than they do now. They viewed, going to prison, for example, **as** the punishment, not going to prison **for** punishment [emphasis added]. As a result, a few years ago, the whole concept of going to prison was to create a normal life for an inmate. The punishment was that he was separated from society. Now, the argument about the death penalty has followed kind

of a similar thing, meaning the argument is over whether or not it is a deterrent, but in fact, I think its first thrust is that it is a punishment. When you commit first degree murder, there is a likelihood that your life is going to be taken. Now, the extent to which it is a deterrent, which is another benefit from it, I think is badly mitigated by the legal system."
EM: "How so?"
BOB: "For a punishment to be effective, it needs to be swift and it needs to be certain and we have such a lengthy appellate process, that it is neither swift nor certain . . ."

While Bob references the potential for deterrence, he is critical of the existing system because of the lengthy process. Somewhat reluctantly, he later offered advice on what improvements might be made, but civil rights concepts such as due process interfere with his ability to do his job quickly and efficiently, and in Bob's view, mitigate any potential deterrent effect.

BOB: "Deterrent impact is another, but I have to admit that it is questionable. It is not questionable because the penalty itself is not a deterrent. It is questionable because of the due process issues associated with it once it is implemented."
EM: "The whole appeals process?"
BOB: "Right."

From Bob's position, the appellate process results in lengthy delays and provides a lawbreaker with an incentive to play a game of chance. Drawing upon psychology, Bob applies a story about two coke machines and the appeal of gambling to illustrate the capricious nature of the system.

BOB: "There is something about human nature. You can set a coke machine so that every time you put 75 cents in it you get a coke, and right next to it, you can set a coke machine that every time you put 75 cents in it, you might get a coke, you might get nothing, and you might get a jackpot. That machine will generate more revenue than the [other] coke machine. Well, extend that concept to murder, and particularly if you come out of an environment where violence isn't a big deal, then not only is it not a big deal, but you are probably not going to get caught at it. And then, if you get caught, you might not get convicted, and then, if you get convicted, you might not get the death penalty, and then, if you get the death penalty, you might never get death."

When asked about the typical length of time from conviction to execution, Bob was able to relate back to his strong belief in the death penalty as a means of punishment. While he argued that timeliness can impact deterrence, he adds that it becomes a moot point regarding the value of punishment.

EM: "How long does it take in [this state], if I get convicted and sentenced to death, how long till I would be executed?"
BOB: "It would range from 15 to 20 years. I think the last one we executed was 17 years."[2]
EM: "That's a lot. People have forgotten the crime then?"
BOB: "Well, people haven't forgotten the crime. What is argued is that the person you are executing now is a changed person, and in fact, from the punishment tenant that would really make no difference."

When the debate in the literature over the impact that the privacy of executions had on deterrence was mentioned, Bob again reminded me of his strong supporting views regardless of any deterrent effect.

> BOB: "Let me tell you something else. You see, anybody who is informed about capital punishment will not detract from the deterrence argument. That is a liberal argument. There are a lot of good and compelling reasons, even if there were no deterrence, to execute somebody. Therefore, just in virtue of the fact that this one group is saying there is no deterrence be careful . . ."

Bob felt strongly that capital punishment is justified. His final reason, the safety of others combined with the cost led him into a critique of the existing system.

> BOB: ". . . there is another very good reason for execution in mind and that relates to the staff of the prison system that is required to maintain those people for the rest of their lives. There are a number of people on death row right now, who've killed while they've been in prison."
>
> EM: "Were they on death row when they killed?"
>
> BOB: "They may have been on death row when they killed or they were in prison for long terms when they killed. And so, for those who argue, well, you can just keep them segregated or incapacitated from society for the rest of their lives, they forget two things. The first thing you forget, is the safety of staff and other inmates who those individuals have to be kept with and secondly, they forget the cost. Now you will see in the literature that there is a lot of stuff about cost and they say it is more expensive to execute than it is to keep them in prison for the rest of their lives. But the reason for that is the process. If there was a limit to that due process, then [that] cost wouldn't occur."

When asked for recommendations to improve the existing system, Bob explained the levels of appeals that death row inmates currently utilize. Somewhat reluctantly he recommended that limitations be placed on the number of times each inmate can appeal through the entire process. He went on to suggest that 'loopholes' be eliminated to "level the playing field."

> BOB: ". . . there just needs to be a limit [on appeals] and I don't know that I am ready to say how they should be structured, because you have to be very careful, but let me tell you what I would say. I would say that technical limitations should be eliminated. So, see I think that would level the playing field for a lot of people because what happens is there is often the argument that it is the poor people that get sentenced to death, not the rich people. What that relates to, I think that everyone would agree that if there is going to be a death penalty, that guilty people should be executed. Right?"
>
> EM: "Yes, makes sense."
>
> BOB: "What's happening is guilty people who can find loopholes don't get executed. Okay? So therefore, if you find a way to close the door or eliminate the technical parts, and make a determination did he or she commit the murder? Well, then that person should be executed. That would mean that poor people, who couldn't afford to find that technical loophole, would be on the same playing field as the rich person who could buy all these attorneys to try to find loopholes. Because in most cases, what's found is a loophole, not necessarily that there was some travesty of justice."

In reviewing Bob's general stance and view of the justice system, there were no surprises. If anything, it appears that he is irritated that some people are not executed because they have the resources to find a 'loophole'. The long appellate process afforded those awaiting execution has been cited by many as a problem and, in Bob's official capacity, it must be frustrating. However, others argue that it is a necessity to ensure that only those most worthy of execution are indeed executed. From 1993 to 1994, 26 states overturned death row sentences of 100 inmates (Hare, 1996). The question remains, without the existing appellate process, would these have been overturned?

WHO HAS ACCESS TO EXECUTIONS?

In the United States, the last public hangings occurred in the 1930s (McClennen, 1992; Bessler, 1993;[13] Madow, 1995[4]). These events were highly public and often conducted before large crowds. This was no new phenomenon as evidence exists to demonstrate that community-involved executions were occurring as early as 3,000 B.C. (Bessler, 1993). It appears that religion has played a predominant role in the legitimizing of executions and Bob references the Judeo-Christian ethic as justification for his duties.[5] However, the Bible upon which the Judeo-Christian ethic is based, proscribes public involvement unlike modern execution practices.

> The appearance of justice can best be provided by allowing people to observe it. . . . People in an open society do not demand infallibility from their institutions, but it is difficult for them to accept what they are prohibited from observing."
>
> Chief Justice Burger, 1984

In Deuteronomy, God mandated the taking of life for certain offenses at the hands of witnesses and others. "The hands of the witnesses shall be first upon him to put him to death, and afterward the hands of all the people. So thou shalt put the evil away from among you."[6] The witnesses were those whose words were used in securing the conviction. Under such a scenario, one would certainly experience the significance of his or her testimony.

Since the 1830s, state executions have become increasingly 'semi-covert' operations. Madow (1995), in a provocative piece chronicling the history of the death penalty in New York, contrasts traditional executions involving public hanging to the modern gas chamber. He identifies three elements that characterize the modern practice: privatization, rationalization, and medicalization. He demonstrates these concepts by describing the shift from public space to prison interiors with stricter control of witnesses, the eradication of ritualistic and religious aspects in favor of 'technical routines', and the development of "purportedly painless methods" to carry out the sentence. In work by Bessler (1993), the media's legal battles to gain expanded access to executions through the use of recording devices is compared to other contemporary legal struggles such as access to trials and prisons and the use of cameras in court rooms. Bessler contends that the original statutes mandating private executions were enacted for 'paternalistic reasons', in response to an abolitionist movement. With these works in mind, I explored Bob's views on who can and should witness a contemporary execution. Initially, I was uncertain whether he would openly discuss the actual process, but was pleasantly surprised with his candor.

> EM: "Now I want to talk to you a little bit about publicity of executions because some of the literature I found suggests that the whole deterrence thing is mitigated by the fact that executions are private. They say that increasing the publicity would ultimately reduce homicides. Some of the literature, not all of it, argues that. But I am wondering what type of publicity you feel is appropriate for executions in [this state]"?
> BOB: "Well let me tell you, I have been struggling with that because there is one part of me says that I want to open up executions for people to see. But, because it is the most profound action that a government takes against an individual citizen, the taking of his or her life, and because we live in a democratic republic, and you ought to be able to view that. But then there is the other side, and let me tell you what that is. I am not sure that the general public, even though they believe it is right by an overwhelming majority, could tolerate observing it, even though now lethal injection is like someone being put to sleep."
> EM: "Like a dog, my dog was just put to sleep?"
> BOB: "No, in fact, they use one of the drugs that they used to put your dog to sleep. Pavulon. But they didn't put your dog to sleep first. What they did, they gave your dog a drug that seized its heart. In our case, we put the person to sleep, just like he is going to have surgery, then when he is asleep, we do something to seize his respiration and his heart."
> EM: "Oh, so he is basically anesthetized like general anesthetic?"
> BOB: "That's right. What you'd see would be somebody on a table, whose eyes would simply close and depending on the position of his head or her head would relax and then a shuddering of the chest, ultimately whether you are on the inhale cycle or the exhale cycle, that has to shudder down. But aside from that, there is not much to it. But I am not sure that our citizenry would have the political will to continue it if they saw it. But yet, they believe in it."

In a few sentences, Bob was able to capture the privatization, rationalization, and medicalization of the contemporary practices supporting Madow's theory and capture the paternalism postulated by Bessler. Uncertainty that the general public could "tolerate observing it" indicates a world-view that says government has a responsibility to protect citizens from discomfort and, at least partially, demonstrates a sense of paternal responsibility. The description of "somebody on a table whose eyes would simply close" implicates the use of technology to ensure a speedy, clean process and assigns value to a rational approach to executions. The reference to a person going "to sleep just like he is going to have surgery" explicitly links the practice of medicine to legalized killing. This direct relationship is remarkable when viewed within its historical context. In 1888, the Gerry Commission was formed to research the most humane and practical method for executing a sentence of death in New York (Madow, 1995). The Commission's report rejected lethal injection on the grounds that it was too closely "associated with the practice of medicine, and as a legitimate means of alleviating human suffering."[7] Today, no similar prohibition exists. With Bob's help, we begin to see a perspective that defines the process as a highly technical, efficient way to terminate the life of a menace to society while making the handler's duties more palatable. We must acknowledge Bob's recognition of execution as "the most profound action that a government takes against an individual citizen" and weigh it against his concern over the public's ability to tolerate observing it. Upon careful consideration, it begs us to ponder what else our government does in our name that we might not tolerate if we knew, saw, or heard. Do we really want or need to know? Seeking more information about who is privy to the sights and sounds of this form of justice, I turned to the inclusion of witnesses.

> EM: "Six states have recently passed laws to allow the victims' family members to witness an execution and I am wondering how you feel on that?"
> BOB: "Right. Well, in [this state], I am required by statute to invite the AG [Attorney General]. I used to be required to invite a physician, but that was eliminated last legislative session, plus witnesses including local law enforcement. Top, and my first priority, is the victims of the crime if they want to come and I do believe that it gives them some closure and I believe that if I believe in punishment, then they are the first that should receive a benefit that is derived from that punishment. And so we always invite the victims and in some cases they want to watch and in some cases they only want to be present, so we may have them present and have a chaplain with them on the grounds."
> EM: "Are there space limitations to the facility?"
> BOB: "It's a room that will hold about twenty-five or thirty people and we don't have them sitting down. They stand."

Bob did not mention any of the witnesses that might be related to the inmate until prompted. Surely, this must be one of the most unpleasant aspects of his duties, the need to interact with the loved ones of the condemned. For them, the duties associated with his position represent the realization of their worst fears. For him, they may be among the few law-abiding citizens who will be most deeply affected and saddened by his obligation to duty. While Bob seems proud to invite the victims, he is more subdued about his duty to extend invitations to those related to the inmate.

> EM: ". . . when I looked at the statute . . ., it is a couple of years old so I didn't know about the physician being eliminated, it said that the person who is being executed can invite someone."

BOB: "Yes, they can invite their attorney, clergy, and I think five friends, family. They can invite a total of seven."

In fact, this state's statute regarding who can be present at an execution requires certain officials to be in attendance. In addition, it requires, at the defendant's discretion, that invitations be offered for up to five "persons, relatives, or friends" and up to two clergy. Moreover, at least twelve reputable citizens are required to witness. These witnesses could theoretically be perceived as a symbolic representation of the public. No mention is made of victims or the media. Apparently both are defined as 'reputable citizens' for the purpose of witnessing. Bob shared that because only one witness room exists, family and friends of the defendant and victim all stand in the same room during the procedure, yet the groups are kept separate through the careful orchestration of the process.

EM: "Were they together with the people that were victims?"
BOB: "No we keep them separate."
EM: ". . . In the same room, but separate?"
BOB: "Right."
EM: "Okay, yeah cos I thought it would be pretty emotional."
BOB: "Yeah."

The visibility of executions is a hotly debated topic with neither proponents nor opponents of capital punishment falling easily on either side of the debate. Retentionist proponents of publicization argue that, by broadcasting executions, society will benefit from the increased deterrent and retributive features (Filbrin, 1992). Abolitionist supporters propose that in a democracy the people have a right to view the actions taken by government on behalf of its people. With the Supreme Court gauging the legality of capital punishment against "the evolving standards of decency that mark the progress of a maturing society," public discourse becomes paramount (Filbrin, 1992).[8] If public attitude is to serve as a marker for legitimacy of an act, abolitionists argue that attitudes formulated about an invisible institution will not reflect an accurate assessment of changing standards of decency.[9] Many opponents of increased publicity view televised executions as decontextualized imagery that amounts to nothing more than government-generated snuff films (Madow, 1995). Both abolitionists and retentionists opposing the televising of executions share the same anxiety, that spectators will interpret what they see and get the wrong message, and depending on the stance, either support or oppose the institution. It is much safer to fight for your side while maintaining the status quo.

The movement to re-publicize executions did not emerge out of concerns generated by the general public. No, they are predominantly comfortable keeping this institution at arm's length, out of their immediate sight and thoughts. Rather, the media has led the charge by presenting legal challenges to the present system which universally prohibits cameras and recording equipment in the witness room (Filbrin, 1992). Likewise, it was the media that fought the historic battle to even be able to report executions once they were moved from public to private sanctums. The press has always felt a duty to report executions. As part of the privatization movement in the 1800s, four states passed laws making it a criminal offense for newspapers to publish any detail of an execution (Bessler, 1993). With the passing of New York's Electrical Execution Act in 1888, it became a crime to publish the details of any execution. The medical community rallied around the prohibition arguing that reporters lacked the medical knowledge necessary to interpret and recite the details of such affairs (Madow, 1995). When newspapers continued printing details by interviewing eyewitnesses (reporters were banished from witnessing too), charges were pressed. It was with glee that newspapers printed headlines such as 'The Herald Indicted for Printing News' (Madow, 1995). Criminal penalties for the publication of execution details were eventually repealed. However, in Minnesota, the Supreme Court upheld the law stating that "the purpose of the private execution act was to surround the execution of criminals with as much secrecy as possible, in order to avoid exciting an unwholesome effect on the public mind" (Bessler, 1993). So far, all challenges to allow the photographing or televising of executions have been defeated.[10] With knowledge of the contemporary debate on increased publicization and an awareness of the statute's failure to mention the media as witnesses, I explored this state's practice.

EM: "What about reporters? We don't have any reporters witness?"
BOB: "Yes we do."
EM: "So, how do you decide who goes as a reporter?"
BOB: "I do that. They don't like it."
EM: "Yeah, I bet."
BOB: "Let me tell you what I do. I rotate it between the print and electronic media in the two major metropolitan areas. And then I invite print media from the outlying areas."
EM: "Okay, so it would be [a certain newspaper] maybe?"
BOB: "And the [name of newspaper], I used to invite the [another newspaper] but they screwed up so I am not going to invite them anymore. Then we rotate among the channels."
EM: "The TV channels, right. I wondered. How many newspapers or media people?"
BOB: "We usually have uh somewhere between 5 and 9."
EM: ". . . Now some states, Minnesota had a law that forbade the printing of anything about the execution, only allowing the time of the execution. Do we have any restrictions like that?"
BOB: "No."
EM: "Then they can describe whatever they want?"

BOB: "Right, they can't take recorders, no books or anything like that with them."
EM: "No videotaping or anything?"
BOB: "Right, they have to agree to be part of a pool that then goes back and reports to the larger media."
EM: "So everyone has access to the information?"
BOB: "Right, at the same time."

Like the involvement of the defendant's families and friends in the execution, Bob's answers to media-related questions were much shorter. This may demonstrate a level of discomfort, or simply that Bob did not have much to say. Certainly working closely with the execution team as Bob does, would make anyone feel some discomfort around those who are least likely to support the duties related to his job. Both of these groups may often represent some form of opposition or the 'other' world-view. His statement that the media are legally entitled to report "whatever they want" may not account for a lot if, in the past, certain media have been banished for 'screwing up'. In retrospect, I regretted not learning more about that particular newspaper's infraction. In fact, later in the interview when discussing the change to lethal injection, Bob's awareness of the media's influence became apparent when he shared his strong belief that a single media person's negative reaction to a gas chamber execution was the driving force behind the change to lethal injection.[11]

BOB: "And let me tell you, the reason we changed to lethal injection was [media person so-and-so]. I mean, he cried on the air and all that cause it was so inhumane to him."
EM: "So it was just an emotional experience."
BOB: "Well, it was for [media person so-and-so], it wasn't for me."
EM: "Well, I saw an article that was published on it in New York Times and that's why I bring it up cause this article said that [Citizens were] horrified by gas chamber execution."
BOB: "Well that was [media person so-and-so]."
EM: "Oh yeah, cause he was the media guy, I gotcha. But, it wasn't horrifying to you?"
BOB: "Well it was a little disconcerting . . ."

It seems evident that Bob does not consider the media a supporter. According to Bob's description, one man's actions resulted in a major change in the standard practices of his job, disrupting his routine. That kind of change can make additional work for anyone. Later in the interview, I asked Bob about access to death row inmates.

EM: "Okay, but how do you feel about these guys who are on death row being interviewed by the media prior to their execution?"
BOB: "No. First, they cannot interview on a face-to-face basis in the [this state's] state prison system."
EM: "Ever?"
BOB: "Ever, I won't allow that."
EM: "Sometimes you see those news shows, but you don't allow any prisoner, not just those on death row?"

BOB: "That's right, not on a face-to-face basis and so I will allow them to be interviewed over the telephone."
EM: "So the telephone?"
BOB: "And we make arrangements for that. Now the reason I won't allow there to be a face-to- face interview and on camera or anything like that is two things that I think are detrimental. The first thing is for any inmate, it makes them notorious within the prison system cause they get a lot of publicity, so that's one. And the second thing is, in the case of death row, it always involves a recapitulation of the crime which then exposes the victims and makes them have to relive it again. So to the extent that I can prevent it I will do that."

Again, Bob reminds us of his responsibility to the victims and, therefore, his choice to exercise his authority to prevent access to prisoners. Near the very end of the interview, I touched upon increasing the publicity of executions. Bob challenged my words.

EM: "The last public execution in the United States was in 1936 or 1937 depending on who you read, and was in Missouri or Kentucky. It was a hanging. Remember I told you before that there are some proponents of the death penalty who are saying we should make executions more public so I wanted to know what your thoughts were on that. You talked a little bit about it with me but . . ."
BOB: "I don't believe there are proponents of that. Let me tell you what I believe. I believe that the Liberals want that to happen because they know what I believe. In other words, they feel that the best way to get rid of the death penalty is to have it public."
EM: "Yes, there is that camp too. But there is the other camp that says that deterrence doesn't work because it is not publicized."
BOB: "And I don't disagree with that, but the larger push is from the liberal side to try to disqualify it based on public opinion. And I struggle with that, in fact, I'll tell you just the other day, maybe a week ago, I was sitting and thinking, well, why not do it, why not? Well, I saw what [mediaperson so and so][12] did, I saw what one man did with the gas chamber. Let me tell you, just between you and me and the gatepost, it ought to be the gas chamber. They ought to hurt and so that is why, I think if I went to the Governor and I said Governor, I'd like to do it open, I'd like to open it to the public or do it televised, then he would do it for me."
EM: "Yeah, there is a huge debate in the law journals over that issue. And there are both sides, there are the liberals who say, I am calling them liberals but I don't know that they are, but they say, if they saw what the government was doing in our name then they wouldn't like it. Then the others say how do you expect this to deter people if they can't see it, if it is private. There are both."

At this point in the interview, Bob summarized the perceived benefits of the death penalty regardless of any deterrent effect (reported earlier). Clearly Bob has a dilemma: He is responsible for an institution that is controversial. On the one hand, he views the institution as very clinical, while on the other, he fears that if 'outsiders' (i.e. the general public) gain access to view the institution, public support will wane and threaten the existence of the institution. As an agent of the public's will, we have put him in a precarious position. With knowledge of who has access to this public – yet private – institution, we turned to Bob's legal duty to attend executions and began to delve more deeply into the actual process.

> The State cannot be permitted, where there is no reason or justification for so doing, to draw an impenetrable veil of secrecy around a public institution being operated by public officials, with public moneys, for the welfare of the public. To permit such a ban on access to a public institution, where there is no need or justification for it, would be to permit arbitrary, capricious, and unreasonable restraints to be placed upon the right of the people to know what their own government is doing. It is inconceivable that this could be permitted in a democratic society.
>
> Texas District Court, *Garrett v. Estelle* 1977.

THE PROCESS

Bob shared freely that he had witnessed five executions, one with lethal gas and four using lethal injection, and gave detailed descriptions of the procedure. His example of how last words are communicated to the public provides an intriguing glimpse at the awesome power of his position. Although he did not openly disclose any discomfort with the proceedings, he had permitted the interview to be recorded and surely recognizes that his position does not allow frivolous displays of emotion even when experienced.

> EM: "Gee, I can't even imagine what the experience is like. Is there anything you can think of that compares to it that I could relate to?"
> BOB: "No, I don't think so. The gas chamber execution was probably the most profound for two reasons. First, it was the first one that we had done in I forget how many years, thirty some years."
> EM: "So there was only one gas chamber execution?"
> BOB: "Yeah, and it took him about 11 minutes to die. And, so it is much more visual than the other one. I mean, you can see involuntary contractions of muscles. Uh, but no, I had been a policeman and I had never seen anybody die, but I had seen a lot of dead people. So, in that sense, it is not real different."
> EM: "But isn't it kind of a somber place? I can't imagine the emotions especially if there are victims and family members of the person who is dying."
> BOB: "No, but it is done very clinically. And what we do is, we keep the victims in one area, we keep the inmates in another area, and we keep the news [media] in another area, and we keep the general witnesses in another area. So they are all separated and they are all briefed, and then they are all brought to the death chamber maybe two minutes before it actually occurs. They then get in and are given numbers, there is a drawing so there is order, and then when they are all there, what happens is just before, there is a big curtain, a picture window, a curtain, before that curtain opens, what happens is, I go in and I ask him for his last words, and then whatever he has to say I leave. Now, that is not broadcast to the witnesses."
> EM: "They don't know what his last words are?"

BOB: "No."

The process is laden with techniques for maintaining control, such as segregating groups of witnesses, drawing numbers to maintain order, keeping the curtain shut until everything is ready, privately obtaining the 'last words', and 'briefing' each group of witnesses beforehand. These combine to illustrate a world-view that allows nothing offensive to be heard or seen and demonstrates a systematic methodology in place to protect and maintain the institution. Certainly, the designers of these procedures recognized that the State's taking of a life is serious business that necessarily must go off without a hitch. Again, the undeniable paternal nature of the procedures leaves one pondering about the government's rightful duty in a democracy. When citizens prefer protection from realities they may find offensive, how far should the government go to grant them their wish? When Bob mentioned that he is the official that asks for the last words, the magnitude of his position became increasingly clear.

EM: "So he says them only to you?"
BOB: "Yeah."

From here Bob changed and went into details of the actual killing and, quite fascinatingly, how the whole process took approximately two minutes from giving the order to proceed to the heart monitor line going straight; two minutes, a mere 120 seconds to legally exterminate a life. When compared to an execution occurring in 1884, a remarkable contrast occurs. Sensibilities may not have changed, but visibility certainly has. In a hanging, reported by the New York Times, two minutes seemed an eternity: "The dying man drew up his legs to his stomach, and spasmodically threw them out and forward, until he began to sway back and forth like a boy in a swing, and his face was a ghastly and pitiful sight as he writhed and struggled in the prolonged torture of one of the most cruel forms of death. The struggle seemed endless to the horrified spectators, and when the arms finally dropped and the limbs hung limp they could hardly believe that the horrible scene had been enacted in less than two minutes."[13] In contemporary times, our rabid need for vengeance, the use of medical-like procedures, and our physical absence from the scene conspire to mediate the possibility of any outrage such as reported in 1884. How provocative that during the information age we prefer to avoid unnecessary details that might invoke uncomfortable emotions. If in attendance, we require that suffering be masked, occurring behind a curtain or mirror or door, where it will not offend our sensibilities.

BOB: "And then, the curtains are opened and the warden goes in and reads the death warrant. Okay?"

EM: "Okay."
BOB: "When the death warrant is read, then I announce to the people, via intercom, the witnesses, that there have been no stays or reprieves and that I am going to order the warden to proceed with the execution. Then I turn to the warden and say, proceed with the execution. At that point, the lethal injection begins and I go watch a heart monitor and when that line goes straight, I note the time and then I tell them. It takes about two minutes."
EM: "Two minutes from the start to the line going straight?"
BOB: "That's right."
EM: "That's not a lot of time, though does it feel like a lot of time? It seems that would be a long two minutes?"
BOB: "Yeah, well you keep watching for that heart line to go straight. And when that heart line goes straight, I say, ladies and gentlemen, the execution is completed at 12:04 hours."
EM: "So it is kind of sanitized?"
BOB: "Oh it is very clinical, very, very clinical for everybody that is involved, then as soon as that happens, and the groups leave, each escort takes them back."

Bob reminds us of the clinical nature of the procedure, highly structured and systematically designed to kill quickly and without making a mess. Perhaps we must perform executions in this manner because systematic procedures enable the formulation of a routine and predictable process, therefore making possible the removal of the emotions and human aspects that might offend or even hurt. The procedures themselves reify the process making it okay because we are "doing it by the book," and in the end, giving society what it has asked for. Our deference to an 'authority' procedure removes any sense of personal responsibility. The interview continues.

EM: "So they leave separately, I mean they stay in their own group?"
BOB: "Right, they leave separately."
EM: ". . . So are you in the same room as these family members and media?"
BOB: "No, I am in the execution room."
EM: "You are in the actual room that the inmate is?"
BOB: "No . . .

As it turns out, Bob and his team are physically and visibly removed from the others. Neither the witnesses nor the inmate can see them and the only communication is via an intercom system. Executions were not always carried out with such a need for privacy. The perceived need for protection and secrecy of death workers is intriguing when compared to similar components in a traditional execution. Madow (1995) recalls the details of a hanging where state officials served as visible displays of power and authority, physical representatives of justice. Moreover, visibility was deemed important enough to merit an elevated platform for the scaffold to enhance exposure. The hangman would hug and kiss the man before pulling down the cap over his face and cutting the rope. The whole event was dramatically staged to remind all of the harsh consequences doled out to those daring to violate God and government. The hangman did not go undercover until after the Civil War when boxes were constructed to hide his identity (Madow, 1995). Today, officials are removed from sight and communication is via technology. The facility, itself, is designed to keep the individual actions private and identities unknown. The subliminal message is one of shame, not pride, for government service. Whether explicit or not, a message is communicated that individual identities should be protected, kept secret, and begs the question "why?" What other law-abiding acts require such secrecy except to protect the security of the nation? The shift from overt to covert in the practice of executions eventually expanded to include the representatives of the state. This intriguing adjustment is a post-Civil War phenomenon demonstrative of the power of changing sensibilities over universal principles of justice.

> BOB: ". . . our death house is set up like this. The gas chamber is set into a wall like this. Then the chair is this way and he faces, he's got his back to the witness room, okay, and then over here is the lethal injection [area]. This is the gurney . . ."

Bob drew a diagram of the death house on a piece of paper before him. Although I did not retain his actual drawing, I have attempted to replicate it in Appendix B. The death house is shaped like a rectangle and on either end are areas for the use of officials. Bob called the area behind the room with the gurney the "special operations area" where the people doing the lethal injection and watching the heart monitor are positioned. The center area is where witnesses stand with windows on two sides. One side is dedicated to viewing the gas chamber which sits between the 'operations' and witness area and the other contains a picture window for observing the room with the gurney used for lethal injection. Behind either execution area is an area for special operations. Regardless of the method, handlers are positioned away from witnesses, and the inmate is alone between the two groups. Two 'death cells' occupy one side of the death house and are used to 'maintain' inmates until execution. Once I had an image of the 'Death House', I wanted to know where Bob was during an execution and what his duties included.

> EM: "So where are you located then? That special operations area?"
> BOB: "First I go in here [points to the room with the gurney] and ask him for his words, then I go out here [to an outside hall] and go in here [the special operations area] and make the announcement to the witnesses."
> EM: "Via intercom?"
> BOB: "Via intercom."
> EM: "So the inmate is between you and the witnesses, that is different. I can't imagine the responsibility on your shoulders."
> BOB: "There is a big window [between the gurney and witness rooms] and then there is a two way mirror here [between special operations area and gurney room] so these people can't see. . . ."

EM: "You?"
BOB: "Yeah, right."

Bob was willing to share the details of how last words are obtained and then communicated with the world. In recent times, a change had occurred in the procedure to ensure that nothing offensive would be transmitted. Previously, last words were broadcast as they were received. They are now spoken to Bob in private, then transmitted, via a public information officer, to the witnesses. The media then are responsible for communicating them to the world. The most recent execution afforded a glimpse of how a dying man must conform to the rigid rules of the penal system. In this case, the inmate had a common-law wife which is not considered a legal union in Bob's state, therefore, he was denied visitation from the woman because his relationship was not officially recognized. Most likely, there are law-abiding citizens in Bob's state who enjoy similar relationships. Of course, those who are incarcerated will not be entitled to the same freedom that others enjoy. How much control should the State have in censoring a condemned man's words or in who is allowed to visit him in his final days? Should we fear that someone may provide him with the means to kill himself and therefore, rob the State of its rightful duty? No one could envy Bob in his duties that require him to answer questions like these on our behalf. As the process unfolds, again, we see evidence of a paternal view that necessitates the censoring of information and adds new elements to the vast control given the State over those occupying death row.

EM: "Well so you ask them for their last words and they say something to you. Do they all say something to you?"
BOB: "No, now I have only done two, the prior ones [the old director] did. The guy in the gas chamber didn't say anything although his last act, he was sitting like this and I was standing here. The gas chamber has these, I was standing right here, um [the Attorney General] was standing right here on the other side and this is a wall. He turned around and looked at [the Attorney General] and then took his hands, you know he's like this and with this hand went like this (flipped a bird). And then the last thing that he did, was he looked at me and the warden and I were standing there and said "Fuck you". That was the last word this guy had."
BOB: "Now the last guy that went, [inmate's name], you know I wouldn't let his common law girlfriend or wife visit him because of the policies, so when I went in to ask him for his last words, he said, "well who are you?" Cause I went in and I said Mr.[inmate's name] do you have any last words? So he says, "Who are you?" And I said, I'm the director. And he said, "You are [Bob]?" And I said yes and he said, "Why wouldn't you let my wife come visit me?" I said because she is not your wife Mr. [inmate's name]. You knew that. He says, "Yes but I wanted her to come visit me." And I said, well I can understand that but you know what the rules were. He had made the decision not to get married because she was receiving social security for her, uh, former husband's passing."
EM: "So she would lose it, it was a money thing?"

BOB: "Right. For her it was, and he says. "Well, I have found Jesus Christ so I'm gonna forgive you for that. Although I understand the need for rules Mr. [Bob], I am going to forgive you." And I said well thank you very much Mr. [inmate's name]. And he says, "Can I shake your hand?" I said yes, so I reached down and I grabbed his hand and he shook my hand and then he wouldn't let go. And he started to say something again about his girlfriend and I said Mr. [inmate's name], do you have any final words? And he looked at me and he said, "Yes, Mr. [Bob], would you please tell everyone that I am sorry." And I said, I will do that Mr. [inmate's name], you go in peace and I walked out of the room."

EM: "You don't broadcast their final words, I mean they don't hear those because there are no speakers in there, but witnesses do they ever hear about those?"

BOB: "The first thing that happens, when I announce that and I come out here, the public information officer meets me right here and says what were his final words? Now the reason I don't, [the old director] used to always do it before they put him on the table. He would do it like right here [points just inside the witness room]. But, he did it in the presence of everyone and the last one he asked told him you can suck my dick, so, I changed it so that I do that in private and I certainly wouldn't want the witnesses to hear that."

EM: "Would the [old director] tell them?"

BOB: "No. He'd just say that he didn't have any last words."

With all the references to the clinical nature of the procedure and the linkages to medicine, I was surprised to learn that no medical personnel were involved. Instead, volunteers from Bob's existing staff are used to perform the procedure.

BOB: "There are no medical people involved in this."

EM: "Wow, there aren't?"

BOB: "No."

EM: "Who administers the lethal injection? I know they used to have it with the electric chair where more than one person would switch on something."

BOB: "We have a team of what we call the special operations team. They are volunteers, they are given psychological exams to make sure that they can tolerate what they are going to do and they select among themselves who is going to do the actual injections. They do the injections and they rotate. Not only that but some of them have to come out here and put the shunts in each arm and all that so there is quite, you know.... They escort him from here to there [the death cell to the gurney] about 45 minutes prior he is put on the gurney. Then they start the IVs and, uh so they kind of, this team kind of rotates. You know."

EM: "How many people are on the team?"

BOB: "Oh, about ten. See the team includes those who escort, those who put on the table, those who start the shunts, and those who actually do the injection and none of those, those are all individual tasks, none of them do two things."

EM: "So when you say a shunt, you mean it is like an IV, where they don't inject it into the person, they inject it into the IV."

BOB: "Right, what they do is they have an IV running into each arm, and they are great big ones, but they are IVs and the injection is done into the IV line. That goes into the shunt."

Despite acceptance that they are performing a just act and insistence that the procedure is not offensive (i.e. like someone going to sleep for surgery), the special operations volunteers are given psychological tests to "make sure they can tolerate doing what they are going to do." This practice acknowledges a potential for distress, even if the procedure itself is 'clinical'. Although I did not anticipate Bob disclosing that he experienced a high-level of anxiety in performing his job, I wanted to at least ask. Again, the need for a smooth operation was emphasized with a reminder that Bob serves society and receives his orders to act from a higher authority, via a red phone. Bob's job and high-status position are contingent upon his ability to perform his duties without negative consequences to the State or the institution's image.

> EM: "Have you ever lost sleep or lost your appetite before an execution? I mean, does it build up to a big thing or is it just part of your routine job?"
> BOB: "Part of my routine job, I have no feeling about it whatsoever. Now the first one affected me, I have to tell you. But after that."
> EM: "But it was the gas chamber and the new ones are lethal injection, so they are much more medical-like."
> BOB: "Right, and the only anxiety I have about it is about the process itself and making sure that it goes as it is supposed to. And then of course, I have a lot of, see I'm the one who takes all the calls on the red phone from the attorneys and like this last one was delayed almost an hour where I have to be sure that before I say proceed with the execution all the appeals are exhausted and I've got the right information and all that."
> EM: "So there is a special phone number that they would call if they needed to stop it or wanted to stop it and you are sitting by that phone."
> BOB: "I am standing by it."
> EM: "I would be anxious."
> BOB: "Well, I, you do get anxious. I probably got 20 calls this last one."
> EM: "On that red phone. And each one they are saying what? Hold off?"
> BOB: "Yeah, hold off, this is happening, hold off, that is happening."
> EM: "Then how did you know to proceed?"
> BOB: "I was called both by the attorneys, the public defender, and the court."
> EM: "All three of them called you and they said . . ."
> BOB: "Yep, it has been exhausted, proceed with the execution."

It helps to remember that Bob is our agent, he works for us and part of his job is to protect us from any aspect of his work that might bother us. In addition, it is our society's systems, institutions, and structures, all flawed to some extent, that direct his actions. His personal feelings, if not in agreement with what we ask, become irrelevant. His position offers prestige, and some power, as long as he publicly articulates the lexicon of the industry he is charged to oversee. Maybe it is a bigger struggle for Bob than he can convey? In the legal flurry to stay the last execution, it was argued that the inmate was mentally unable to comprehend between right or wrong and what was happening to him, Bob could not possibly agree.

> BOB: ". . . it was a big deal and then they contended that he was, uh, retarded and all that. That is just like his conversation with me, you know one of the things he said to me was,

"Well I understand why there has to be rules, but I still wanted my wife to be with me." Well you know you wouldn't expect someone who was retarded to say well I understand why there has to be rules, but I want you to understand that I wanted my wife and that kind of stuff so you know . . ."
EM: "Yeah, I wasn't very familiar with that case at all. But, wow, I just can't even imagine. I think it would be very hard to have your job."
BOB: It is interesting. I was at a meeting where one of my peers stood up and said that I don't believe in the death penalty, but yet I do it. I thought, you know, how can you do that?"

Most people exposed to death find it a difficult experience. For Bob, death confronts him as part of his job and he must remain composed and in charge. I wondered whether participation in state executions served as his initiation into such events.

EM: "Was the first time you witnessed a death, at an execution, in your life?"
BOB: "Actual death?"
EM: "Yes, where you were present at the death, I guess. I know in law enforcement, you probably saw dead bodies but . . . not the actual dying."
BOB: "Yes it was, yeah. Well, no it is not, I'm sorry. I saw people die in Vietnam."

Apparently, Bob was in the military during the Vietnam era and witnessed death. It surprised me that he originally did not remember this. Apparently, the differences between the deaths he has witnessed are very different to him as he did not immediately recognize that executions were not his first exposure to death. In exploring the process used to dispense this form of justice, the stress-laden nature of Bob's position became apparent and I could begin to see how his livelihood is irretrievably linked to a smooth killing operation. Therefore, logistics are important to Bob.

LOGISTICS

If your job were to carry out a death sentence for the government, your focus would be on doing your job with minimal disruptions and unpleasantry, regardless of any personal feelings. Your success would be directly tied to your ability to pull off each killing with smooth precision. Bob knows his job well and was comfortable discussing the logistical problems. The major concern was the lack of medical expertise available for training of personnel. This results in a number of problems.

EM: "Now I wanted to talk to you about the logistics, but some of them you have already told me about, like the kinds of problems that the corrections system has in getting ready for an execution? It must a logistical nightmare."
BOB: "Oh it is. Let me tell you about some of them. The biggest logistical nightmare is finding someone from a medical background to train personnel in doing it because the American Medical Association is against the death penalty"

EM: "I didn't know that."
BOB: "Yeah, so they try to sanction doctors who participate in it."
EM: "So what do you do then?"
BOB: "Well we find somebody who has medical training. Now it turns out that the people who insert the shunts are by and large former corpsmen or army medics and stuff like that. But that's one issue. Another issue is, you know, ensuring that everybody is practiced and that you have the correct amount of drugs because those are controlled substances and they have to be very tightly inventoried."
EM: "How do you get the drugs? If a doctor doesn't write you a prescription, do you go to Mexico?"
BOB: "No we obtain them from a pharmaceutical supply and you don't need a prescription because we are using them for a legal means."
EM: "Oh, I see."
BOB: "One of the issues that we have been concerned with is what's called the cutdown where their veins are so abused that you can't get a shot in there and so you have to find another way into a major artery."
EM: "From drug abuse?"
BOB: "Yeah."
EM: "What do you do if they don't have arteries or veins?"
BOB: "You just have to find another one, it might be a carotid artery or it might be another artery. So that is a logistical issue, let me tell you though, there are a lot of logistical issues associated with staging the execution. I talked a little bit about keeping all the groups separate, making sure they all arrive on time, You know there is the issue of notifying them and making sure that they are going to be there. There is the issue of controlling the traffic access to the prison and around the prison."
EM: "Is that usually a problem?"
BOB: "No, but we also are responsible for the pro and con death penalty death watch."

The staging of an execution requires coordination to ensure the timely arrival of all parties and control of the media and other onlookers. Since the privatization movement of the 19th century, all these activities take place in the early hours of the morning when executions are typically held. Bessler (1993) cites the shift to hours between midnight and 6:00 am as part of the strategy initiated to prevent the general public from hearing the details of executions.

BOB: "And then we have to make sure that the coroner is there on time and make sure that we have a debriefing psychologist for our people right afterwards. So there is just an inordinate amount of preparation and logistical planning to make sure.... Then we have media. We are constantly having to deal with them.. I've had them mount cameras that were inappropriate to take pictures of the special operations people coming out. The last time I was going to make them disassemble, they have a boom that has a microwave link on it and one of the channels also has a minicam on it so then they couldn't disassemble it and all that so I made them put a bag over it. And then they got all mad and I said 'well that's too damn bad'."
EM: "Yeah, because you really have to protect the privacy."
BOB: "The privacy, yeah, that's exactly right."
EM: "Yeah, that's been an issue I guess, allowing even the videotaping of the executions."

BOB: "The Supreme Court ruled that no one has a right to view that except who is authorized by statute or by the prison officials. So the media has not liked that at all but that's fine'
EM: "I guess its the media that have brought charges all the way to the Supreme Court?"
BOB: "Right, right"
EM: "Okay, I wondered about that, I didn't know. Um, you mentioned before the special operations team are debriefed after. Can you tell me about that because in California I found an article where before and after an execution there is a lot of counseling and stress management that all staff members go through. So I am assuming you have something similar in place for your staff too."
BOB: "Only the ones that participate in the execution."
EM: "Okay, is that only the special operations team?"
BOB: "Right"
EM: "The other people are really just on the periphery, like other staff?"
BOB: "Yes, well first, they are volunteers, secondly we evaluate them to see if they have the psychological stamina to be able to carry out this task and then we have at least two debriefings, or we have a debriefing and a later counseling session. Because, as you know, in post-traumatic incidences, the syndrome may not manifest itself immediately. It may manifest itself a few weeks down the road so we have one where they immediately deal with it and then they have a subsequent one where if there is any manifestation of it down the line, that could be dealt with as well."

A huge weight is on Bob's shoulders for he must make sure that his staff are protected from any breech in the privacy of their identities while keeping the press at bay, coordinating all the witnesses, monitoring the red phone for last minute reprieves, and ensuring the precision of actual killing. All this, while protecting witnesses from any offensive sights or sounds.

FREEFORM RESPONSES TO EXISTING PRO AND CON STATEMENTS

During the last part of the interview sheets of paper containing quotes from students either for or against the death penalty were handed to Bob and then read. Bob was asked to comment. Four were from students opposed to the death penalty, while the remaining two were supporters. All quotes were taken from a study on the effects of a death penalty class at a university in northeastern Alabama (Bohm, 1989). Bob's responses felt somewhat less rehearsed.

QUOTE 1. I feel that with all the arbitrariness and discrimination that there is, I don't feel it's right. I feel that if there was some way to kill them all, then I might be for it. I just don't think it's fair to kill one and not kill the other. Also, I don't know if life without parole just might be a stricter punishment
RESPONSE
BOB: "Well life without parole isn't a stricter punishment."

EM: "It is not?"
BOB: "No, because you get to live and you have certain constitutional guarantees and things you can't take away from people. Um, given the due process that is afforded someone on death row, I can't imagine, this is obviously an uninformed person."
EM: "These are students."
BOB: "Talks about arbitrariness. As far as discrimination, I have some real concerns about that. I think there are a larger percentage of minorities in prison and on death row because minorities commit more crimes. Now said another way, is there a socio-economic difference, I doubt it. Because the vast majority of poor people are neither in prison or on death row."
EM: "I guess that could be true."
BOB: "It is true, let me assure you. Now what I would say is it's true that they are less able to avail themselves of the loopholes. That is absolutely true. But that's not the discrimination, that's simply until we get an egalitarian society where everyone has the same amount of wealth and that's not going to happen."
EM: "I don't think that will ever happen."
BOB: "That's right."

Bob responded to the first statement arguing that the system does not discriminate. Instead, some are better equipped to afford the costs involved in finding loopholes. This seems to fly counter to his opinion that over-representation of minorities in prison is a direct result of minorities committing more crimes, when minorities are more likely the ones with limited resources. Perhaps both provide a partial explanation.

QUOTE 2. (1) so many murders are committed and only a handful of those involved are caught and prosecuted; (2) under the present dispensation most people found committing the offense and sentenced after trial are the minority and poor; and (3) death as a punishment does not make an offender sorry for the wrong; instead it helps them to ease the torture of remaining in prison without joining the outside world. The aim of punishment is to make the offender know and realize the effect of his action and to give him a chance to repent. That chance is not there in the case of death.
RESPONSE
BOB: "Only a handful of the perpetrators are actually apprehended and sentenced and convicted, so, I am not sure that that has any logic in its argument whatsoever. The fact is there are minority and poor that are in prison. But the fact is that as far as poor, until we can equally distribute wealth that doesn't mean that if you are poor and you commit a crime you shouldn't be sanctioned. Unfortunately, we have a system where if you are rich and you commit a crime, you might not be sanctioned."
EM. "OJ Simpson?"
BOB: "Yeah. That is exactly right. That is an issue of the deficiency of our justice system. It has nothing to do with the sanction imposed whether its burglary. The same thing happens in burglary, the same thing happens in rape, the same thing happens in theft."
EM: "It isn't unique to the death penalty, that is what you are saying."
BOB: "That's right, exactly, right. And thirdly, I don't know if there is a moment more intense in a human being's life than the last seconds where he has to choose whether he or she is going to be remorseful or not and so I think this last one is patently ridiculous. [Inmate's name] completely denied it up until the moment before his execution. Said he

was innocent but then when it came time for him to go see Jesus, he said tell everyone I'm sorry."
EM: "So, is that common in your experience when you hear their last words?"
BOB: "Yes."
EM: "So they really do have a chance to repent?"
BOB: "Yep, that's exactly right. Now there are two of them that I know of that didn't. And there are a bunch of them that go quietly."
EM: "They never said anything?"
BOB: "Right, but I have to believe in my heart that inside, in fact, there's only two that I know of that didn't pray in their last moments."

Again, Bob references the deficiency of the justice system and acknowledges that if you are a rich criminal, you might not be sanctioned. It seemed that Bob fervently needs to believe that repentance is part of capital punishment, more than other penalties, and its implementation effectively pushes inmates to that point.

QUOTE 3. I am strongly opposed to the death penalty. It's about people, human lives, not numbers and statistics, but human beings. If you value life, any life, then you cannot condone the destruction of life in a cold-blooded manner such as the death penalty.
RESPONSE:
BOB: "That's a belief and that's hard to argue. An individual can believe in the sanctity of life. Now where that value comes from I wonder. Because I am not aware of a formal religion that doesn't believe that taking a life costs a life. I mean the Muslim religion does, Jewish religion does, the Christian religion does, um, I don't know about eastern religions but I think by and large they do so, um, that is founded in intellect and belief and not necessarily in any deep theological or philosophical or ethical argument."
EM: "Right."
BOB: "So, when someone believes that, you have to give them the right to believe that. But, uh, again, I think that our laws are founded on the Christian-Judeo ethics and that's why the outcome of the death penalty."

I found it surprising that Bob referenced intellect and belief as different than theology, philosophy, or ethics. It appears that these 'schools of thought' represent external sources of truth for Bob. Again, the justification of capital punishment and its relationship to Judeo-Christian ethics was cited in addition to other world religions. While this may be enough evidence for certain citizens, many counter-arguments can and have been made.

QUOTE 4. I feel that no human being has the right to say that another human being's life is of no value. We are all children of God and he loves us all and for us to kill somebody regardless of the circumstances is taking on the role of God. I am also against the death penalty because of the unjust way it is applied in the U.S. I also feel that it is cruel and unusual punishment.
RESPONSE:
BOB: "Well you know, even if you accept that we are children of God, God gives us a set of rules in which he said thou shalt not kill. Um, now, depends on how you interpret God's

views, right? Because, again if they came to you from Christianity where we are taught an eye for an eye and so on and so forth. If you interpret through Jewish religion, it is essentially the same. If you interpret it through Muslim religion it is essentially the same."

EM: "I know, they cut your hand off if you steal anything."

BOB: "That's right. So, it is interesting. Yeah, this is a good one because on the one hand, if you use that argument, then it's a violation of God's rules to kill that person, and then all of the prophets that were supposedly the most expert in interpreting what God's meaning on earth all said the same thing. So, but that's a belief."

EM: "And it is really hard to respond to . . ."

BOB: "It is unjustly applied. There is no question about that, but that's not the death penalty's problem, that's the justice system's problem. Okay. It's because we've had attorneys and they've screwed it up like everything else they've gotten involved in. And uh you know cruel and unusual is interesting because you know I really feel like the, uh, I mean I haven't died in the electric chair but that's cruel and unusual so I don't think once, once.... Let me think you are going to love this. The eighth amendment is about life. Okay, the death penalty is about death so you can't apply the rules of life to death. So once you've made the determination that that person is going to die, then in my mind the cruel and unusual doesn't apply."

EM: "What you're saying is it's different if you were electrocuted in the electric chair versus. . . ."

BOB: "Right, someone who has been condemned to die. That is only a method so it doesn't seem to me to apply logically, I mean . I understand what they are trying to say but you know when you've decided that that person is going to die it doesn't really matter if it is a shot in the head or the electric chair or lethal gas or lethal injection or hanging."

Bob continued to reference external authorities as the determiners of the rightness of the death penalty. If capital punishment is against God's will, then it is because all the prophets have been wrong. Whether the inmate suffers becomes irrelevant once a mandate for death comes from the heavens (or statute).

From here we moved to the two quotes that supported capital punishment, for Bob's final comments.

QUOTE 5. I do not believe there is a harsh enough punishment for a person who takes another's life. But under the control of law and order this is the most strict and harsh punishment I believe can be given.

RESPONSE:

BOB: "Okay, this one is saying that I do not believe there is a harsh enough punishment. . . ."

EM: "Which is sort of what you were just saying."

BOB: Yes, that's right, but there my belief is that the death shouldn't match, in other words, the method of death shouldn't try to somehow offset the terrible thing that the person did.. Let me use this analogy. The analogy that I use is about going to prison as punishment rather than going to prison for punishment. Now, the liberal approach to it is that you go to prison as punishment. That's one thing. I believe that you go to prison for punishment. It should be a place that you don't like to go to, you don't want to come back to. It should be safe and humane and a lot of other things but you shouldn't want to come back to it. In

the case of death, for me, the death penalty, that's it. You shouldn't be executed in order to somehow, um, in order to somehow offset how that happened. So I guess I'm kind of arguing both ways because this guy is arguing exactly what I'm arguing about prison."
QUOTE 6. I am very strongly in favor of the death penalty. I am a strong believer in the law. If we have laws and someone breaks these laws, there is a punishment. Aggravated murder is the ultimate crime. Therefore, it should have the ultimate punishment – death, not life imprisonment. That is no punishment.
RESPONSE:
BOB: "That's exactly how I feel."
EM: "She actually says that life in prison is no punishment."
BOB: "Well that's not true, but see I guess what I would say to this is that it's founded in, you know, its founded in, its just not a law. A lot of laws are enacted that have some basis in logic and some of them don't even have that. Some don't and we deal with those everyday, so but these laws are founded in a lot of things like they're founded in the Christian Judeo ethics, they're founded in the sanctity of human life. There is a very rich underpinning of those laws and it is truly interesting that in the hierarchy of laws if you start down here with organizational policies and make your way all the way up here that a lot of people will say all of these should be followed. But when you get to that last one, it gets somehow different."
EM: "The exception. What about traffic laws, people don't feel the same about following traffic laws?"
BOB: "You know the whole issue of full enforcement where you know it is impossible to have full enforcement of any given law. And so you get to the law about the death penalty and you say well we can't arrest everybody and prosecute everybody and execute everybody that commits murder, so we can't any law, you know?"
BOB: "You know you often hear someone trying to make this statement in a positive sense that says we are a country of laws. That's true. We have become such a country of laws that in that we've lost common sense. You know we used to be tried by a jury of our peers. That was the people around us that knew us and they knew when we were telling the truth or they knew how much weight to put on what we said because they knew us. Now our juries are such that we can't allow them to know the person at all."
EM: "Well they are contrived too because there is a whole industry around jury selection. There is no justice."
BOB: "There is no such thing as an objective justice but there is a justice that you accept and there is a justice that I will accept and in democratic society out there is this form, if you will, kind of to look at it in terms as the platonic ideal of justice that most of us believe and will accept. What we're dealing with right now because of the mass media and the O. J. Simpson trial is we are coming to realize that what we believed was justice all these years is what we saw in the O. J. trial and that's why our system isn't working. Let me tell you, I think we are coming to a real crisis point. You know, because we have absolutely lost common sense. And that is what law is supposed to be all about."

Even though Bob agreed with the last two students' quotes he partially disagreed and was willing to comment. Bob referenced justice as an ideal that was being challenged by our loss of common sense as a nation. He recognized the fluid nature of laws and perceptual distinctions made between them. Most importantly, he acknowledged that no objective justice exists, that justice is relative, therefore punishments that are considered just today, such as the death penalty, may not be deemed just tomorrow. In the present, Bob must do his job, dispensing justice in the manner and method deemed appropriate by our society.

SUMMARY

Bob works for us and faithfully does what we have asked him to do. "Please handle these heinous felons for us, Bob. Do it neatly and without a lot of negative press for we don't want to appear merciless or inhumane. We don't want to know the gory details, Bob. Just handle it." Everett C. Hughes (1962) provides an historical context for our expectations of Bob by reviving the case of the Nazi *Mein Kampf* (final solution) in an effort to remind us of "dangers which lurk in our midst." Hughes relates our attitudes toward criminals as potentially analogous to the Nazi Germany view of Jews. "The German, pushed the Jews firmly into an out-group: they were dirty, lousy, and unscrupulous. Having dissociated himself clearly from these people, and having declared them a problem, he apparently was willing to let someone else do to them the dirty work which he himself would not do and for which he expressed shame."[14] He speaks of a collective unwillingness to know unpleasant facts that might threaten our conception of ourselves as a people, parallel to our desire to remain ignorant of how the problem is handled. Like Nazi Germany, we are able to find agents who will handle the details for us, people who will spare us the indignation that comes with knowing the details. These people are not that different from us really, perhaps more dedicated to serving society's wishes, if anything. We sleep better because someone else willingly acts on our behalf so we never have to directly participate, thereby exonerating us in the event that something ghastly happens. Bob performs well, working diligently to protect us from the unpleasant details of the duty we have bestowed upon him. Bob handles our business and carries out our wishes.

This chapter serves as an exploratory tool – arguing to expand the discourse on capital punishment to include human dimensions in the conversation. Capital punishment is not a universal truth, accepted by everyone, as eleven states currently do not legalize this form of justice. Therefore, we can draw from no universally accepted authority to proclaim capital punishment as 'correct'. Seeking an external authority to justify our nation's institutions is a poor practice, intolerable in a democracy. We, the people, and we alone, decide what will be the codified practices and acceptable behaviors of our nation. Likewise, we determine how and under what circumstances we will sanction offenders. As citizens, we must not defer to any external authority, whether religious or statutory, for the responsibility that rightfully belongs on our shoulders.

According to a United States Supreme Court precedent, the goals of the death penalty are deterrence of capital crimes and retribution (Bessler, 1993). Furthermore, the Supreme Court has held that the legitimacy of capital punishment is gauged by "evolving standards of decency" within society. Consequently, the responsibility for capital punishment resides with the citizens. We are the ultimate source of authority on what will be considered right and wrong in our society. If we choose to shelter ourselves from the details of capital punishment or any other institutions because awareness makes us uncomfortable, then we are falling down on our responsibility. Prior to granting our support, the burden falls upon each of us individually to participate in conversations exploring the realities associated with capital punishment. If we decide in favor of the taking of life, questions must be answered with deliberation. How paternal should the institutions associated with the practice be? Exactly how much do we want and need to know about the actions taken on our behalf? Once we agree to consciously take a life, what, if any, actions will we consider as too extreme? Under what circumstances will we evoke this penalty? Already we have seen the fluid nature of crimes defined as capital offenses changing over time as demonstrated by Bob's state.

Capital punishment involves much more than the final two minutes it takes to extinguish a life. The average age of a person sentenced to death is 35 (Hare, 1996), therefore, offering society numerous chances to intervene before the capital offense occurred. When products are manufactured and come off the assembly line with defects, a savvy businessman would not just destroy the faulty merchandise. He would walk the line to discover the root cause of the defect and take action to prevent it in the future. Correspondingly, when socialization fails and humans do not internalize our definition of morality or the enticements of rewards associated with conformity, incarceration, or death threats will never address the root cause. Do we dispose of socially unacceptable human beings like faulty products? The enormous complexity of social problems requires complex solutions. Perhaps we need to consider our own culpability when social mechanisms fail? There are many horrors associated with capital punishment, and if we agree to its practice, it seems unfair to ask for protection from the horrors. On the other hand, if we concur to practice some other form of justice, I am confident that Bob will follow our wishes and do that duty equally well. Whatever the decision, it must be made only after we are armed with an understanding of the consequences and a grasp of the human dimensions involved in capital punishment.

We know that there are those who walk among us who seemingly are not of us. And because some of them rob and hate and kill and throw away their lives we call them children of the Devil. Quite often they regard themselves as such. But we, as well as they, are wrong, and when we wreak blind vengeance upon them we do a futile and a tragic thing. Unwittingly we seek to propitiate a malignant god whose goal is to rob us of our humanity.

Caryl Chessman, Death Row, San Quentin, 1954

NOTES

1. This information was contained in a packet of information that Bob's staff prepared for the researcher. (McClennen, 1992).

2. Between 1977 and 1994, the average time spent between the imposition of the most recent sentence received and execution was slightly more than eight years (Hare, 1996). Apparently, Bob's state has a longer than average time.

3. Bessler, p. 365. Bessler reports that the last public execution was carried out in Galena Missouri on May 21, 1937.

4. Madow, p. 465. Madow reports one of the last public hangings as in Owensboro, Kentucky during 1936 where many of a group of ten thousand onlookers stormed the gallows and tore off swatches of the dead man's clothes for souvenirs.

5. The Bible's first reference to legitimate killing occurs in Genesis, the first book of the Bible. "Whoso sheddeth man's blood, by man shall his blood be shed. . . ." Genesis 9:6.

6. Deuteronomy 17:7. The capital crime being idolatry here, but references to other capital offenses are described including breaking the Sabbath and harlotry.

7. Report of the Commission to Investigate and Report the Most Humane and Practical Method of Carrying into Effect the Sentence of Death in Capital Cases, N.Y. Senate Report No. 17, at 81 (1888). Respondents were divided as follows: 80 favored retention of hanging; 87 urged the adoption of electrocution. Poison attracted 8 votes, the guillotine 5, the garrote 4, etc. One recommended that hanging be retained for men, but that women be electrocuted. (Reported in Madow, 1995).

8. Filbrin, p. 150.

9. For a comprehensive look at the debate and legal struggles over publicization of executions, read Bessler 1993; Madow, 1995; and Filbrin, 1992.

10. For a comprehensive review of media cases related to access to executions, read Filbrin, 1992.

11. As of December 31, 1994, lethal injection is the predominant method of execution with 27 states authorizing its use. (Bureau of Justice Statistics, February, 1996).

12. Bob references the media person who cried on the air, which he feels resulted in the shift from gas chamber to lethal injection.

13. *A Horrible Death Scene*, N.Y. Times, August 2, 1884. (Madow, 1995, page 531.)

14. Hughes, Everett C. (1962) p. 7.

REFERENCES

Athens, L. H. (1992). *The Creation of Dangerous Violent Criminals.* Chicago, Illinois: University of Illinois Press.

Bessler, J. D. (1993). Televised executions and the constitution: Recognizing a first amendment right of access to state executions. *Federal Communications Law Journal, 45*(3), 355.

Bohm, R. M. (1989). The effects of classroom instruction and discussion on death penalty opinions: A teaching note. *Journal of Criminal Justice, 17,* 123.

Chessman, C. (1954). *Cell 2455 Death Row: A condemned man's own story.* New York: Prentice-Hall, Inc.

Dobrin, A., Wiersema, B., Loftin, C., & McDowall, D. (1996). *Statistical Handbook on Violence in America.* Phoenix, Arizona: Oryx Press.

Filbrin, P. D. (1992). Pictures at an execution. *Thomas M. Cooley Law Review, 9*(1), 137.

Grisham, J. (1994) *The Chamber.* New York: Dell Publishing.

Hare, G.n F. (1996) *Death Row, 6.* Carlsbad, CA: Glenn Hare Publications.

Hughes, E. C. (1962). Good people and dirty work. *Social Problems, 10*(1), 3.

Madow, M. (1995). Forbidden spectacle: Executions, the public and the press in nineteenth century New York. *Buffalo Law Review, 43*(2), 461–562.

McClennen, C. (1992). *Capital punishment in Arizona: Past, present and future. Arizona Attorney, 17.*

Norris, J. (1988). *Serial Killers.* New York: Anchor Books.

Prejean, H. Sister (1993). *Dead Man Walking: An eyewitness account of the death penalty in the United States.*

APPENDIX A

General Stance

I understand you support the death penalty. In your opinion, under what circumstances should the death penalty be imposed?:

murder rape treason child abuse

What are your thoughts on the benefits derived from the death penalty?

The System

How do you feel about the current justice system and its use of the death penalty? Is the system fair? How could it be improved? What do you feel is the biggest problem with the current system?

Publicity of Executions

Some studies suggest that increasing the publicity of executions would reduce homicides.

What kind of publicity do you feel is appropriate for state executions?

How do you feel about family members of victims witnessing executions?

How do you feel about reporters and the media witnessing executions?

How do you feel about the video-taping of executions?

Should there be restrictions on who can witness an execution? and why?

Duty to Witness

I understand that your position requires that you witness executions.

How many executions have you witnessed in your career?

Are there any other life experiences that compare with the experience to help me get your perspective?

What kinds of emotions do you experience during an execution?

Do you ever lose sleep or your appetite before or after an execution?

Was the experience different with gas chamber vs. lethal injection.

Was your first experience being present at someone's death as a witness at a state execution?

Logistical Issues

What kinds of problems does the corrections system encounter in carrying out executions?

Who is allowed to witness an Arizona execution?

How do you feel about death row inmates being interviewed prior to their executions?

In California, staff members involved in executions undergo special training to assist them in dealing with stress. Does Arizona provide any special training for staff before or after executions?

In response to the gas chamber execution on April 6, 1992, Arizona law changed to require death by lethal injection. As a witness to both types of executions, can you describe the differences between the two methods.

Historically, executions were public events. The last public execution was in 1936 in Missouri. Some proponents argue that the death penalty would have a greater deterrent effect if executions were less private. What are your thoughts on this? How do you think the public would react if executions were public?

Opponents and proponents to the death penalty cite several reasons for their opinions. I would like to share with you some statements from surveyed students and hear what you think of what they are saying.

APPENDIX B

The Death House

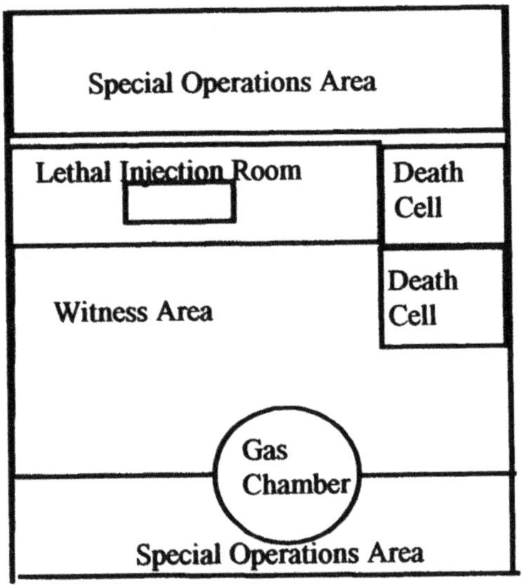

DISCURSIVE ANALYSES OF LEGAL AND SOCIAL CONTROL PROCESSES

LEGITIMATING MURDER? AN ANALYSIS OF NEWSPAPER COVERAGE OF VIOLENCE AT ABORTION CLINICS

Morgan Blake Ward Doran and Gray Cavender

ABSTRACT

Since 1993, extremists have murdered five individuals linked to the abortion profession: Dr. David Gunn, Dr. Bayard Britton, Lieutenant Colonel James H. Barrett, Ms. Shannon Lowney, and Ms. Leanne Nichols. Other victims have been wounded. We analyze the content of 151 articles from The New York Times, The Los Angeles Times, *and* USA Today *subsequent to each incident. Because these crimes satisfy multiple dimensions of newsworthiness, coverage of the 'abortion violence' has been extensive. However, in contrast to traditional crime coverage that condemns the crime and the criminal, the newspapers depicted these incidents as part of an ongoing political debate. To understand why these crimes were covered in this manner, we draw on the literature that deals with media frames.*

INTRODUCTION

On March 9, 1993, Dr. David Gunn was shot to death outside his Pensacola, Florida clinic by an anti-abortion extremist. Dr. Gunn was murdered because he

Sociology of Crime, Law and Deviance, Volume 2, pages 169–189.

ISBN: 0-7623-0680-7

performed abortions. Since then, Dr. John B. Britton, Lt. Col. James Herman Barrett, Shannon Lowney and Leanne Nichols have also been murdered. Dr. George Tiller, Ms. June Barrett, and other victims have been wounded.

Public understanding of these events is, in large part, informed by their representation in the news media. Indeed, media coverage of what have been called 'the abortion killings' or 'the abortion violence' has been extensive. This is not surprising given the media's affinity for violence and especially murder (Chermak 1995; Lotz, 1992). Media coverage of crime and violence has become so pervasive as to constitute a predominant theme in journalism (Altheide, 1997, p. 648). In addition to their inherently violent nature, the abortion killings possess other characteristics coveted by the news media: political protest, moral dilemma, and social controversy (Yale, 1993). These provocative attributes, coupled with the dramatic nature of the events and their connection to the contentious issue of abortion, made the abortion killings newsworthy.

Because murder is so sensational, media coverage of such crimes is usually intense. Murder stories, responding to the public injustice of the act, invoke formats that condemn the crime and the criminal. Crime news is predicated upon an almost Durkheimian (1964) sense of crime as a social wrong (Ericson et al., 1991). This Durkheimian penchant is so fundamental that the condemnatory nature of crime news is not seen as a challenge to journalistic notions of objectivity.

Nevertheless, some scholars characterize crime coverage as overly-simplistic (Ericson et al., 1991). Crime news emphasizes the sensational event and the individual responsible for it and de-emphasizes the social context in which the act and the actor exist (Ericson et al., 1991, p. 8). The focus of this criticism is not so much the lack of balance in coverage, but rather the degree to which crime news de-politicizes criminal events.

In contrast, in the case of the abortion killings, media coverage did emphasize the crimes' context: the political debate over abortion. The abortion controversy has been increasingly newsworthy over the past ten years; media coverage has focused on the pro-life and pro-choice debate (Grindstaff, 1994). However, a focus on political debates in stories about violence at clinics potentially shifts attention away from the criminal dimension of the violence.

The newspapers could have covered the abortion violence in several ways. We argue that the newspapers invoked a frame that resembled political news more than crime news. To understand why the abortion violence was covered in this manner and to assess the implications of such coverage, we draw on the literature that deals with media frames, and with how social movements seek to influence the media.

In this chapter, we offer a case study of three newspapers' reporting of the abortion violence. We focus on how these newspapers depicted the crimes, the accused criminals and victims, and the impact that the sources who were cited had on the representations.

The News: Politics, Abortion, and Crime

Analogous to the ideal of positivism in the hard sciences, the ideal of journalism defines the news media as mirrors that reflect and describe social reality. Like laws in the normative sciences, the news tends to achieve a state of assumed validity. This effect is maintained through a ritualistic adherence to the strategies that convey objectivity (Tuchman, 1978).

However, media scholars, like scholars in the sociology of science, question an objective reality. Patricia Ewick notes that we cannot know the world in a culturally unmediated way; the identification of a fact reflects the theories and assumptions and beliefs of the historical period (1994, p. 93). Media scholars make a similar point: reporters do not simply report facts; they rely on news frames which are selection principles that determine what gets reported and how, and that inform public perceptions of issues (Tuchman, 1978).

News frames are a specific instance of Erving Goffman's (1974, p. 21) general statement of frames as "schemata of interpretation" that allow people to make sense of the occurrences they encounter. Frames, by making occurrences meaningful, "function to organize experience and guide action" (Snow et al., 1986, p. 464). Similarly, news frames organize the news and allow journalists "to cast an incident as an instance of something" (Fishman, 1978, p. 534).

Of course, on most public issues competing frames are available (Gamson & Modigliani, 1989). Indeed, according to scholars, competing social movement organizations actively work to have their world view, i.e. their own particular frames, accepted by the media and by the larger society (Meyer & Staggenborg, 1996). Often such social movements are organized to provide the media with information supportive of their perspective (Gamson & Meyer, 1996).

The issue then becomes how news media select a news frame. The framing process that transforms an event into news is influenced by prior frames, organizational requirements, external factors (such as public information efforts by organizations), and by ideology (Gitlin, 1980). In print news, for example, organizational exigencies such as space limitations and deadlines require concise, timely narratives. Newspapers respond by presenting headlines and lead paragraphs that contain the important details of a story (Ericson et al., 1991). Journalists also make use of predictable sources, like government officials or public information spokespersons, who have information at hand and impart objectivity (Fishman, 1980).

There is a symbiotic bond between reporters and sources. The media legitimate the sources from whom they seek information, and sources, especially official sources, confer legitimacy to news. There is an ideological dimension to this bond and to news frames generally. Journalists in the mainstream media tend to share with the mass audience core cultural values, and to use a media discourse that resonates with those values (Ericson et al., 1989, p. 14; Gamson, 1992, p. 24). Thus, there is a political dimension to journalists' selection of news frames: they choose frames that reflect and reinforce dominant cultural values, and, in so doing, constitute a common system of meanings (Gamson, 1992, p. 27).

To enhance legitimacy, the news media generally undertake "fair and objective" reporting. Fair and objective reporting is practiced through the presentation of "both sides" of a story. Reporters ensure balanced-looking reports by structuring stories in dyadic formats that quote sources on both sides of an issue (Bennett, 1996). For example, citing Democrat and Republican sources creates a sense of balanced, objective political coverage.

Such coverage, when presented as a conflict between opposing sides, creates a narrative tension that makes stories dramatic and newsworthy. A conflict focus is especially apt for political elections: reporters dwell upon the motives and strategies of opposing candidates (Cappella & Jamieson, 1997, p. 9). Because they give a sense of balance while making stories newsworthy, themes of conflict and strategic intent increasingly dominate not only elections, but coverage of other public issues as well (Cappella & Jamieson, 1997, p. 33).

Abortion coverage represents a particular instance of "fair and objective" reporting. Drama and conflict make abortion the perfect media issue, but journalists face a dilemma. Opinion polls reflect that the public favors a woman's right to choose abortion, but a powerful minority, one that holds great sway over electoral politics, opposes freedom of choice (Goggin, 1993; Grindstaff, 1994; Luker, 1984). As a result, the media have no clear populist indicator for abortion coverage. Reporters therefore cite sources on both sides of the abortion debate who put forward their world view – pro-choice or pro-life. As in political coverage, the conflict between the opposing sides, not the underlying issues that might offend either side, becomes the story. Reporters use descriptive warfare metaphors, and "right to life" and "right to choose" become master narratives or motifs in the discourse of abortion coverage (Grindstaff, 1994, p. 65, 74–75). Through this dyadic coverage, reporters tap the drama and conflict that make abortion newsworthy, but appear to be balanced in their coverage.

While objectivity is a media ideal, not all news coverage pursues the standard of fair and objective reporting to the same degree. Crime news employs a unique news frame that diverges dramatically from this balanced style of coverage. Crime news is framed through a recurring pattern of coverage that reaffirms moral boundaries, imputes responsibility, individualizes blame, and promotes resolution (Cavender & Mulcahy, 1998). For instance, murder is not presented in a neutral manner, but rather is condemned as a deplorable act.

In keeping with this condemnatory approach, the alleged murderer usually is named and blamed in language edged with accusatory tones (Lotz, 1991). Depictions of criminals lack the motive-based justifications that are common in political coverage (Ericson et al., 1991). There are exceptions. For example, extremely short, synopsis-like crime reports contain little detail beyond the basics of who did what to whom (Chermak, 1994). Moreover, certain types of crime stories may lack the denunciations that are characteristic of the crime news frame. A good example of such coverage occurs in stories about spousal violence or homicide. When media cover incidents wherein men murder women mates, the men are depicted as "having snapped." The implication is that these men somehow lack the blame that inures to non-spousal violence (Jurik & Gregware, 1992; Websdale & Alaverz, 1998). More commonly, however, the media resist efforts on the part of those charged with violent acts to justify their conduct or to portray themselves as legitimate, even when such interpretations are feasible. For example, British and U.S. newspapers refused to recognize IRA members as legitimate political prisoners, instead portraying them as criminals and terrorists (Mulcahy, 1995).

In contrast, crime victims appear sympathetic and innocent in crime coverage (Chermak, 1995). Media portrayal of criminals and victims demonstrates the dyadic format of crime news. Unlike political news, however, the two sides represent poles, not balance: criminals are evil and victims are good. There are exceptions – again, in situations wherein men murder women mates. Women victims in such coverage sometimes either appear less than sympathetically or perhaps as contributing to their victimization (Websdale & Alaverz, 1998).

Crime news, like the news generally, fragments and personalizes coverage (Bennett, 1996; Ericson et al., 1991). Crime stories tend to individualize blame while also de-contextualizing and de-politicizing crime (Chermak, 1995). There are some exceptions. Stories about juvenile offenders sometimes situate them within a social context, e.g. gangs, especially ethnic gangs. However, such coverage often deals in caricatures. For the most part, the social conditions that are crime's milieu are elided altogether or are reduced to stereotypes within the crime news frame. Even when stories contain an obvious political dimension, crime news de-politicize the acts and the actors and focus instead on individual responsibility (Cavender et al., 1993).

Some theorists draw upon a Durkheimian model to explain how the media cover crime and why crime coverage differs from other news. This model suggests that the news media frame crime as a breach of the community's social values (Alexander, 1988). Media portrayals of crime generally, and murder in particular, symbolically condemn deviant actors and their acts, thereby reaffirming the social order (Dayan & Katz, 1992).

In this paper, we assess how three newspapers covered a particular series of crime stories that were thematically united as abortion violence. Specifically, we focus upon how the coverage framed the abortion violence more as political conflict than as crime news. An important effect of such coverage is its potential to portray the violence as a political strategy rather as acts to be condemned. We offer some theoretically grounded speculations about why the abortion violence was covered in this manner.

To better understand this potential, our analysis focuses on how the newspapers presented the crimes, the alleged criminals, the victims, and who they cited as sources. Before turning to these analytic dimensions, we first present our methodology.

METHODS

Our analysis is based upon newspaper coverage of the five deaths and multiple injuries that occurred during four incidents of violence that were thematically combined in the news as Abortion Violence. The articles were drawn from *The New York Times* (NYT), *The Los Angeles Times* (LAT), and *USA Today* (USA) between March 10, 1993 and January 14, 1995. Respectively, these dates represent the time of the first murder, Dr. Gunn, and a date two weeks after the last murders of Shannon Lowney and Leanne Nichols.

We selected NYT and LAT because they are among the prestige press, newspapers thought to represent the best in reporting. As such, each newspaper is frequently analyzed in media research. We included USA because it is a national publication with a diverse audience, and yet is often overlooked in media research.

To obtain our data we searched the Nexis news service. We generated an initial data set by entering the term 'Abortion and (Two Week Date Restriction)' into the database for each incident. Combining the results for the four searches yielded 610 stories. We deleted articles that did not reference the abortion violence, which left 221 articles. Because we were interested in actual news coverage, we discarded 70 stories that were either a news summary or an opinion/editorial piece. Our final data set comprised 151 articles: 59 were from NYT; 54 were from LAT; 38 were from USA.

A searchable word processing program and a searchable database program allowed us to sift through the data. By segmenting the database program into sections we were able to tailor our searches to pertinent areas of interest. Each section represented a category that would facilitate organization such as 'Headline', 'Date of Article', and 'Newspaper'. Additional thematic sections such as 'Sources Referenced', 'Sources Cited', and 'Motivation Cited' assisted in the analysis. Each of the 151 articles was summarized on a single database sheet.

Document retrieval with Nexis allowed us to translate the articles into a searchable word processing program. This permitted key word searches within the text of each article. Search terms were chosen after we established a repertoire of the vocabulary used by the newspapers through an initial reading of the entire data. The word processing format facilitated searches for organizations such as Planned Parenthood, Operation Rescue, and the Police. In addition to the master document that held all 151 articles, each event was given a separate word processing document so that we could conduct temporally specific searches.

We traced each incident's coverage for two weeks after it was first reported. Using the two search engines, we were able to obtain numerical representations of the three newspapers' coverage. Our analysis includes computations of these data as well as a qualitative assessment of the articles.

We present our analysis in three sections: The Crime; The Criminal and the Victim; The Context of Violence.

THE CRIME

Consistent with past research (Grindstaff, 1994) of thematically linked events, we observed a cumulative pattern in our data. Each succeeding event generated a greater number of articles. Dr. David Gunn's death on March 10, 1993 was reported in 35 articles. Dr. John Britton's and his escort Lt. Col. James Barrett's deaths on July 29, 1994 generated 43 articles. Leanne Nichols' and Shannon Lowney's deaths on December 30, 1994 generated 57 articles. Thus, the escalation in violence at abortion clinics was matched by an escalation in news coverage.

However, the depictions of the abortion violence in the three newspapers diverged from traditional crime coverage. The articles focused less on the acts as crimes and more on their political implications. Specifically, the language, structure, and placement of the crimes were subordinate to the greater abortion issue. The abortion conflict and the strategies and motives of the opposing sides were important themes in the coverage.

We first considered headlines, which are designed to quickly convey a story's thematic focus. Our findings indicate that the headlines addressed the abortion debate more than the violence. As Table 1 demonstrates, criminal event-driven terms such as 'murder', 'slaying', and 'shooting', appeared in 31% of the headlines, while politically-driven terms such as 'abortion', 'activist', and 'war' appeared in 76% of them. The frequency of politically-driven terms increases to 97% if the term 'kill', which appeared in 21% of the headlines, is coded as a political term. According to Merriam-Webster (1991), 'kill' is a neutral term that merely states the fact of a death, while 'murder' implies motive and premeditation, and thus entails moral responsibility. These definitions agree with the common usage of the terms in newspapers, so we coded 'kill' as a neutral term throughout the analysis.

Other, clearly political terms such as 'war' and 'battle' constitute a warfare metaphor that, when used in these headlines, emphasized the conflict between opposing factions. For example, a headline in a story about Dr. Gunn's death

Table 1. Frequency of Terms in Headlines

TERM	LAT	NYT	USA	TOTAL	PERCENT
Politically Driven					
Abortion	40	26	17	83	55.0
Abortion Doctor	5	5	1	11	7.3
Abortionist	0	1	0	1	0.7
Activist	6	0	1	7	4.6
Battle	0	1	1	2	1.3
Fight	1	1	3	5	3.3
War	1	2	3	6	4.0
TOTAL	115				76.2
Kill	13	18	1	32	21.2
TOTAL				147	97.4
Event Driven					
Murder	4	0	1	5	3.3
Shoot	12	7	3	22	14.6
Slay	10	7	3	20	13.3
TOTAL				47	31.1

exclaimed that there had been a 'Slaying in Pensacola', but then invoked the warfare metaphor by adding, "A doctor pays with his life in the war over abortion" (NYT 3/14/93, p. 4:2). A headline which noted that a Florida city, "is (the) unwitting vortex of abortion war" (LAT 7/31/94, p. A1) emphasized an embattled place. In the coverage, the violence of the crime was co-existent with the political stance that initiated it.

Similarly, within the text of the articles, the crime was dominated by the political debate. Crime stories usually open with the important information about the incident, and then "let details. . . dribble out in later paragraphs" (Lotz, 1991, p. 156). In the articles that we examined a quick reference to the crime was attended by a more extensive discussion of strategy or other aspects of the abortion issue. A story that began with a report about the "killings of two workers at abortion clinics in Massachusetts," immediately re-focused on abortion strategy by observing that the crime had "touched off a debate among . . . prominent Catholic and Evangelical Protestant leaders over whether opponents of abortion should pull back from sidewalk protests and turn instead to prayer within church walls" (NYT 1/9/95, p. A1). The article, and others like it re-addressed the crime, but deeper within the text. In the articles, political, theological, and ideological themes dominated the narratives.

Specific terms also convey an article's thematic focus. The term 'murder' was referenced an average of 1.5 times (N = 222) per story (see Table 2). However, the term 'kill', which we coded as a neutral, non-condemnatory term, averaged 3.2 references (N = 484) per article. Although these terms convey unique meanings in a sentence, their uniqueness was often diminished by conjoining references to abortion or abortion clinics. 'Abortion' was the most frequently used term averaging 11.5 (N = 1731) references per article. Even sentences that contained violent terms tended to emphasize abortion. By stating that, "the killing was the first of its kind since abortion was legalized in the US more than 20 years ago" (NYT 3/12/93, p. A17), the political issue is as much a focus of the sentence as the murder.

While the crime was thematically subordinate in some articles, in others it was elided altogether or was used as a referent for an issue-driven narrative. A progression from the crime theme to a political theme emerges in the following story:

> For women's rights leaders, the slaying of Gunn and the departure of the two Melbourne doctors are a result of . . . an intensified campaign by anti-abortion forces to stop abortion through the harassment, humiliation and intimidation of doctors.
>
> Until last year, the anti-abortion movement had hopes that its influence on the Republican White House . . . and the U.S. Supreme Court would win a reversal of the . . . Roe vs. Wade decision and end the right to legal abortion. But the court's ruling affirming

> . . . abortion . . ., followed by the election of Democratic President Clinton, dashed those hopes. Undaunted, the most extreme factions of the anti-abortion movement have simply changed their focus, turning up the pressure on abortion providers instead. And their campaign seems to be working (LAT 3/14/93, p. A1).

Initially, the article references the crime, but each subsequent paragraph moves incrementally away from the murder to other issues: the operation of abortion clinics, the anti-abortion movement's links to the Republicans, the role of the Supreme Court, President Clinton's election, and anti-abortion strategies and their effects. In doing so, the article shifts focus from a politically motivated murder to the politics of abortion. To a degree, the article treats violence as a political strategy.

Another example of politicizing the articles is seen in the recurring use of the term 'justifiable homicide'. Crime stories typically do not offer justifications

Table 2. Frequency of Term Usage in the Coverage of Abortion Violence in Three Newspapers (N = 151 Articles)

TERM	Incident4 12/30/94 n = 57	Incident3 7/29/94 n = 43	Incident2 8/19/93 n = 16	Incident1 3/10/93 n = 35	Total Usage of Term	Average Use Per Article
Violence Related						
Attack	115	34	11	18	178	1.18
Assassin	1	1	0	3	5	0.03
Battle	5	10	5	15	35	0.23
Fight	9	5	4	8	26	0.17
Gunman	61	0	0	4	65	0.43
Homicide	14	20	6	1	41	0.27
Kill	170	176	37	101	484	3.21
Murder	76	71	27	48	222	1.47
Shoot	169	96	64	60	389	2.58
Slain	12	18	0	11	41	0.27
Slay	11	26	6	19	62	0.41
Slaughter	10	2	0	0	12	0.08
Violence	149	131	40	74	394	2.61
War	21	12	1	6	40	0.27
Abortion Related						
Abortion	577	426	227	501	1731	11.46
Abortion Opponent	20	13	4	7	44	0.29
Abortion Protest	23	26	12	29	90	0.60
Activist	16	22	0	42	80	0.53
Anti-Abortion	103	113	47	84	347	2.30
Protest	115	97	49	99	360	2.38

for murder, while political coverage speculates about motives (Ericson et al., 1991; Capella & Jamieson, 1997, p. 19). Although the majority of the articles (58%) followed the traditional crime pattern, a large minority (42%) discussed motivation and justification. The justifiable homicide defense accounted for 19% of the motives proffered. Paul Hill, who first invoked the term on the 'Donahue' television program, unsuccessfully employed the defense in his conviction for the deaths of Dr. John Britton and Lt. Col James Barrett. The term became notorious when prominent pro-life leaders signed a declaration advocating violent force to save unborn lives (NYT 8/4/94, p. A1).

'Justifiable homicide' was used as an exculpatory term, although mainly by pro-life extremists cited in the articles. Even so, the legitimacy of the defense was countered only by pro-choice activists, which tended to render it more viable. One week after violence in Brookline, Massachusetts, an article stated that, "the shootings represented a turning point in the fight over abortion rights . . . that signaled a shift from a political, moral and intellectual debate to violence" (LAT 1/7/95, p. B4). By presenting violence in this manner, the murders were depicted as a strategy in an ongoing political conflict. Further, by introducing politically exculpatory motivations for violence, the articles "carry excuses and justifications for behavior(s)" that are otherwise not subject to dispute (Ericson et al., 1991, p. 269).

The headlines, the texts, and the terms in these articles de-emphasized the crime in deference to the abortion conflict. The political and strategic ramifications of the deaths, not the condemnation characteristic of murder stories, were the focus of the articles. This political format is also apparent in the coverage of the accused and the victims, to which we now turn.

THE CRIMINAL AND THE VICTIM

News coverage of political issues features polemical discourse that dramatizes conflict while affecting neutrality (Cappella & Jamieson, 1997). A story appears to be balanced when both sides receive equal coverage. In contrast, crime news, and especially murder stories, condemns the criminal for killing the sympathetically portrayed victim.

Unlike traditional crime news characterizations, the articles that we examined presented the accused and the victim in a confrontational framework. They were cast as proponents of polemic political agendas. Stories portrayed defendants as committed anti-abortion extremists and victims as strident advocates of women's rights (LAT 7/30/94, p. A1). Commensurate with political formats, these individuals were defined, not so much by their criminal-victim relationship, as by their 'abortionality', that is, by their position, pro or con, on the abortion issue. These descriptive markers invoked a master status for the individuals who were portrayed in this manner.

As a further clue to the defendant's and victim's mediated identities, we examined their sequential identification in the articles. Because reporters convey important facts early in an article, we examined how the accused and the victim were first identified. Crime stories usually name the defendant and the victim early in an article (Chermak, 1995, p. 59). In 68% of our articles, the first reference to the accused failed to name him/her; 45% of the initial references neglected to name the victim. Instead, these articles first identified the subject according to her or his political position, e.g. as "a committed abortion-rights advocate" (NYT 7/31/94, pp. 1,26).

The trend toward such identification was amplified as stories progressed. After the accused or the victim was identified, ideological 'tags' were attached that positioned them in the greater abortion conflict. Thus, stories identified the accused as "being obsessed by religion and abortion" (LAT 1/2/95, p. A1), or as "a Scripture-quoting student hairdresser" (NYT 1/10/95, p. A12). While these and other characterizations of the defendants were unflattering, they lacked the condemnatory language that we expected in coverage of murder. Moreover, these identification tags were oriented toward the abortionality of the accused, rather than criminality. References such as, "An Oregon woman who treated abortion protesting like a full-time job goes to court today, accused of shooting a Kansas doctor at his clinic" (USA 8/23/93, News, p. A3), depicted criminality as an extension of her abortionality.

Victims were identified in a similar manner. Dr. David Gunn was frequently referred to as "a local abortion doctor" (USA 3/16/93, p. A7). His and other victims' identities were conflated with their abortion-related occupations. As Table 3 illustrates, the term 'abortionist' was used in 23% (N = 34) of the articles. Together with the phrase 'abortion doctor', these characterizations occurred in 91% (N = 138) of the articles. This usage is contrasted with descriptors such as Gynecologist (N = 3, 2%), Ob-Gyn (N = 0, 0%), or Obstetrician (N = 3, 2%), terms which more aptly describe these doctors' specializations. What is missing from these articles is the sympathy-evoking language that usually characterizes murder victims and their families. Instead, abortion-related identifications appeared that may diminish their aura of innocence. As the "only doctor regularly performing abortions in this area" (NYT 3/13/93, p. 1, 6), Dr. Gunn's death was presaged.

Although identification based upon a victim's abortionality has clear political undertones, the most potent examples of political framing occurred when the accused or the victim was not named at all. In these articles (Victim, N = 10, 7%; Accused, N = 32, 21%), the crime narrative quickly gave way to the politics of abortion. For example, the lead paragraph of one article noted that, "an abortion doctor and his volunteer escort were shotgunned to death Friday outside an abortion clinic, 16 months after a similar fatal attack here prompted federal legislation banning violence against such facilities" (LAT 7/30/94, p. A1). The victims were mentioned, although not by name, but their identities were largely an abstraction. The abstraction was subsumed by a quick shift from the specific event to the broad abortion controversy.

By presenting abortion, not the clinics, as the site of conflict, the coverage instated a confrontational tone that privileged warfare metaphors over individual losses. The notion of conflict is a recurring pattern in political stories in general and abortion coverage in particular (Grindstaff, 1994; Cappella & Jamieson, 1997). For example, one story referenced a sign held by anti-abortion protesters that read, "John Salvi – Prisoner of War" (LAT 1/2/95, p. A1). Warfare metaphors add to an article's tension by reproducing the opposition that defines the two factions. Another article noted that Mr. Salvi was, "the latest hero in a battle that once was waged only with words" (USA

Table 3. Frequency of Medical and Political Terms in the Text of the Coverage of Abortion Violence in Three Newspapers (N = 151 Articles)

TERM	Incident1 12/30/94 n = 57	Incident2 7/29/94 n = 43	Incident3 8/19/93 n = 16	Incident4 3/10/93 n = 35	Total Usage of Term	Average Use Per Article
Doctor Related						
Abortion Doctor	13	57	28	6	104	0.69
Abortionist	5	10	7	12	34	0.23
Gynecologist	1	0	0	2	3	0.02
OB-Gyn	0	0	0	0	0	0
Obstetrician	0	0	0	3	3	0.02
Political Organizations						
Abortion & Reproduction	3	4	0	0	7	0.05
ACLU	1	0	0	1	2	0.01
Feminist Majority	10	10	1	3	24	0.16
Nat.Abortion Federation	2	5	1	4	12	0.08
NOW	5	3	1	7	16	0.11
Operation Rescue	25	23	13	45	106	0.70
Planned Parenthood	99	6	2	20	127	0.84
Pro-Life Groups	26	21	6	24	77	0.51
Rescue America	1	1	1	15	18	0.12
Right to Life	11	8	7	11	37	0.25

1/3/95, p. A3). This language portrayed the participants as combatants. In such a context, violent acts may become elements of an insurgent mentality.

When the accused and victim are depicted as disputants in a battle, coverage of the outcome of their actions is politicized as well. One article specifically addressed the political ramifications of the murder of Dr. Gunn.

> Most anti-abortion activists concede that their momentum has ground to a crawl. The killing of . . . abortionist Dr. David Gunn and subsequent pro-choice fund-raising ads ("They shot him in the back because he stood up for our right to choose") are just the most recent setbacks. A combination of court decisions and Clinton Administration reversals had already driven the movement to its knees (LAT 3/22/93, p. E1).

In such a framework, the accused's actions were assessed primarily according to their effect on the political plane rather than on the victim. The victims' deaths were re-shaped as unfortunate casualties of war.

Such depictions of the accused and the victims positioned the coverage as political rather than as crime news. Of course, reports did not specifically exculpate the accused, but, unlike traditional murder stories, they presented ideological motivations for the crimes. The portrayal of the victims, particularly the emphasis on their abortion-related professions, denied them the status of innocent victims and perhaps rendered them partially responsible for their victimization. In contrast to typical murder stories, the coverage did not offer the symbolic condemnation that creates a sense of social wrong. One important aspect of condemnation arises from the choice of sources who are cited in a story. We now turn to this dimension.

THE CONTEXT OF VIOLENCE

Reporters provide a thematic context for the stories they cover. This context organizes reality for them and for the news consumer (Altheide, 1997); it also determines whose perspective will reach the audience. Sources are an important aspect of a story's context because they symbolically relate an article's thematic focus. Through their choice of sources, the newspapers that we examined contextualized the abortion killings more as instances of political contention than as criminal acts.

For reasons of efficiency and legitimacy, the news media typically cite authoritative sources. In crime coverage, sources are usually criminal justice officials (Fishman, 1980). Stephen Chermak (1995) found that 49% of the sources cited in newspaper crime coverage were police or court officials. This number was much smaller in our study.

Although a decline in criminal justice sources for highly newsworthy incidents is consistent with past research (Chermak, 1995), the extent of this

Table 4. Sources Cited in Text in the Coverage of Abortion Violence in Three Newspapers (N = 151 Articles)

REFERENT	Total	Percent
Abortion Organization	149	30.9
Abortion Provider	52	10.8
Pro-Life Activist (Non-Affiliated)	41	8.5
Pro-Choice Activist (Non-Affiliated)	5	1.0
Court Official	12	2.5
Misc. Government	14	2.9
Police	18	3.7
Politician	51	10.6
Defendant	17	3.5
Defendant Acquaintance	26	5.4
Victim	13	2.7
Victim Acquaintance	21	4.4
Church	28	5.8
Witness	11	2.3
Citizen	15	3.1
Media	2	0.0
Other	8	1.7
TOTAL	483	99.8

reduction far exceeded expected levels. As illustrated in Table 4, police and court officials accounted for only 6% of the sources cited. Further, only 20% of the total number of sources were accounted for even when all government sources (police, courts, government leaders, and miscellaneous government officials), were combined. Clearly, these newspapers' use of government, especially criminal justice sources, was unusually low for murder stories.

In an attempt to provide clarification for the lack of official sources, we added the category 'abortion groups' to the sources that we analyzed. Abortion groups are defined as individuals and organizations who were referred to as pro-life or pro-choice activists, spokespersons, or workers (e.g. doctors who performed abortions or clinic administrators). This category included 36 distinct political organizations that were cited as factions, pro and con, in the abortion debate.

Abortion groups accounted for 51% of all sources cited across the four events covered. As shown in Table 4, this category overshadowed all other sources. Within this category an almost equal distribution of pro-choice and pro-life sources were cited, a strategy which appears balanced, and thus shoresup objectivity as a media ideal (Tuchman 1978). Two prominent organizations, Planned Parenthood and Operation Rescue, were cited 1.7 and 1.8 times more than the police, respectively. In many cases, all of the sources cited in an article were from abortion groups.

As we have noted, murder stories usually denounce the crime and the criminal. Because the police arrest criminals and solve crimes, their use as a source invokes an inherent condemnation (Ericson et al., 1991, p. 182). However, in part because they were cited more often, and, in part, because of the debate format adopted in the articles, condemnation of the violence most often came from pro-choice groups. The following excerpt from a LAT article illustrates both the "two sides of coverage" political frame, and the tendency for pro-choice spokespersons, not criminal justice sources, to be used as critics of the violence.

> A Torrance obstetrician who occasionally performs abortions was rudely surprised . . . when he saw his photo . . . on an anti-abortion flyer titled: "Wanted! For Killing unborn babies. . . . 'I don't know how crazy these people are, so I'd rather not be singled out', said the Torrance doctor, declining to be quoted by name. 'I have a relatively secure, easy life that I'd like to finish off in the time God's allotted me'."

Following this pro-choice oriented quote, the article shifted to a pro-life concern: "Abortion foes say they are non-violent. They say they are being wrongly portrayed as fanatics and terrorists." Next, the article shifted to a pro-choice response that condemned pro-life violence: "Abortion rights advocates express outrage at such campaigns, asserting that the tactics turn doctors into targets." Finally, the pro-life side is allotted a response:

> Abortion foes counter that their opponents are attempting to capitalize on the Florida death. At the Pro-Life Coalition, Finn was particularly critical of New York Times columnist Anthony Lewis' piece last week characterizing most anti-abortion activists as religious fanatics (LAT 3/21/93, p. B3).

The article oscillates between the two sides of the abortion debate, a technique that provides the sense of balanced coverage characteristic of political news (Cappella & Jamieson, 1997). When violence is condemned, the pro-choice groups condemn it, but the pro-life faction is then afforded a response. The sort of information about the crime that is typically provided by criminal justice sources is not imparted.

This tendency is further demonstrated in our analysis of the group referents made in the articles. In contrast to a source citation which delineates an actual quote by a person or group, a referent category is simply the number of times a group's name appeared in the text of these articles. We found that, similar to the source statistics, the largest number of referents were to abortion groups (see Table 3). The top ten (of 36) abortion affiliations were referenced 426 times, with Planned Parenthood (N = 127) and Operation Rescue (N = 106) heading the list. The coverage was again informed by the political frame, which gives equal time to both sides in a controversy. In contrast, the term 'police' occurred only 306 times. While police were quoted as sources only 18 times, the term 'police' was used more often, although considerably less often than were the abortion groups.

Authoritative sources legitimate news coverage because they have official information and, with political news, because they provide a sense of balance (Cappella & Jamieson, 1997). In the articles that we examined, pro-life and pro-choice groups were given the status of authority with seemingly equally viable perspectives. Consistent with political coverage, each side was afforded an opportunity to present its world view. And, as social movement organizations, each side had a developed world view that either attempted to legitimate or condemn the violence. The discussion of facts by police, which inherently condemns violent crime, was displaced. In such a context, the abortion murders were not condemned by governmental authority but were, instead, discussed within a political news frame as a strategy.

CONCLUSION

Our analysis is a case study of three newspapers' treatment of the violence that has erupted at clinics that provide abortion services. Specifically, we examined the thematic depictions of the newspapers' coverage of these murders. Our findings conform with many of the tenets of media research.

Reporters used an existing news frame, the abortion controversy, to organize the four events into a thematic whole: abortion violence. Consistent with recent coverage, the reports emphasized the political confrontation between the two sides of the abortion controversy. This dyadic format accentuated the dramatic conflict that makes events newsworthy and also invoked the rituals that help legitimate the media and their ideal of fair and objective reporting (Tuchman, 1978). The newspapers addressed debate, strategy, and the efficacy of different tactics (see Cappella & Jamieson, 1997).

While there usually are several competing frames available to the news media on public issues (Gamson & Modigliani, 1989), what is interesting about this coverage is that the newspapers invoked a political news frame in stories of murder and assault. Certainly there is a political dimension to the abortion violence reported in these stories. Indeed, there is a political dimension to crime generally; but mainstream media rarely discuss it. What is significant in the coverage that we analyzed is that the political frame so completely overwhelmed the crime news frame.

Of course, the coverage that we analyzed did not condone, much less actually legitimate these murders. Nevertheless, the politics of abortion, not crime, was the focus of these articles. This focus was apparent in the language that was used to describe the four events. War metaphors and political terms like 'protest' predominated over value-ladened words like 'murder'. In place of the usual 'good and evil' dyad employed in crime stories, alleged criminals and victims were defined by their abortionality. As a result, these individuals' position in the abortion debate was presented as more important than their relationship to the crime.

Neither the crimes nor the criminals received the condemnation that tends to occur in crime news. When condemnation did occur, it came from pro-choice sources, not criminal justice officials. Thus, the coverage lacked the categorization of crime as a social wrong and the reaffirmation of the social consensus that a Durkheimian model of crime news would suggest (see Ericson et al., 1991, p. 110; Alexander, 1988). Similarly, the newspapers did not portray the crime victims in the sympathetic light that usually occurs in crime news. As a result, the coverage tended to normalize their victimization rather than treat it as an injustice.

In addition to the politicizing language and the focus on abortionality rather than the criminal/victim relationship, the sources cited in the articles were more likely to be Planned Parenthood or Operation Rescue than the traditional crime news source: the police. The coverage differed markedly from what might be expected in reports of a series of murders.

News frames are designed to organize reality for news workers and consumers. They signal to the public how events are to be understood (Savelsberg, 1994). Embedding the abortion violence in a political frame potentially has problematic legitimating effects. First, it offers authority to pro-life extremists who appear as news sources, an inherently legitimating role; it also allows them to attempt to justify the violence. Second, the 'balanced' presentation of a political frame suggests that the opposing arguments over the violence are merely differences of political opinion: each as acceptable as the other (see Meyer & Staggenborg, 1996). Finally, imposing a politically grounded warfare metaphor upon the murders suggests that, although extremists murdered five people, the violence was precipitated by, and arguably was a result of, the greater war against abortion.

Violence at four clinics that resulted in five deaths made these stories newsworthy. Because the victims provided abortions at these clinics, the abortion controversy is a viable frame of coverage. Even so, to frame these events as political news is unusual when reporting acts of violence and murder. The media resist efforts to portray political violence in a positive manner

(Mulcahy, 1995). Yet, a news frame was chosen in these articles that did not condemn the violent acts in the usual fashion, but covered them with a sense of balance reserved for legitimate political conflict.

The obvious question is why did the newspapers choose a political rather than a crime news frame. We suggest several interrelated answers. First, abortion is a contentious issue that lacks a clear ideological position for mainstream coverage. Lacking such a safe position, the newspapers fell back on the traditions of fair and objective reporting: they covered both sides of the controversy. Second, the pro-life and pro-choice sides were well organized social movements, each with their own clearly established position and with the resources available to respond to media inquiries when sources were needed. Third, by letting each side say its piece, the media avoided controversy (and possibly pressures from these social movement organizations) and, as is standard in political coverage, made the conflict the story. Fourth, criminals seldom receive favorable coverage: their actions violate social norms and they rarely are members of legitimate social movements.

Ultimately, however, the decision to frame the abortion violence in this manner must be seen as a political decision: one that is consistent with the media ideal of objectivity and balance, and, at the same time, seeks the safe ideological ground in terms of cultural values. The media, by choosing a political frame over a crime news frame, demonstrates its power to encourage or discourage notions of injustice. In this case, the newspapers failed to evince the sense of injustice that generally attends media coverage of the murder of innocent victims (see Snow et al., 1986).

The newspapers that we examined opted for a political frame of coverage in these stories. The use of Planned Parenthood and Operation Rescue as sources produced the sense of 'balance' that passes for objectivity in reporting. By simply reporting the rhetoric of the opposing factions, these newspapers avoided the controversy that accompanies the abortion issue. In this way, these newspapers remained above the fray and appeared to be objective. However, by politicizing the four events, this brand of news coverage surely understates the deaths of Dr. Gunn, Dr. Britton, Lt. Col. Barrett, Ms. Nichols, and Ms. Lowney.

ACKNOWLEDGEMENT

We appreciate comments of Nancy Jurik and Michael Musheno on earlier versions of this manuscript.

REFERENCES

Alexander, J. (1988). Culture and Political Crisis: Watergate and Durkheimian Sociology. In: J. Alexander (Ed.), *Durkheimian Sociology: Cultural Studies*. New York: Cambridge University Press, pp. 187–224.

Altheide, D. (1997). The News Media, The Problem Frame, and the Production of Fear. *The Sociological Quarterly*, *38*, 647–668.

Bennett, L. (1996). *News: The Politics of Illusion*, 3rd Ed. White Plains:Longman Publishers.

Cappella, J., & Hall Jamieson, K. (1997). *Spiral of Cynicism: the Press and the Public Good*. New York: Oxford University Press.

Cavender, G., & Mulcahy, A. (1998). Trial By Fire: Media Constructions of Corporate Deviance. *Justice Quarterly*, *15*, 501–521.

Cavender, G.,Jurik, N. C., & Cohen, A. K. (1993). The Baffling Case of the Smoking Gun: The Social Ecology of Political Accounts in the Iran-Contra Affair. *Social Problems*, *40*, 152–166.

Chermak, S. (1995). *Victims in the News: Crime and the American News Media*. Boulder:Westview Press.

Dayan, D., & Katz, E. (1992). *Media Events: The Live Broadcasting of History*. Cambridge:Harvard University Press.

Durkheim, E. (1964). *The Division of Labor in Society*. New York:The Free Press.

Ericson, R., Baranek, P., & Chan, J. (1989). *Negotiating Control: A Study of News Sources*. Toronto:University of Toronto Press.

Ericson, R., Baranek, P., & Chan, J. (1991). *Representing Order: Crime, Law and Justice in the News Media*. Toronto:University of Toronto Press.

Ewick, P. (1994). Integrating Feminist Epistemologies in Undergraduate Research Methods. *Gender & Society*, *8*, 92–108.

Fishman, M. (1978). Crime Waves and Ideology. *Social Problems*, *25*, 531–543.

Fishman, M. (1980). *Manufacturing The News*. Austin:University of Texas Press.

Gitlin, T. (1980). *The Whole World Is Watching*. Berkeley: University of California Press.

Gamson, W. (1992). *Talking Politics*. Cambridge:Cambridge University Press.

Gamson, W., & Meyer, D. (1996). Framing Political Opportunity. In: D. McAdam, J. McCarthy, & M. Zald (Eds), *Comparative Perspectives on Social Movements: Political Opportunities, Mobilizing Structures, and Cultural Framings* (pp. 275–290). Cambridge:Cambridge University Press.

Gamson, W., & Modigliani, A. (1989). Media Discourse and Public Opinion on Nuclear Power: A Constructionist Approach. *American Journal of Sociology*, *95*, 1–37.

Goffman, E. (1974). *Frame Analysis*. Cambridge: Harvard University.

Goggin, M. (1993). A Framework for Understanding the New Politics of Abortion. In: M. Goggin (Ed.), *Understanding the New Politics of Abortion* (pp. 1–18). Thousand Oaks: Sage Publications.

Grindstaff, L. (1994). Abortion and the Popular Press: Mapping Media Discourse From Roe to Webster In: T. Jelen & M. Chandler (Eds), *Abortion Politics in the United States and Canada* (pp. 57–88). New York: Praeger.

Jurik, N. C., & Gregware, P. (1992). A Method for Murder: The Study of Homicides by Women. In: J. Holstein & G. Miller (Eds), *Perspectives on Social Problems, Vol. V* (pp. 179–201). Greenwich, CT: JAI Press.

The Los Angeles Times (1993). Selected articles, March 10, 1993–January 14, 1995.

Lotz, R. (1991). *Crime and the American Press.* New York: Praeger.
Luker, K. (1984). *Abortion and the Politics of Motherhood.* Berkeley: University of California Press.
Meyer, D., & Staggenborg, S. (1996). Movements, Countermovements, and the Structure of Political Opportunity. *American Journal of Sociology, 101*, 1628–1660.
Mulcahy, A. (1995). Claims-making and the Construction of Legitimacy: Press Coverage of the 1981 Northern Irish Hunger Strike. *Social Problems, 42*, 449–467.
The New York Times (1993). Selected articles, March 10, 1993–January 14, 1995.
Savelsberg, J. (1994). Knowledge, Domination, and Criminal Punishment. *American Journal of Sociology, 99*, 911–943.
Snow, D., Rochford, B. Jr. Worden, S., & Benford, R. (1986). Frame Alignment Processes, Micromobilization, and Movement Participation. *American Sociological Review, 51*, 464–481.
Tuchman, G. (1978). *Making News.* New York:The Free Press.
USA Today (1993). Selected articles, March 10, 1993–January 14, 1995.
Websdale, N., & Alvarez, A. (1998). Forensic Journalism as Patriarchal Ideology: The Newspaper Construction of Homicide-Suicide. In: F. Bailey & D. Hale (Eds), *Popular Culture, Crime and Justice* (pp. 123–141). Belmont, CA: Wadsworth.
Yale, M. (1993). Abortion, Elections, and the Media. In: M. Goggin (Ed.), *Understanding the New Politics of Abortion* (pp. 134–153). Thousand Oaks:Sage Publications.

READING PRISONS: A METAPHORIC-ORGANIZATIONAL APPROACH

Bruce A. Arrigo and Christopher R. Williams

ABSTRACT

Current research on prison organizations has not developed a comprehensive framework for how to interpret these complex networks. This is surprising given that learning how to 'read' the behavior of correctional facilities can have an enormous impact on program development and policy formation. The present study endeavors to fill this gap in the literature. The authors utilize Morgan's (1997) organizational template on metaphorical analysis and apply it to the configuration of modern prisons. To facilitate this investigation, the authors rely upon an eclectic mix of existing corrections literature to demonstrate how the particular metaphor in question can be used to analyze the prison as an organization. This perspective offers a more systematic explanation of how these metaphors routinely shape correctional research, practice, and policy. Eight images are identified and examined, including the prison as: (1) culture, (2) organism, (3) brain, (4) instrument of domination, (5) political system, (6) machine, (7) psychic confinement, and (8) flux and transformation. The authors conclude by commenting on the viability of adopting the proposed organizational model for future prison studies.

INTRODUCTION

Understanding how prisons behave requires a comprehensive examination of the diversity of underlying operational practices which are present in most

Sociology of Crime, Law and Deviance, Volume 2, pages 191–231.

ISBN: 0-7623-0680-7

correctional facilities across the country. This analysis of 'reading' prison organizations allows for an assessment of both strengths and weaknesses existing within the system. Modifications based on such assessments are fervently pursued by the public, government, and law enforcement, and are necessary for the evolution of this complex network.

Identifying how decisions and relationships internal and external to the prison are affected by confinement philosophy and practice is the first step in the process. Organizational theory facilitates this process by suggesting the method and means. The existing prison literature can further contribute to the analysis by consolidating organizational principles with sociological, criminological, legal, and psychological perspectives. This collaboration fosters an integrated and comprehensive understanding of organizational theory which shapes the daily behaviors of correctional facilities.

Accordingly, the following analysis will attempt to examine the more common practices and behaviors in the prison system from an organizational perspective, informed by the traditional theoretical and applied prison literature. The study will approach such behaviors in the context of eight metaphors as outlined by Gareth Morgan (1997). The use of metaphor in assessing organizational behavior enables us to appreciate more systematically the prominent features of the prison milieu. Through metaphoric inquiry, we are able to acknowledge a number of themes which present themselves in the day-to-day operation of a prison. Though the metaphors are treated largely independently for the purpose of this assessment, it is not difficult to understand how correctional facilities assume properties of many of these eight metaphors in practice.

On The Meaning of Metaphors in Organizations and Some Comments on Method

Morgan (1997) establishes a template, or framework for addressing organizational issues through the use of metaphor (see also, e.g. Morgan, 1980, 1981, 1983, 1993, 1996). While the suitability and applicability of his model has not been tested in the prison context, Morgan (1997: 3–8) contends that his model is useful for understanding all managed networks. He concedes that the selected metaphors are by no means exhaustive, though, they adequately represent the broad range of ideas and perspectives that exist regarding organizational theory and practice.

According to Morgan (1997: 8), the metaphor is 'central to the way we read, understand, and shape organizational life'. The use of metaphor establishes a "*way of thinking*" about and a "*way of seeing*" through our everyday experiences (Ibid.: 4). In this context, metaphor implies a certain limited perspective in which understanding unfolds. This is not so much a criticism of metaphorical analysis as much as it is a recognition that the process both reveals something quite unique about the world in which we live while, at the same time, concealing several of its distinctive characteristics.

Central to Morgan's (1980) analytic scheme is a conviction that it is possible to uncover and interpret the core assumptions that 'characterize and define any given world view' regarding the behavior of an organization (p. 607). Typically, these core assumptions are situated within a paradigm (Kuhn, 1970) that inform how reality is constructed for the organization or institution in question. Morgan (1980) identifies four such paradigms: the functionalist; the interpretive; the radical-humanist; and the radical-structuralist.[1] What is important about these paradigms or world views in relation to Morgan's theoretical approach is that each one embodies certain precepts about the nature of science and the nature of society (Burrell and Morgan, 1979). Moreover, each paradigm is linked to favored metaphors in accord with preferred views of reality, demonstrating how 'puzzle solving' activities in organizational analysis are fundamentally a product of the social construction of scientific knowledge (Morgan, 1980: 607).

Morgan's conceptual scheme on paradigms is a framework or 'blueprint' for metaphorical analysis. Where the individual paradigms represent preferred ways of approaching and assessing a shared reality (e.g. the organization and behavior of prisons), metaphor is the foundation upon which social inquiry in organizational analysis unfolds. Metaphor is what makes the paradigm in use concrete by giving it identifiable form (Ibid.: 610). Indeed, the naming of 'concepts, ideas, facts, observations, etc., do not so much denote external 'things', as conceptions of things activated in the mind by a selective and meaningful form of noticing the world which may be shared with others' (Ibid.).

Methodologically, the application of metaphorical analysis to reading the behavior of an organization (including the modern prison) entails a series of intellectual steps. The approach begins by accepting the premise that metaphor is a basic mode of symbolism and that all theory is constructed as a symbolic form. In this regard, the use of metaphor in organizational analysis requires one to interactively cross images (e.g. the prison as a machine; the prison as culture, the prison as an organism); that is, to selectively create comparisons or substitutions for the purpose of assessing how such linkages generate new understandings or interpretations (Black, 1962). In this process of crossing images, it is important that there be some degree of difference between the subjects involved in the metaphorical process. This difference is important because the investigator's task is to focus **only** on shared likenesses, representing partial truths about that which is metaphorically examined. In other words, this model of organizational behavior deliberately emphasizes certain common features in the selective comparison while suspending judgement on certain others that are less common.[2] If the degree of difference between the subjects compared is perceived to be too dissimilar, or too identical, then the 'metaphorical process produces either nonsensical or weak imagery' (Morgan, 1980: 611). The best fit is one in which the articulated difference is substantial but not total. Thus, the metaphorical method of inquiry is a creative, intuitive, and subjective endeavor in which constructive falsehoods are fashioned as a means by which to emancipate the imagination. This emancipation is essential as the method stresses, indeed encourages, thoughtful, critical inquiry as a vehicle to transcend favored or orthodox perspectives in the sense making process.

One example where the application of Morgan's metaphorical analysis is particularly useful is in an assessment of the military apparatus.[3] For example, if we assess the organization of this institution from within a functionalist paradigm, certain metaphorical images emerge. Examining its behavior, as defined through enlisted staff roles and rights, organizational values and norms, and departmental tasks and responsibilities, tells us something quite distinctive about the prevailing prisms (metaphors) through which the organization operates, shapes its identity, and defines rules for individual participation or membership. These processes are a way of thinking about the military apparatus and a way of seeing it in relation to its personnel, other organizational networks, and society in general.

Consistent with Morgan (1997), our position is that theories of prison organization are based on tacit images or metaphors, informed by identifiable world views with their core assumptions, that allow us to see, interpret, respond, understand, and manage the correctional milieu in particular but *always* incomplete ways. This is because all theory is metaphorical, including theories of prison organization and management. Thus, valuable insight is certainly possible through the metaphoric prism but such insight comes at the expense of provisional and fragmented understanding.[4]

The significance of metaphor in criminology, psychology, criminal justice, and law, most recently has been alluded to by Arrigo (1996a: 3–22, 1993: 44–58; see also Manning, 1979, 1992 on *organizational discourse*, Manning, 1988 on *symbolic communication and the police*; and Thomas and O'Maochatha, 1989 on the *'critical' metaphor of justice*). For example, in addressing issues of psychiatric justice, Arrigo (1996a) creates an intellectual environment in which the use of metaphor facilitates an understanding of confinement practices, currently accounted for with significant deficiency in the literature. Certainly the concept of metaphor is not new. The application of metaphor to the criminal justice system, particularly the prison setting, though, is lacking. Morgan's (1997) collection of organizational metaphors was not intended specifically to address the prison environment; however, they provide a solid framework for the review of prison organizational practices.

This article, then, explores whether and how Morgan's metaphors can be applied to the prison system.[5] If the analysis is persuasive, it indirectly demonstrates how this system benefits from a metaphoric-organizational approach, informed by the existing corrections literature. To facilitate this more indirect assessment, several summary observations are presented in the conclusion section. These comments address how the conditional and partial understanding of prisons, based on metaphorical analysis, produce certain tangible insights for correctional research, practice, and policy.

APPLYING MORGAN'S METAPHORS TO PRISON ORGANIZATIONS

Morgan (1997) identifies eight metaphors for better comprehending the behavior of organizations. As applied to correctional facilities, these metaphors include the prison as: (1) culture, (2) organism, (3) brain, (4) instrument of domination, (5) political system, (6) machine, (7) psychic confinement, and (8) flux and transformation. Each section begins with an overview of Morgan's position on the respective metaphor under consideration[6] and why it directly applies to prisons. This is followed by an identification of where and how the relevant corrections literature supports the existence of the metaphor in question, thereby informing our understanding of prison organizational behavior. We conclude each section by exploring the implications of Morgan's 'frame' for research, practice, and policy.

The Prison as Culture

1. Morgan's Organizational Analysis of the Culture Metaphor

When we discuss culture we are usually referring to a knowledge system (e.g. ideas, values, laws), and day-to-day rituals from a societal perspective (Morgan, 1997: 120). Interestingly, culture need not be defined as such. Morgan (Ibid.: 138), for example, interprets it as a process of reality construction which allows people, through shared meaning, belief, etc., to understand the various aspects of their lives in characteristic ways (also see, Schutz, 1967; Berger and Luckmann, 1967 on the *social construction of reality*, and Gergen, 1985, 1991 on the *psychological dynamics* of this perspective). Morgan (1997: 129) further describes organizations as 'mini-societies' with "distinctive patterns of culture and subculture." Subcultures generate a unique group identity which serves to separate them from the overall culture in which they exist (Ott, 1989: 97–8). Therefore, examining the prison organization through the culture prism allows us to assess the often unstated and socially constructed reality operating within this milieu and to explore the patterned behavior of its diverse constituencies (e.g. inmates, guards, administrative personnel).

Historically speaking, however, the concept of culture has acquired various meanings. From Durkheim's (1934) work on industrialization and culture (also see, Scott & Christensen, 1995 on the *institutional theory approach*), to Garfinkel's (1967) analysis on how realities are 'accomplished' (also see, Weick, 1979, 1995 on *enactment and retrospective sense-making*), to Manning's (1992) critique of organizational communication (also see, Arrigo, 1993; Henry & Milovanovic, 1996 on how *language "shapes the reality" of law and crime*), the culture construct has, explicitly and implicitly, undergone considerable refinement.

Most recently, the concept of culture within an organizational context has been defined as a "persistent, patterned way of thinking about central tasks of and human relationships within an organization . . . [and] like human culture it is passed on from one generation to the next" (Fleisher, 1996: 82; Ott, 1989). Bolman & Deal (1991: 268) extend this concept in a symbolic sense, stating that "distinctive beliefs and patterns over time" are unconsciously processed, and illustrated in myths, rituals, stories, ceremonies, and other symbolic forms (also see, Arrigo, 1992: 24–27; Cornell, 1991: 147–152; for applications to *imaginative discourse, feminist jurisprudence, and the psychoanalysis of law*).

By applying Fleisher's definition of an organizational culture to the prison, we can explore the idea that a patterned way of thinking about the tasks and relationships within a correctional facility clearly exists. Further, Morgan's definition of culture provides insight into the construction of a prison reality (pervasive in both inmate and staff subcultures), and indicates the necessity for such a creation. Aspects of Bolman and Deal's symbolic representation of culture may also be applied to the correctional setting.

2. *Prison Research on the Culture Metaphor*

The culture of prisons has inspired an abundance of research, as sociologists, criminologists, and psychologists have endeavored to understand and explain the everyday reality which exists in such institutions. From Clemmer's (1958) *prisonization* thesis, to Goffman's (1961) *total institution* argument (also see,

Sykes, 1958; Sykes & Messinger, 1960 on the *deprivation model*), to Irwin and Cressey's (1962) *importation model*, to Thomas' (1970, 1972) *integrationist* perspective, reading prison culture has been a source of considerable controversy (Hunt, Riegal, Morales & Waldorf, 1993; Grapendaal, 1990). The overall culture and subcultures of the prison represent a very distinct aspect of society, yet similarities can be drawn to the societal formations observed outside of the prison milieu.

A principal basis for the existence of a prison culture is the deprivation and confinement of life behind the correctional walls (Clemmer, 1958; Goffman, 1961), creating a need for certain relationships and items deemed necessary for comfortable existence during one's incarceration. The limited reality of prison life, however, constrains that accessibility. While the environment of the prison is much more detached from that of the external society, basic human desires and needs remain the same. Resulting cultural equivalents surface in various forms, including an underground economy involving exchanges of goods and services, sexual and other interpersonal relations and affiliations, entertainment, crime, and language, etc., (Silberman, 1995: 54–54, 181–188).

Abadinsky & Winfree (1992: 523) discuss an economic system catering to the 'high demand' needs of inmates. "Drugs and other contraband, sex, and virtually anything else an inmate might want" are available for barter (Ibid.). Transactions often involve exchanges of goods, sex, or substantial amounts of cash, a product of the illegality of goods and services. As in the outside world, exchanges of goods, money, sex, and the like are subject to criminal interference. As money is often the object of theft, extortion, robbery, and blackmail in society, the 'convict code' (Silberman, 1995: 34–7) offers much the same. Sexual acts are often analogous to prostitution, making victimization and exploitation relevant issues.

Language is another aspect of culture which is observable within the prison. Wittenberg (1996: 45) notes that "the prison society has always had its own unique language . . . [which] . . . has evolved over the years." Distinctive terms are often used by inmates and staff to describe the various aspects of prison life. The evolution of a distinct language is common to all cultures, and is one of the primary ways of distinguishing one culture from another. Thus, language serves as another example of a cultural parallel which is operational within the prison.

An additional characteristic of most cultures is the existence of several subcultural forms. Similar to cultures outside of the prison, the diversity of racial, ethnic, religious, and criminal orientations necessitate the formation of subcultures within the overall prison milieu. As Irwin (1977: 32) noted ". . . there is no single inmate culture or inmate social system. . .. [T]he variety of cultural and subcultural orientations [ethnic, class, and criminal], the variety of pre-prison experiences, and the . . . hostility between segments of the prison population preclude this."

Shover & Einstadter (1988: 85) describe inmate relationships as occurring in the context of "small friendship groups formed [along] racial or ethnic lines, with prisoners restricting most of their informal interactions to members of these groups." Groups may also be formed on the basis of mutual interest, such as criminal history, religious orientation, adjustment issues, and need for protection. Subculture inmate groups generate a collective identity, allowing inmates to achieve more significant power in pursuit of goals or to confront issues within the prison.

Another dimension to inmate subculture includes the presence of political entities within confinement systems. The 1960s and 1970s produced more contentious inmate groups than in previous decades (Shover & Einstadter, 1988), as civil rights issues were of increasing significance. Modern inmate groups have assumed a different form than their predecessors as the more dangerous and destructive prison gangs have emerged (Silberman, 1995: 42–47). These gangs differ from traditional inmate subculture in that they are organized principally along racial and ethnic lines, routinely in search of institutional power and control (Abadinsky & Winfree, 1992: 527). The modern prison gang does serve a similar purpose to that of the traditional inmate group (e.g. cohesion, solidarity), though, with (political) differences attributable to the changing world outside of the prison. The prison gang, then, signifies the importation model (Irwin & Cressey, 1962) of prison organization and culture from a modern perspective.

3. Implications for Research, Practice, and Policy

The examples used for purposes of our inquiry (i.e. economy, crime, language, subgroups), offer evidence of a prison culture. The intent of the present analysis is not to explain or understand the culture of prisons per se. Instead, the metaphoric analysis of the prison as a culture attempts only to affirm the existence of such a phenomenon, and identify the similarities between the prison environment (its behavior, norms, values, etc.), and a traditional conception of culture. The examples offered serve their primary purpose. However, acknowledging that a prison culture exists is not sufficient. Understanding how the prison *behaves as a culture* can create a vastly different perspective from which to address a multitude of research, practice, and policy issues in corrections. As this article will demonstrate, the application of Morgan's (1997) additional organizational metaphors substantially moves us in this direction.

The Prison as Organism and Brain

1. Morgan's Organizational Analysis of the Organism and Brain Metaphors

In this section two of Morgan's (1997) metaphors are linked given their obvious interdependence.[7] The prison as organism refers to its capacity to function as a living, vital system (Ibid.: 33–34; also see Burrell & Morgan, 1979 on the *biology of social and organizational theory*; Wilson, 1975 on *sociobiology*; and Kauffman, 1993; Lewin, 1992 on the study of *complexity theory and nonlinear dynamics*). Metaphorically, understanding the internal social ecology of the individuals, groups, and sub-systems composing a prison can tell us a great deal about the structure, operation, purpose, and effectiveness of this complex network. Further, all living systems exist in a wider environment. Typically, this wider environment contributes to the sustenance of the organism of which it is a part, otherwise the organism itself becomes extinct. Thus, understanding the impact of the external ecology on a prison helps to explain how some of its needs are met through the wider environment on which this living system depends (Morgan, 1997: 34–39).

The metaphorical analysis of the prison as a brain is based upon learning system theory (Ibid: 78–82; also see, Belden, Hyatt & Ackley, 1993; Senge, 1990; Pedler, Burgoyne & Boydell, 1991 on *learning organizations*). A learning system is a type of open system (Miller, 1978) which emphasizes the processing of information in an exchange with the environment to effectively self-regulate (Morgan, 1997: 79; Argyris & Schon, 1978). The learning approach requires that organizations scan and anticipate changes in their milieu, relate such information to current operations, detect deviations, and initiate corrective measures. Current norms and assumptions about organizational operations must be challenged to allow appropriate patterns to emerge. Thus, Morgan's brain metaphor allows us to read (process) how the prison, as a living system, behaves and makes decisions over time (Henry, 1991; Ford & Gioia, 1995).

2. Prison Research on the Organism and Brain Metaphors

Goffman (1961) identified the prison as a 'total institution'. This conception has been, until recently, a widely shared belief within legal and criminological psychology (e.g. Farrington, 1992). The characteristics of the total institution have also been identified through public perception. The prison is commonly regarded as impermeable and secure, aspiring to incapacitate society's criminal element (Ibid.). Consequently, Goffman's total institution phenomenon remains an integral component of prison behavior in modern America (Irwin & Austin, 1997).

Goffman's classification of the prison was based on the principle that this institution was a "place of residence and work where a large number of like-situated individuals, cut off from the wider society for an appreciable period of time, together lead an enclosed, formally administered round of life" (Goffman, 1961: xiii). The total institution model assumes that the organization retains complete control over all aspects of inmate and correctional staff matters (Grapendaal, 1990: 5–7), and that the organization is essentially dissociated from the effects of societal, cultural, economic, and political factors.

As a result, external influences on prison organization have received little attention (McCorkle, Miethe & Drass, 1995). Farrington (1992: 6), however, notes a "relatively stable and ongoing network of diverse transactions, exchanges, and relationships, which connect and bind together (a) the prison, (b) its immediate host community, and (c) society more generally." Farrington & McCorkle et al., both allude to a more recent conceptualization of the prison as a 'not-so-total' institution. The prison, impacted by the wider society in both highly visible and readily comprehensive ways (Ibid.: 7–10), has recently incited theoretical reformulations of its practices and its influences (e.g. Grapendaal, 1990; Hunt, Riegel & Morales, 1993).

Goffman's reference to the prison as existing in a state of separation and isolation from the wider society, is somewhat synonymous with the organizational concept of a 'closed system'; one which is self-contained, self-referential, and exhibits unchanging parameters (also see, Maturana & Varela, 1980; Kickert, 1993; Luhmann, 1995 on *autopoiesis*; Touraine, 1977, 1988 on the *self-production of society*; and Arrigo, 1995: 452–5 for *applications to law and criminology*). The prison may, however, be organizationally conceived of as an 'open' and 'learning' system, capable of adapting to external inputs, processing this information, and learning for the sake of accommodating such influences (Grapendaal, 1990).

3. *Implications for Research, Practice, and Policy*

Morgan's metaphoric-organizational framework of prisons as organisms and learning systems is particularly useful in reconstituting research, practice, and policy in the field. The parameters of the prison organization have been assumed to be constant (Irwin & Austin, 1997). Accordingly, scant research has focused on the prison as an open system, or one with changing boundaries. Further, the concept of the learning organization is related to the idea of open systems (Bateson, 1972). It suggests that the prison could be viewed as a complex network capable of learning from its environment and, subsequently, changing its internal structure to attain more effectively its desired goals.

Open systems theory describes organizations as dependent on the wider environment for various kinds of sustenance; that is, as living systems which must achieve an appropriate relationship with that environment for continued survival (Katz & Kahn, 1978). Though initially the prison may not seem to embody this property, considering it as such provides meaningful insight. An open system must be organized with the environment in which it co-exists. Its interaction with various external sources is essential to its long-term viability. The prison must concern itself with such influences as labor unions, government agencies, the court system, the police, current correctional practices, the characteristics of inmates upon entry into the prison, and the manner in which parolees are assimilated back into society. The impact of these influences on the prison is substantial and dictates, to some degree, the internal structure and changes within the system itself. The prison must be able to anticipate changes occurring externally and manage itself internally to respond to those changes.

Another key issue in open systems theory is the concept of interrelated subsystems which function within the larger organizational matrix (Morgan, 1997: 42). Such subsystems as industrial crews, medical and psychological personnel, custodial staff, correctional officers, and administrative workers must function collectively within the complete structure. Intra-organizational relations between technical, social, managerial and strategic components are necessary for internal preservation as each is dependent on the other. Identifying these subsystem patterns and their potentially dysfunctional behaviors, is essential for actualizing the organism's objectives. We see, then, that the various subsystems within the prison are analogous to a living entity or the human body: each part is mutually dependent on the others for overall self-maintenance and reciprocal sustainability.

While, in general, the prison is certainly not an organization which focuses its attention on challenging existing internal operations, examples of such behavior do exist. Inmates must learn to cope in order to reduce the stress associated with prison life. Correctional officers must learn to adjust to ambiguous or conflicting directives in order to find the most appropriate action. Management personnel must decide on which outside groups to invite to the prison (e.g. potential funders, prison rights advocates, politicians) in order to promote system goals. Further, as the philosophy of prisons change, staff adjust to different ways of treating and counseling inmates (Lester, Braswell & Van Voorhis, 1992: 3–22). Different styles of managing staff and inmates must be employed in response to changes in correctional philosophy (Klofas, Stojkovic & Kalinich, 1990: 292–296). The system as a whole learns to adapt and process information as social ideas about society and punishment evolve. The same is true for changing legislative statutes, police practices, political tactics, fiscal constraints, and other variables that affect the input and output of the prison organization. Following Morgan's metaphoric schema, we contend that research protocols, practice studies, and policy analyses entertaining these, and related matters, could considerably inform our understanding of the modern day correctional facility.

The Prison as an Instrument of Domination

1. Morgan's Organizational Analysis of the Domination Metaphor

The basis of this metaphor is the notion that organizational behavior is grounded in processes of domination and exploitation (Morgan, 1997: 303–304). Concepts central to the organizational metaphor of domination have roots extending back to, and inspired by, the writings of Max Weber (1946; also see McNeil, 1978; Mouzelis, 1979; Salaman, 1978 for applications to *modern organizations*), and Karl Marx (1976; also see Burrell & Morgan, 1979). Where Weber (1946: 228) viewed bureaucracy as an iron cage, Marx (1976) believed that domination was engendered in the pursuit of capital accumulation. These fundamental insights produced enormous interest into the role that domination assumes in many facets of contemporary society and its organizational networks.

Following the insights of Marx (1976), radical organizational theory links social domination with the everyday behavior of complex systems (e.g. McNeil, 1978; Mouzelis, 1979; Salman, 1978). In this model, the relationships between organizational life, class, and control are sharply defined. Organizational reality is seen as ideology which physically and emotionally bankrupts the labor force (e.g. Bridges, 1994; Howells & Wood, 1993; Rifkin, 1995; Shostak, 1996). Brutal management methods are consistently employed (Eberle, 1996), and occupational hazards are routinely identified (Pauchant & Mitroff, 1992; Frost, Mitchell & Nord, 1992).

Notwithstanding the radical organizational perspective, it was Weber (1946) who defined the various forms of social domination characterizing societies. The most significant for our purposes is institutional power, understood as rational-legal authority. In this process, will is imposed on others because of the perception that one has an unquestionable right to do so. This expression of domination is legitimized, normalized, and regarded as socially acceptable (Morgan, 1997: 304), through the instrumentation of laws, rules, regulations, and procedures. This form of authority is most relevant for understanding the metaphor of domination in prison organization and behavior.

2. Prison Research on the Domination Metaphor

The processes of domination and exploitation are evident in every prison (Silberman, 1995). They are intrinsic to the system's organization and function as a framework from which goals are formulated (Jacobs, 1977; Duncan, 1996). The prison's population consists of a majority of inmates working for and complying with the orders of their supervisors. Their welfare and desires often have limited value, as the organization must maintain its goal of control above all else (Irwin & Austin, 1997) .

In analyzing the exploitative practices of the prison, Goffman's (1961: 17; 1963) concept of the total institution includes descriptions of the means by which individuals within the institution undergo "rites of passage." He refers to this rite as a ritual, upon entry to the institution, which reinforces within the individual the belief that he or she is the property of the institution. The rite additionally serves to distance the individual from those performing the rite (or within the prison), creates a separation between inmates and prison officials, and fortifies the power structure (see also, Richards, 1998).

Goffman likens this process to the practices of the armed forces, in which the recruit's isolation from society and sense of impoverished power within the military complex are most pronounced. In comparison to other institutions, however, the prison's rites of passage are much more extreme and intensive (Toch, 1998). Goffman (1961) refers to the inmate's introduction to institutional officials as 'obedience tests', or 'will-breaking contests', serving to establish the prisoner's sense of 'belonging to' the correctional facility. Defiance results in increasing degrees of punishment until the inmate assumes his or her dictated role.

The intensity of this process is illustrated in the dehumanization which is often an integral part of the inmate's introduction to the prison. The inmate is stripped of all possessions and issued institutional apparel. This process essentially reduces the prisoner to a 'non-person', an object or, simply, a convict (Silberman, 1995: 15–19). The dehumanization of inmates upon entry to the prison, creates the foundation for exploitation and domination.

Exploitation is perhaps most evident in the industrial aspect of the prison (Institute for Economic and Policy Studies, 1988). While inmates are often not only expected but forced to perform hard labor, they are granted little in return (Kurshan, 1992). Prison labor is intended to act as a means of discipline or custodial control over the inmates (Sellin, 1976). It is considered to be a consequence of their punishment (Sykes, 1958). While this may be true, the prison does benefit in various ways from inmate labor and shares little of the benefit with those who perform it.

Domination (the process of individuals or groups imposing their will on others) is embedded in the philosophy of prison organization (Adams, 1992). Prison staff are able to control and dictate inmate behavior, with little or no consequence of action (Marquart, 1983; Toch, 1998). Prison staff are perceived by the public and the inmate population as having the right to impose their will upon the inmates. The general consensus of the public is that more emphasis should be placed on the punishment of criminals, and that inmates should expect to have few or no amenities (Wittenberg, 1996b). The 'punishment' philosophy, in turn, reinforces and legitimizes the domination and exploitation of inmates. This domination is justified in tradition and in the laws and rules (policies) of the organization (Durham, 1994: 22–23). Power is legitimized in this form and inmates and lower-level staff are socialized to abide by it (Marquart, 1983).

Evidence of a distinct class structure is also evident in the prison organization (Irwin & Cressey, 1962). Upper-level staff assume the identity of the ruling class, inmates are forced to comprise the lower class by necessary intention, and lower-level staff are left to fill the middle-class gap. The concept of organizational management within the prison becomes a significant factor. The large inmate population must be supervised by staff, who must then be supervised to ensure adequate performance of responsibilities. With the upper echelon of staff not directly involved in the supervision of inmates, there is a need for organizational management to play the role of intermediaries. This class division within the prison creates a distinct organizational power structure and, thus, creates the necessary environment under which domination and exploitation operate (Wittenberg, 1996b).

3. *Implications for Research, Practice, and Policy*

Morgan's domination metaphor allows us to see where and how exploitation operates within complex networks as a part of their everyday behavior and practices. Prisons are a representative example of a complex institution. Existing correctional literature draws attention to where and how the exercise of individual and/or group power is used and legitimized. The prison milieu, then, can be characterized as an exploitative environment where will or force is employed to chorale conformity from others. Ensuring compliance in this controlling fashion can produce deleterious consequences, and these consequences serve as the basis for research, practice, and policy.

The existence of hazardous conditions within the prison is one troublesome area contributing to the exercise of institutional power and exploitation. Correctional facilities currently suffer from extreme overcrowding and often display living conditions for the inmates that would be unacceptable under most other circumstances (Durham, 1994: 31–57). Ekland-Olson (1991: 389–391) suggests that any realistic attempt to improve prison conditions must take full account of population and capacity (also see, Cox, Paulus & McCain, 1984: 1148).

We recognize that overpopulation is an issue which has certainly received a substantial amount of attention (e.g. Bureau of Justice Statistics, 1994). However, its detrimental effects persist (e.g. *Castello vs. Wainwright*, 1975; Territo, Halsted & Bromley, 1998: 483–485). These effects, informed by Morgan's domination metaphor, are protean areas of investigation.[8]

In addition, while the humaneness of prison conditions and treatment have come under much scrutiny (Toch, 1998), and while legislative (1990) and policy (McCarthy, 1988) action have ensued, the general state of the prison continues to be criticized.[9] Although many in the general public would argue that prison conditions are far too pleasant for inmates (Wittenberg, 1996b), reality suggests that the prison environment fosters significant physical maladies (Durham, 1994: 67–97). Tuberculosis, AIDS, mental illness, and the like, are all health-related problems that, when linked to the metaphor of domination, are suggestive for the contemporary organization of prisons. These, too, are matters requiring further examination. The correctional system must weigh the cost of enhancing conditions against the consequences of foregoing this undertaking. As limited regard for inmate needs often exists, cost containment issues resulting from lack of funding have generally been given more attention than prisoner welfare (e.g. Zedlewski, 1987: 5). In addition, with living conditions seriously suspect (Durham, 1994: 31–57), and adequate medical/psychological care often unavailable, inmates are prone to sustained physical and mental ailments (Steadman, McCarty & Morrissey, 1989), especially where the initiative or opportunity to alter such hazards are infrequent at best. Examining these circumstances, given the operation of institutional domination, could shed greater light on the modern day administration of correctional justice.

The Prison as a Political System

1. Morgan's Organizational Analysis of the Political System Metaphor

Related to the domination metaphor is the notion that prisons are political systems. Organizations variably function as loose or tight systems of government, informed by, *inter alia*, autocratic (Michels, 1949), bureaucratic (Weber, 1947), technocratic (Galbraith, 1967), and industrial-democratic (Vanek, 1975) principles. Domination is often embedded in the means by which complex networks conduct their affairs. These practices are typically expressed through the politics of organizational members (e.g. Pfeffer, 1981; Pettigrew, 1973; Frost & Egri, 1991). The political dynamics are often in the form of self-interest (Frost, 1987), interpersonal, interdepartmental, and subsystem conflicts (e.g. Burrell & Morgan, 1979; Follett, 1973; Putnam & Poole, 1987), and displays of exploitative power (Clegg, 1989). Thus, politics in organizational settings, through specific system features, is rather commonplace (Morgan, 1997: 307–314).

Most prisons can be regarded as systems of government, employing political principles (Kurshan, 1992). The disciplinary order present within the prison is much akin to the political systems observable outside the prison (Bright, 1996). When one thinks of politics and its relation to correctional facilities, external political influence on the prison system is often primary. There exists, however, a separate political sphere which operates solely within the prison. This sphere is constructed around fundamental ideological principles, such as power, control, and domination (Irwin & Austin, 1997). The resulting effect is an exercise of politically shaped governance and disciplinary procedures, bringing order to the world which lies behind the physical walls of confinement systems.

Marxist thought describes prisons as 'instruments of power', expressing ruling class ideologies, policing priorities, and imposing upon the poor and the powerless the discipline of the powerful (Bright, 1996: 1). Bright suggests that, beyond Marxist thought, power may be understood not simply as its result, but as a strategy and subsequent practice employed by those within a sphere of power relations. The explanation of correctional goals begins with strategy, and evolves as a prescription for normative standards and discipline. Within the prison, this strategy and practice is most especially observable on a prison to inmate and staff to staff basis.[10]

2. *Prison Research on the Political System Metaphor*

Power is an issue which has been more than adequately discussed in the political sociological literature (e.g. Habermas, 1975; Adler & Longhurst, 1994). Further, the issue of power and its relation to the control of knowledge and discursive practices has been equally well established (e.g. Foucault, 1972; Garland, 1990; also see Henry & Milovanovic, 1996 on *power and the constitutive dynamics of crime, law, and penology*). The prison offers an example of an institution which displays power in terms of combinations, rivalries, and alliances where more power is afforded to some than to others (Bright, 1996). The control of power further establishes ongoing discursive practices, thus, reinforcing the system. The power established by control of discourse (Foucault, 1972, 1977; also see, Arrigo, 1993, 1996a for applications to *psychiatric confinement*), for example, is notable in that there is an intended control of communicated knowledge and information. This leaves those who control such information with the most power and the most 'truth'.

Power is thus largely bureaucratic and technical, as it is firmly established in the hands of the upper echelon of the system (Foucault, 1977). Control of all resources is delegated to a select few, and the control of all property rights is taken away from the inmate upon entry. These practices further the concept of the bureaucratic prison. Most staff and inmate behavior within the facility is subject to strict rules and regulations set forth by policy. The law-like administration leaves little room for discretionary practices or deviations from established routine.[11] Though a variety of administrative forms are seen in prisons throughout the country producing different political configurations, most prisons appear to display authoritarian, bureaucratic, and technical tendencies (Bright, 1996; cf. Scharf's, 1977 *just community* concept and Murton's, 1976 *participatory management model*).

As previously noted, the most salient concepts for the political metaphor are self-interest, conflict, and power (Morgan, 1997: 161–198). Each is observable within the organization of prison practices and behaviors. Differing goals and desires are certainly of importance within the system, particularly when distinguishing between staff and inmates. When considering the goals of staff, certain proclivities arise. While the hierarchical structure of the correctional facility leaves little legitimate power in the hands of lower-level personnel, a desire for career advancement is often present. As with any organization embodying a similar power structure, the interests of lower-level and higher-level employees are not always in accord. This is perhaps most evident in the differing attitudes and values among staff. For example, prison employees have distinct reasons for pursuing their careers. Money, job security, justice, and convenience are only a few of the possible motives for working in a correctional setting. As these motives differ, interests and means chosen to pursue them vary as well. As a result, arriving at consensus on appropriate organizational practice and policy is not easily attainable. Workers may seek out others with similar interests to form an informal coalition. Their shared interests allow them to work together and to achieve (hopefully) their organizational goals. This process consolidates power.

Conflict within the prison organization arises when the diversity of interests clash. Conflict is embedded within the structure of the prison, with the clear distinction between roles and a strict hierarchical system imposed on its subjects (Useem, et al., 1995). Differing objectives within the organizational sub-units of the prison create contrasting attitudes about the operation of the organization. Policy makers and line staff (guards) have different perspectives on the organization, as neither communicates directly with one another, and neither can completely understand the responsibilities of the other (Kaufman, 1988). Further, the hierarchical structure acts as a sort of career ladder, with those at the lower end struggling to ascend. These conflicts become institutionalized within the setting, as witnessed in the various attitudes, stereotypes, and rituals that may exist. Without conscious processing, inmates and different levels of staff find themselves adopting certain persona, etc., which often accompany their position in the organization. As these conflicts are part of the organization and the people within it, they are often difficult to identify and to resolve. The primary means of controlling such conflicts of interest is the use of power and authority within the prison organization (Marquart, 1983).

Sources of power within the prison setting that are most prominent include formal authority and the control of knowledge and communication by those in superior positions.[12] Formal authority within the prison system is the most recognized form of power. The strict divisions between staff and inmates, and between lower- and upper-level staff generally goes unquestioned in the prison (Kaufman, 1988). The authority that is exercised within the prison is a power given to appointed officials within the setting (Owen, 1988). This power has been earned through training, experience, and leadership ability. Formal authority in the prison is based on rights and privileges that accompany the appointed position, and it is the foundation of the hierarchical structure seen in the correctional milieu.

Control of knowledge and communication (discourse) is a second source of power in the prison. By allocating knowledge necessary to maintain operations in the confinement setting to specific individuals, the prison authority is able to control the amount and content of information that is received by all members of the organization (Zupan, 1992; Irwin, 1980). This allows those who are currently in positions of power, to maintain their status by only letting certain types of knowledge filter down to the other levels. When other, lower-level personnel are left either uninformed or inadequately notified, they are in no legitimate power position to question the actions or decisions of their superiors (Kaufman, 1988). The information that is received by lower-level staff is often limited, consisting only of information that is necessary to perform their role within the organization. This censorship and suppression restricts a pattern of discourse from which criticism may arise. By suppressing such discourse, potentially constructive ideas are left unexplored, and the possibility for a diffusion of power is abated (Henry & Milovanovic, 1996: 222–223).

3 Implication for Research, Practice, and Policy
Recognizing that prisons are organized and behave as political systems directs our attention to the type of authority, knowledge, and communication underscoring their daily administration. The often technical, authoritarian, and bureaucratic composition of prisons as political entities informs our understanding of the various organizational relationships (e.g. staff to staff, inmate to staff) occurring in these confinement settings. These relationships are based upon power differentials that serve to maximize the interests of some correctional constituencies while minimizing those of others. Given these realities, researchers, practitioners, and policy analysts would do well to assess whether and how the hierarchical and control-driven framework underscoring the political systems metaphor of the modern day prison advances the administration of correctional justice. Indeed, if one legitimate aim of incarceration is to reduce prospects for recidivism, then exploring the extent to which a more technical and bureaucratic organizational approach readies, if at all, prisoners for adjustment to prison life and adjustment to post conviction release (Richards, 1995) is essential. On this point, we contend that Morgan's metaphorical framework offers an organizational blueprint by which this and related questions can be entertained.

The Prison as a Machine

1 Morgan's Organizational Analysis of the Machine Metaphor
Morgan's (1997) machine metaphor is an invitation to explore the bureaucratic elements of complex systems or, metaphorically, the ordered array of organizational 'parts' which, collectively, behave in a mechanistic method (Ibid.: 12–13; also see Fayol, 1949; Mooney & Reiley, 1931; Gulick & Urwick, 1937 on *classical management theory*). The machine metaphor assumes the figurative existence of a set of clearly defined and ordered parts which work to ensure productive relations and outcomes (Morgan, 1997: 13). The success of these relations and outcomes can be anticipated, modified, sustained, or maximized through principles of scientific management (Taylor, 1911, Gilbreth, 1911; Sward, 1948). The organizational aim is total efficiency and optimal precision. Utilizing Morgan's machine metaphor allows us to assess how, and to what degree, prisons behave as a collection of parts whose ultimate goal is correctional efficiency.

Recently, numerous criticisms of the mechanistic approach to organizations and management have surfaced, particularly in relation to the deskilling of jobs (e.g. Shostak, 1996; Scarbrough & Corbett, 1992; Knights & Wilmott, 1989; cf. Hammer & Champy, 1993 on the *re-engineering movement*). In essence, these assessments question the modern work process as stifling and homogenizing the more humane aspects of existence (e.g. see Ritzer, 1996 on the *McDonaldization of society*). Notwithstanding, the pervasive image of organizational life as machine-like remains undisputed (Morgan, 1997: 12).

The conceptual basis for this metaphoric exploration is derived from Weber's (1946, 1947) classic sociological and organizational portrait of the bureaucracy. The bureaucratic representation of the prison or correctional organization is by no means a novel depiction (e.g. Snarr & Wolford, 1985; Shover & Einstadter, 1988; Abadinsky & Winfree, 1992). Nevertheless, the examination of the mechanistic quality of the prison is extremely important. Consistent with the previous metaphoric analysis, exploring this perspective allows the researcher to extricate implications for preparation and response to the changing correctional environment.

Morgan's (1997: 15–17) treatment of the origins of bureaucratic patterns in organizations draws attention to the system's militaristic qualities. He likens this behavior to the leadership behavior of Frederick the Great of Prussia. The military, in its automated assembly and execution, emerged as the prototype of the bureaucratic organization (Ibid. 1997: 15). Frederick the Great's military reformations included, inter alia, the introduction of ranks, uniforms, and standardization of regulation. These mechanistic characteristics of a mid-1700s military regimen, clearly invoke images of the modern prison in its everyday conception.

Weber (1946, 1947) described the characteristics of what he termed the bureaucratic organization. Included within Weber's portrait are formalized division of labor, a formal hierarchy of authority, and established channels of communication between superiors and subordinates. The mechanistic aspects of bureaucracy were emphasized by Weber, as he analogized the bureaucracy's routinization of the administrative process to the machine's routinization of production (Morgan, 1997: 17). The formalization and routinization of organization, as described by Weber, primarily intends to promote organizational regularity, reliability, efficiency, and precision. While Frederick the Great's objective was to routinize his military, and while Weber's analysis focused on the production-oriented organization, the philosophy of each has contributed significantly to other organizational forms. The prison is one such complex network.

The paramilitary analogy often applied to the prison (Snarr & Wolford, 1985; Shover & Einstadter, 1988) is justified by numerous aspects of prison philosophy and behavior. Examples which may immediately come to mind include the separation of prison staff into distinct categories based on rank and title, the allocation of distinctive uniforms to both staff and inmate populations, and the separation (classification) of inmates based on personal characteristics (Archambeault & Fenwick, 1988). Weber's analysis of the bureaucratic organization and Frederick the Great's militaristic philosophy contain all of these key features. Thus, these characteristics, as applied to the prison, will be analyzed with this in mind.

2. *Prison Research on the Machine Metaphor*

The prison (and correctional systems in general) has historically assigned distinct titles to its employees (Foucault, 1977). While the specific titles have undergone minor variations over the years (Abadinsky & Winfree, 1992), there remains a strict division in employee rank which is signified by title. The delegation of rank and title to employees creates a formal hierarchy of authority, with equally formalized division of power, influence, and knowledge. As previously discussed in the political metaphor, the control of knowledge within the prison fortifies the hierarchy of power (Milovanovic & Henry, 1991; for a feminist analysis see, Howe, 1994). This control of knowledge exemplifies what Weber discussed as formalized channels of communication. Weber's concept of division of labor is also observable, as assignment of titles to employees also delegates and defines the quality and quantity of labor which that employee will perform.

An additional example of the militaristic quality of the prison, is the issuance of uniforms. In maximum security prisons, inmates are required to wear institutional clothing at all times (Abadinsky & Winfree, 1992: 511). Prison staff are also required to attire themselves based on specific, formal guidelines. The uniform requirement further dictates status and role within the prison organization, labels inmates as such, and reinforces the hierarchical structure of staff authority.

Given these observations, it is clear that the prison borrows heavily from the militaristic, bureaucratic organizational style that has been evident for many years in other segments of society. Elements of Frederick the Great of Prussia's legacy, and Weber's organizational analysis both persist in clear ways. Beyond these earlier influences, other, more modern theories and philosophies have equally contributed to the mechanistic organizational foundations of the prison.

The advent of classical management theory contributed to the bureaucratic organizational structure, some aspects of which are still observed in the prison setting (Snarr & Wolford, 1985). Classical management theory brought added emphasis to the process of planning, organizational command, and coordination and control in management (Morgan, 1997: 18). A number of principles were set forth in an effort to promote this agenda. Some of the principles of classical management theory that pertain to the organization of the prison include unity of command, scalar chain, staff and line, division of work, authority, responsibility and discipline.

Unity of command suggests that an employee should receive orders from only one superior (Morgan, 1997: 19). As previously discussed, the hierarchical nature of organization is prominently featured in the prison, with marked superiors supervising and commanding a defined group of employees (Archambeault & Fenwick, 1988; Marquart, 1983). The typical maximum security prison is headed by a warden or superintendent who maintains ultimate authority. Below this person are middle managers, security staff (correctional officers), and treatment personnel. Each assumes his or her position within the hierarchy (Clemmer, 1958). Along with that defined position comes a predetermined allocation of power. Each member of the organization, including the warden or superintendent, must report and/or is directly governed by a superior. As one ascends the chain of command, the number of supervisors lessens, yet the structure of authority provides for employees to receive orders from a specific supervisor or superior (Kauffman, 1988).

The scalar chain is a resulting hierarchical structure in command, which is a derivative of the unity of command principle (Morgan, 1997:19). The top-to-bottom organization of superior/subordinate relationships in the prison is the basis for communication and decision making. The top of the organizational chain controls communication and, ultimately, the decision making within the prison (Owen, 1988).

Staff and line suggests that personnel have limited amounts of input, yet must not violate the line authority (Morgan, 1997: 19). This practice can be observed within the prison, especially with the limited discretionary privileges correctional officers receive (Shover & Einstadter, 1988). While some discretion is valuable in that it allows for individuality in judgment when faced with ambiguous situations, the line of authority is still firm (Archambeault & Fenwick, 1988). Lower-level staff are often granted a limited amount of situational input, yet are always subject to the boundaries of their own authority (Marquart, 1983: 15–25).

Division of work entails efficient attainment of organizational goals through specialization of tasks (Morgan, 1997: 19). The specialization and segmenting of tasks is a valuable tool for assuring order and command within the prison. Correctional officers, supervisors, treatment staff, and industry personnel all have specific functions they are expected to perform or maintain. An absence of this organizational division of duties would result in less efficient and effective production, treatment, inmate supervision, etc., and would open the door for a breakdown in the corrections process (Owen, 1988).

Authority, responsibility, and discipline are additional principles which guide the process of maintaining order within the prison (Marquart, 1983; Irwin & Austin, 1997). All are integrated to achieve the necessary pattern of obedience to authority and respect for those higher up in the chain of command (Morgan, 1997: 19). Authority allows for prison employees to have a necessary degree of command to ensure that their tasks are executed. Prison policy often acts as an established system of rules which governs the delegation of that authority, and provides a basis for the respect of that control (Zimmer, 1989). Similarly, policy and authority also ensure that inmates display the necessary level of respect for staff. Inmates are disciplined to respect such decision making, and are rewarded in various ways for assuming responsibility in such relations (Sykes & Messinger, 1960).

3. Implications for Research, Practice, and Policy

All of the above principles are commonly observed in modern organizations. The prison provides a setting in which they can be examined in near perfect form (Archambeault & Fenwick, 1988). Further, the bureaucratic foundation on which many prisons operate serves an essential need in the correctional setting. It establishes distinct rules, regulations, command bases, delegated responsibilities, and control of authority and power within the prison. Thus, the machine metaphor gives us insight into the ways in which a prison is able to maintain order in a setting where the absence of it might well result in chaos. Morgan's machine framework offers us a unique perspective by which to interpret and assess the relative efficiency of modern day correctional facilities. Research, administration, and policy developed along these lines could potentially ensure the smooth, safe, and effective operation of the prison milieu and the parts that constitute the institution itself.

Prisons as Psychic Confinement

1. Morgan's Organizational Analysis of the Psychic Confinement Metaphor

Morgan (1997: 406–407) contends that there are two important aspects to the psychic confinement metaphor. One is associated with the unconscious and the other addresses the role of idealogy. Given our exposition on prisons as both metaphors of domination and power, consistent with *ideologically-based reality construction*,[13] we focus upon only the more psychological dimension of the psychic prison metaphor.

There are many ways in which 'cognitive maps' shape how organizations think and behave (e.g. Morgan, 1993; Schon, 1963). These maps or mind sets need to be challenged so that paradigm shifts (Kuhn, 1970) can occur, and adjustment to postmodern (Henry & Milovanovic, 1996), postcapitalist (Drucker, 1993) society can follow. The aim here is to avoid the 'traps' of group thinking or taken-for-granted interaction. Thus, various critical (e.g. Mason & Mitroff, 1981; Mitroff & Linstone, 1993), dialectical (Morgan, 1989) and paradoxical (Harvey, 1995; Zohar, 1995) modes of imaging are essential. By identifying and exposing the prevailing cognitive maps operating within the prison setting, it becomes possible to critique the behavior of the correctional institution and, to some meaningful degree, implement changes in its essential functioning.

It is also important to examine the unconscious dynamics at work within and throughout any complex system. The unconscious is a repository for repressed impulses such as sexuality (e.g. Freud, 1953, Vol. 7; Marcuse, 1955, and for applications to *sexual repression and the rise of formal organizations* see Burrell, 1984, 1992). It is also the locus of other desires including narcissism (e.g. Lasch, 1979; Schwartz, 1990); patriarchy (Fromm, 1971; Gherardi, 1995); death and immortality (Freud, 1953 Vol. 18; Becker, 1973; Denhardt, 1981). These impulses, when inappropriately quelled within an organization, can produce anxiety, giving rise to unconscious defenses (Argyris, 1993, 1994; Hirschhorn, 1988). Additional insight may be gleaned about the behavior of an organization by examining its relationship to unconscious, archetypal roles assumed by occupational members or the entity itself (Jung, 1971; Bowles, 1990, 1991; Bradshaw & Newell, 1993). This depth-oriented approach offers important information about the 'shadow' side of human nature and its influence on everyday, organizational life (Zweig & Abrams, 1991). Thus, by exploring the subliminal, archetypal forces activated within a correctional setting, we can obtain invaluable data on the structure of behavior underscoring institutional decision making.

Morgan's (1997) position on the metaphor of psychic confinement resonates with prison organizations. An excellent illustration of this point comes from Plato's (1941), *The Republic*. In Book VII, Plato provides us with a portrayal of an imprisoned collective psyche. In this tale, a group of people are chained within an underground cave in such a way that they cannot move. They are forced to recognize reality only as it appears on the wall of the cave in front of them. The wall is illuminated by a fire, such that the shadows of figures and objects external to the cave appear on the wall. Thus, the confined individuals interpret the shadows as reality, naming and discussing them, and equating sounds to movements on the wall. Reality for the prisoners is defined by the shadows, as their predicament restricts them to such identifications.

Plato suggests that if one of the prisoners were able to leave the cave, s/he would realize that the shadows were simply reflections of a much more intricate reality, and that the group's understanding of this reality was considerably erroneous. If the prisoner were to return to the cave and disclose these findings to the others, this person would be chastised. The group member would be unable to return to the old way of living, and fellow prisoners would fear the outside world of which this person spoke.

Plato's allegory of the cave illustrates a familiar characteristic of organizational, group, and individual behavior. The psychic prison metaphor utilizes the essence of this allegory, and attempts to understand the processes by which organizations become imprisoned or confined by ideas, actions, and thought processes. Patterned ways of thinking and behaving form a web or cycle upon which organizations come to rely. In doing so, organizations and individuals within it, limit the perspectives from which a situation may be construed, addressed, and reconciled. While cognitive maps may eliminate the possibility of undesirable consequences caused by unconventional action, they also negate the increased efficiency and effectiveness that might stem from initiating less orthodox, more novel responses. Organizations can be caught in a vicious cycle of conscious and unconscious patterned thoughts and behaviors, affecting situations and relationships within the organization. The prison environment illustrates this tendency in several ways.

2. Prison Research on the Psychic Confinement Metaphor

The "trap of favored ways of thinking" (Morgan, 1997: 216), may be most evident and influential in the stereotypes which exist, in both prison staff and inmate populations. For inmates, the cognitive map often includes the negative characteristics of correctional officers, past on through tradition (Jacobs, 1977). Robert Johnson (1987: 120), for example, notes that the symbolic image of the "subhuman and senseless brutal custodian" lives on, often used as a collective inspiration for the inmate identity. The traditional image of the correctional officer leaves inmates susceptible to the 'trap' into which s/he may succumb. This trap is accurately and vividly depicted in one inmate's narrative:

For a prisoner . . . a guard is possibly the lowest imaginable form of humanoid life . . . The intriguing aspect of this view of guards, is that no inmate I've ever met came by it through his own experience . . . It's an opinion a prisoner automatically picks up at the door along with his issue of prison clothes . . . and from that point on he simply looks for incidents to confirm the view' (Schroeder, 1976: 151–152).

Self-righteous inmates often provoke the incidents which result in violence by correctional officers, thus confirming the stereotype in their own eyes and those of other prisoners (Toch, Adams & Grant, 1989; Johnson, 1987: 120). Over the years this stereotype has been almost equally applied to the correctional officer by prison analysts and society in general (Johnson, 1987: 122; Kurshan, 1992). These observations provide examples of how generalizations enter the world of the prison, creating a reliance on traditional, favored ways of thinking about groups and individuals within it. What is most intriguing is that the stereotype often has little, if any, basis in reality (Marquart, 1988). As noted above in the inmate narrative, prisoners adopt an unflattering image of correctional officers, based entirely, at that point, in myth (Schroeder, 1976: 152).

Stereotypical images of correctional personnel may be further augmented by the importation of cultural beliefs. Inmates often possess experience and knowledge of the criminal justice system, either directly or indirectly, through the environments in which they existed prior to incarceration (Fleisher, 1989; Sykes & Messinger, 1960). Through friends, family, stories, etc., they are led to perceive prison officials, custodial staff, authority figures, and prison life in a generalized fashion. Inmates who carry such preconceived views into the prison setting often experience the negative consequences of their cognitive maps (Bondeson, 1989). These patterned ways of thinking and behaving by inmates, both consciously and unconsciously, affect the nature of their existence in the correctional milieu. By accepting and maintaining particularly limiting mind sets, relations and practices within the prison are hindered (Jacobs, 1983). This undoubtedly compromises the inmate's ability for correctional adaptation and long-term success while incarcerated (Toch, Adams & Grant, 1989; Heffernan, 1972; Irwin, 1980).

Similar cognitive dispositions are evident in correctional employees, especially with regard to their often unfounded perspectives on offender characteristics (Kauffman, 1988). Swigert & Farrell (1976), for example, investigated the evaluation and processing of homicide defendants by staff at a diagnostic clinic. They found that the clinic staff generally regarded homicide offenders as 'normal primitives', or as having cultural predispositions toward violent behavior and an aggressive response to any situation of personal challenge (Ibid.: 101). Clinic staff employed this conception as a basis for recommendations to the courts.

It is not difficult to understand how similar practices exist in the prison. Most prisons have been operational long enough to have developed specific treatment dispositions for inmates. Often these philosophies remain intact even though staff change or move on (Owen, 1988; Arrigo, 1996a: 95–128 for applications to the *psychiatric confinement of criminals*). Cognitive distortions such as those just described can impede the capacity of prison personnel to adequately meet the physical, psychological, and related needs of inmates, significantly jeapordizing rehabilitiative prospects.

3. Implication for Research, Practice, and Policy

As a society, we are encouraged and taught to maintain certain attitudes and prejudices toward prisoner and criminal populations. These attitudes are, arguably, embodied in the collective psyche of our country. Personal beliefs about race, ethnicity, gender, and morality all carry with them certain consequences when brought into the correctional environment. As we are introduced to such attitudes, through the intended and unintended teachings of others, we reach a point where viewing the inmate population from any other perspective is difficult if not impossible. These attitudes and prejudices are often present in both administrative and custodial prison staff from the moment their employment commences. Further, these attitudes are certainly present in and accommodated by the legal and judicial system, the legislature, and other institutions that directly impact the prison (e.g. Hamm. 1990).[14] As a result, the employees of the prison bring with them certain predispositions that will, in turn, effect their treatment of and relations with inmates. Those relationships with inmates, on all levels, will determine, in part, the organizational success and effectiveness of the prison. Correctional research, practice and policy must expose and investigate the cognitive maps and the prevailing imagery underpinning the administration of modern day prisons. Indeed, the preferred ways of thinking about and behaving in confinement settings may ultimately be what undermines or energizes prospects for organizational change and institutional advancement.

Prisons as Flux and Transformation

1. Morgan's Organizational Analysis of the Flux and Transformation Metaphor

The metaphor of flux and transformation invites us to consider the organization as it exists in a continuously flowing state, embodying constants as well as change (Lewin, 1992; Kauffman, 1993). This mix of order and disorder is essential to understanding and mapping the behavior of all systems (for recent *applications to crime and justice* see Milovanovic, 1997; Arrigo & Williams, 1999; Arrigo & Schehr, 1998), including the organization of prisons. As 'pockets' of instability and flux are generated from within and without, the 'whole' is simultaneously symbolized by turbulent local movement and patterned global activity (Prigogine, 1984). This is what is meant by the notion of system chaos. It is a sort of 'predictable unpredictability' or order out of disorder thesis (Arrigo, 1995. 452–454).

Morgan (1997: 253–298) explores several features related to the science of chaos theory.[15] The most relevant for purposes of this inquiry include the notions of flux and transformation. Until recently, little attention has been given to chaology by social science researchers (Butz, 1997). Since its modern inception, chaos theory has typically been reserved for analytic use in the natural sciences (e.g. Elliott & Kiel, 1996). Chaos theory or dynamical systems theory has only recently seen any attempt by scholars to incorporate its notions of nonlinear dynamics and complexity, with some legitimacy, into social science methodology (e.g. Young, 1992, Weinstein & Weinstein, 1993; Hayles, 1990).

Researchers are now discovering the inherent value of nonlinear analysis or orderly disorder practices in their application to the unfolding logics of system change (e.g. applications to *individual, couples, and family therapy*, Butz, 1997: 103–174). Arrigo (1996a: 1992–200), for example, has successfully applied such a methodology to his postmodern analysis of the institutionalized psychiatric subject. Morgan (1997: 261–265, 414–415), too, notes some general principles of chaos and complexity as they relate to the development of organizational networks. The nature of such analyses supports the use of chaology in the organizational study of an institution, including the modern prison.

Chaology is the result of discoveries involving nonlinear dynamics. Nonlinear dynamics refer to the study of temporal (phase-space) evolution within dynamical systems (e.g. the unpredictable movement tendencies within prisons). A nonlinear system is one which displays behavioral dynamics revealing instability in relationships (e.g. Stewart, 1989; Butz, 1992a; 1992b; Briggs & Peat, 1989). Chaos theory, then, in its general thesis, holds that the social world and, thus, the organizational world, is unpredictable. The initial randomness of events at the micro-level, however, is not sustained at the macro-level. Nonlinear systems, despite the surface, everyday, interactional flux, display an emergent order. In other words, the chaotic system appears to be devoid of situational pattern or regularity, but does, over time, embody global order. Both internal dynamics and external disturbances produce routine uncertainties. A consequent process of change (transformation) results from the 'shocks' (flux) which a system endures. As the shocks are indeterminable and unpredictable, the system leaves itself open to a vast and complex array of possible outcomes (Elliott & Kiel, 1996). On a macro-level, however, each seemingly random event and its effects generate an overall emergent (dis)order; that is, a state of chaos. Given these observations, the internal and external stimuli impacting the prison system are of considerable consequence. Understanding these organizational shocks can tell us a great deal about the relationship between prison management officials, correctional officers, and the prisoner population. In addition, the metaphor of flux and transformation may help deepen our appreciation for the nature of correctional policy formation in the face of ongoing, micro-level perturbations (inputs) affecting the prison milieu.

2 Prison Research on the Flux and Transformation Metaphor

Arrigo's (1996a, 1997) analysis of the psychiatric citizen and criminal/civil confinement suggests a degree of similarity with the dynamics of prison behavior and practice. The correctional system assumes that life within its institutional parameters is predictable or can be made this way (Clemmer, 1958). The routinization of *all* activity by prison staff and inmates alike confirms this notion (Owen, 1988). Arrigo (1996a: 195–6) notes, however that this prescription for management ignores or conceals the differences, inconsistencies, anomalies, and absurdities of general living and human social interaction. These are essential to how organizations behave.

Similar to Arrigo's (1996a) assessment of institutional life in psychiatric confinement systems, the prison management assumption applies a linear, reductionistic, and controlled perspective to the actions of correctional administrators, staff, and inmates (Bondeson, 1989). The experience of the subject within this system is, however, much more individualized and chaotic. There is always some unpredictability, indeterminability, and uncertainty in one's experiences. These nonlinear behaviors within the system, as manifestations of inevitable 'shocks', can either receive system-wide accommodation or be subject to organizational ignorance, avoidance, or oppression (Morgan & Zohar, 1995).

Within the prison, the unpredictability may materialize in the form of inmate riots, inmate adaptation to incarceration, institutional violence (Toch, Adams & Grant, 1989), the changing personal agendas of correctional officers and administrators (Marquart, 1983), and the ebb and flow of drugs or other contraband through the underground economy (Durham, 1994). Prison policy, by definition, presumes that each 'shock' to the system can be alleviated through established procedures (Owen, 1988). The dynamics of correctional policy create a continuing paradigm of order, as individual experiences are left to be allayed by rigidly prescribed, tightly controlled methods (Kaufman, 1988). These experiences seldom receive the necessary degree of mutable and fluid attention they deserve from the formal system. Indeed, their essential complexity and instability receive, at best, practical and established solutions for abating the consequences of organizational shocks (Bondeson, 1989). Thus, innovative interventions are rarely part of the correctional equation of system behavior.

Finn (1996: 35) provides us with a current example of a movement-in-flux within the penal system. In recent years, correctional administrators have eliminated or reduced the amenities and privileges granted to inmates. The 'no-frills' movement illustrates a shock to the correctional system, resulting in apparent short-term, situational transformations. External forces (e.g. outcry from the public, legislators concerned with re-election, prison rights lobbyists appalled by such inmate treatment), as well as internal dialogue among correctional administrators have created a temporary state of uncertainty; one which will result in some system change. Chaology regards this process as unpredictable (Butz, 1997: 8–26). The impact of the no-frills proposal will undoubtedly invoke 'ripple' effects throughout the correctional system, with indeterminable and immeasurable consequences. The point is that the no-frills movement, in addition to creating modifications in correctional policy, as well as criminal and judicial practices, substantially contributes to the chaotic behavior of the prison organization understood as a nonlinear system. It is a system in regular flux undergoing constant transformations. However, as Arrigo (1996a: 198) reminds us, the cornerstone of chaology is 'orderly disorder'. The correctional system, if examined on a macro-level, despite the apparent chaos within its behavioral regime of eliminating inmate amenities, will, over time, display an underlying structure representative of order, sustainability, boundedness.

3. Implications for Research, Practice, and Policy

Acknowledging that all natural and social systems are constituted by their mix of both constants and dynamics, implies that the behavior of any organization will always be, at some appreciable level, unpredictable. The complex functioning of the modern day prison embodies this sentiment. Precisely because there are internal and external pockets of instability and uncertainty at work within the correctional setting means that its operation can and will never be entirely subjected to tight, rigid control. Researchers, practitioners, and policy experts, embracing Morgan's flux and transformation metaphor, need to examine the implications these insights hold for the future administration of correctional justice.

For example, what are the lessons of prison violence, as a state of disequilibrium or local disorder, given the inescapable presence of chaos in the confinement institution? How should prison administrators respond to riots, inmate abuse, sexual exploitation, the underground economy, and other facets of 'dysfunctional' or maladaptive prison behavior (Toch, Adams & Grant, 1989), given that seemingly random events can trigger a state of chaos within the correctional milieu. Why is it better to impose lock-downs, sequister prisoners, enforce punitive sanctions for non-compliance, perform body cavity searches, or otherwise further pathologize and criminalize those confined, all in the name of system control, when the nature of human existence ineluctably resists (deviantly or otherwise) efforts at total homogeniety and complete conformity? We contend that Morgan's organizational analysis of the flux and transformation metaphor requires a substantial re-examiniation of these and similar correctional matters through ongoing research, management studies, and policy evaluations.

SUMMARY AND CONCLUSION

Metaphoric analysis of prison behavior yields a series of distinct prisms through which to better read organizational behaviors and practices. As Morgan (1997: 5) accurately asserts, all theory is metaphor. The foundation of any analysis, be it empirical or philosophical, is theory. Thus, the adept researcher must be aware of the impact and benefit of employing metaphor in any prison inquiry. The eight metaphors outlined by Morgan are by no means exhaustive; however, they do provide a solid foundation for assessing organizational settings.

The primary objective of the present study was to ascertain whether or not metaphorical analysis is applicable to the prison. Each of the eight metaphors has provided useful insight into the various behaviors of this complex system. While the metaphors employed here assume different frames of reference in their application, each is prominently featured within various elements of the correctional milieu.

The diversity in organizational philosophies and practices across prisons creates obvious difficulties in formulating one all-encompassing, integrated theory of prison metaphoric-organizational life. It would be equally futile to apply a single metaphor to the behavior of the confinement system, arguing that it was exclusively or even primarily representative of how the system always functioned. While normative practices are generally infrequent, some common features in prison organization are, however, notable.

Most prisons have similar objectives, yet the means by which they choose to achieve and maintain these objectives are variable. The application of organizational metaphors to the correctional milieu accomplishes several goals. It accentuates some of the organizational practices and conflicts that arise in pursuit of goals, and provides insight into how organizational philosophy and policy influence these pursuits. More importantly, it equips the researcher with a framework from which to analyze the behaviors prominent in various prisons.

Each and every outlined metaphor, when applied to the prison, exposes the diversity of philosophy and behavior that shapes the organization of the prison. As metaphors are akin to theory, no prison displays all metaphors in their entirety, and no prison is devoid of all aspects of any particular organizational metaphor. Correctional facilities tend to present an organizational framework which incorporates components of all metaphors in practice. The primary difference between prison organizations across the country, in the world, and throughout history, is which metaphors are most conspicuously featured. Analysis of any one prison would reveal the presence of a variety of different organizational metaphors. Understanding these metaphors and how they operate and contribute to the practices of the prison, can provide a means for critiquing and changing this seemingly eternal institution within different societies.

Metaphoric analysis of the prison thus concludes by establishing itself as an integral component of organizational observation. There is little doubt that prison policy and practice will continue to undergo transformation. This transformation will be linked to the conceptual prisms through which correctional facilities are organized and operate. In a society where crime, criminals, and social attitudes require the continuous re-assessment of criminal behavior, the formulation of effective and efficient methods for implementing and sustaining change becomes imperative. This formula necessarily includes accurately, or more completely, reading the behavior of modern penal institutions. Clearly in this endeavor, the metaphoric-organizational framework, as identified here, is one model that could have profound and long-lasting significance.

NOTES

1. Morgan's (1980: 607–609) four paradigms can be briefly summarized. The *functionalist* paradigm assumes that society has a concrete, real existence in which order, regulation, and control are paramount. It emphasizes social theory in which the tangible, lived, and concrete human affairs of individuals are assessed through the value-neutral, essentialistic method of positive science. The *interpretive* paradigm presupposes that the nature of society is precarious and uncertain. Reality does not exist in any ontologically concrete sense; rather, it is a by-product of ongoing human social interaction and the subjective and inter-subjective processes associated with it. Society has meaning only in the context of the social scientist participating in its constitution. The *radical-humanist* paradigm acknowledges that reality is both socially constructed and sustained; however, it recognizes that these very manifestations of reality bound and imprison cognitive structuring and behavior processes. The result is social alienation, psychic despair, and ideological domination. The *radical structuralist* paradigm agrees, consistent with its humanist counterpart, that society is a potentially dominant force and

foe. However, it is wedded to a materialist conception of social reality, anchored by concrete, everyday structures. Thus, reality exists independent of individual or intersubjective perceptions in which the tensions and struggles of opposing segments are lived out through the use of power and control.

2. On this point, Morgan (1980: 611) describes a boxer as a tiger in the ring. The effect of crossing these images works to the extent that "we conjure up specific impressions of a fierce animal, moving at times with grace, stealth, power, strength and speed in aggressive acts directed at its prey" (Ibid.). The comparison is impactful because the metaphor suggests that the boxer possesses these attributes and is a formidable opponent when fighting in the ring. In this process, however, we dismiss certain other comparisons that could be made through the use of metaphor. For example, drawing attention to the tiger's claws, coloring, size, fur, etc., are all ignored. Again, the emphasis is on a selective comparison in which certain shared features are assessed.

3. The bulk of Morgan's practice work in metaphorical analysis is linked to numerous institutions broadly encompassing the domains of business, industry, and commerce with an emphasis on the administration and management of such systems (e.g. Morgan, 1989, 1993). The operation of the military is one example of a complex network consistent with Morgan's general thesis on the relevance of metaphor in organizational analysis.

4. We note that the proceeding analysis draws attention to a detailed application of well chosen metaphors pertaining to the organization of prisons rather than to a sustained analysis of world views or paradigms that frame the metaphorical imagery of correctional facilities. Our strategy allows for a more careful assessment of metaphorical inquiry as a bona fide method by which to consolidate the disparate, yet familiar, prison literature described in the article. The related task of explaining the relationship between the four paradigms and the metaphorical images each evokes in correctional settings, although certainly worthwhile, is decidedly beyond the scope of the present work. While our strategy could be fairly criticized as lacking sufficient explanation regarding the conceptual utility of the metaphorical template, we contend that the model is specifically designed to demonstrate how metaphorical imagery helps consolidate the way in which prison behavior can be understood.

5. To the extent that Morgan's metaphoric template has not been systematically applied to the organization of correctional facilities is itself worth noting, especially given the breadth of his model's apparent utility in other disciplines. In order to fairly assess the applicability of his paradigm to the prison milieu, the existing corrections literature will be closely scrutinized.

6. When reviewing the specific metaphors, our explication of Morgan's thesis is substantiated by reference to other relevant source material. The intent here is to draw attention to the development of thought related to the 'frame' under consideration, and, thus, to demonstrate indirectly how Morgan's more metaphoric analysis falls within an established history of organizational theory and practice.

7. As we subsequently demonstrate, what makes this interdependence obvious is that Morgan's 'organism' frame refers to how the (prison) system lives on the basis of both internal and external stimuli, while the 'brain' metaphor refers to how this vital organism learns, adapts, and behaves, given such environmental inputs.

8. We note that attempts have been made to address the physical environment of prisons. However, as Shover & Einstadter (1988: 158) noted, the "psychological

consequences of correctional processing have not yet been systematically litigated." We contend that the physical and psychological consequences of the prison environment must both be confronted, as each contributes to the presence of exploitation.

9. Criticisms of the modern prison run the gambit from overcrowding and riots, to special population inmates (e.g. AIDS, tuberculosis, elderly), to sexual exploitation, to gangs. For an overview of these and related matters see, Durham (1994); Henderson (1990); Silverman & Vega (1996); and Camp & Camp (1985) respectively.

10. We recognize that there are other configurations in which power manifests itself in prisons (e.g. inmate to staff, staff to prison, inmate to inmate). However, our purpose is in highlighting the operation of each metaphor as a way of reading the behavior of correctional organizations. For some additional commentary on the relationship between power as expressed in other staff and inmate prison arrangements, see our discussion on the metaphors of culture and domination.

11. This observation is a bit of an over-generalization. Indeed, there is a great deal of complexity to the power, authority, and control mechanisms at work in modern prisons. This complexity is much more apparent when one considers the differential social organizations found along the traditional custody-treatment continuum of correctional institutions. This differentiation is further complicated when one evaluates male to female as well as adult to juvenile facilities. The point, however, is that in the context of prison to inmate and staff to staff relationships, official (political) authority significantly quashes or neutralizes prospects for deviation from established parameters of bureaucratic decision making (Garland, 1990).

12. The use of detailed rules and regulations (policies) and certain interpersonal collectives observable in prison guard and inmate alliances are also sources of power. We allude to the former in the context of the machine metaphor, and discuss the latter in the context of prison culture.

13. This notion can also be understood on the basis of *alienation* (see e.g. Marx, 1975; Marx & Engels, 1846; Fromm, 1961), *the deep structure of power* (e.g. Foucault, 1972, 1977), on *knowledge and human interests* (e.g. Habermas, 1970a, 1970b), and *critical theory and the Frankfurt School* (e.g. Horkheimer, 1972, Horkheimer & Adorno, 1973).

14. For example, in the wake of the public acrimony following the not-guilty-by-reason-of-insanity (NGRI) verdict in the John Hinckley Jr. case, Congress moved quickly to enact legislation that substantially restricted the use of the insanity defense, which gave rise to the guilty-but-mentally-ill (GBMI) verdict (Steadman, et al., 1993; Arrigo, 1996b). This disposition essentially ensured that defendants could be found criminally culpable (bad) and psychologically disordered (mad) at the same time. Further, when defendants are adjudicated NGRI, they can languish in mental health care settings for greater periods of time than if the defendant had been found guilty in a court of law (*Jones vs. U.S.*, 1983).

15. These include four chaotic processes which he refers to as "logics of change". These processes can be further defined as (1) autopoiesis, (2) shifting attractors, (3) mutual causality, and (4) the logic of dialectical change. Embedded within these logics are insights propagated by the 'new' sciences of chaology and complexity (for applications to *criminal confinement for the mentally ill* see, Arrigo, 1994; for applications to *chaos and law* see Milovanovic, 1992).

REFERENCES

Abadinsky, H., & Winfree, L. T. Jr. (1992). *Crime and justice: An introduction.* 2nd ed. Chicago: Nelson-Hall.

Adams, K. (1992). Adjusting to Prison Life. *Crime and justice: A review of research, 16,* 275–359.

Adler, M., & Longhurst, B. (1994). *Discourse, power and justice: Towards a new sociology of imprisonment.* New York: Routledge.

Archambeault, W. G., & Fenwick, C. (1988). A comparative analysis of culture, safety, an organizational management factors in Japan and U.S. prisoners. *Prison Journal, 68,* 3–23.

Argyris, C. (1993). *Knowledge for action.* San Francisco: Jossey-Bass.

Argyris, C. (1994). Good Communication that Blocks Learning. *Harvard Business Review, 43,* 77–85.

Argyris, C., & Schon. D. A. (1978). *Organizational learning: A theory of action perspective.* Reading, MA: Addison-Wesley.

Arrigo, B,. A. (1992). Deconstructing jurisprudence: An experiential feminist critique. *The Journal of Human Justice, 4*(1), 13–30.

Arrigo, B. A. (1993). *Madness, language and law.* Albany, N.Y.: Harrow and Heston.

Arrigo, B. A. (1994). Legal discourse and the disordered criminal defendant: contributions from psychoanalytic semiotics and chaos theory. *Legal Studies Forum, 18*(1), 93–112.

Arrigo, B. A. (1995). The Peripheral core of law and criminology: On postmodern social theory and conceptual integration. *Justice Quarterly, 12*(3), 447–472.

Arrigo, B. A. (1996a). *The Contours of psychiatric justice: A postmodern critique of mental illness, criminal insanity, and the law.* New York/London: Garland.

Arrigo, B. A. (1996b). The Behavior of law and psychiatry: rethinking knowledge construction and the guilty-but-mentally ill verdict. *Criminal Justice and Behavior, 23*(4), 572–592.

Arrigo, B. (1997). Transcarceration: Notes on a psychoanalytically-informed theory of social practice in the criminal justice and mental health systems. *Crime, Law, and Social Change: An International Journal, 27*(1), 31–48

Arrigo, B., & Schehr, B. (1998). Restoring justice for juveniles: A critical analysis of victim offender mediation. *Justice Quarterly, 15*(4), 629–666.

Arrigo, B. A., & Williams, C. R. (1999). Chaos theory and the social control thesis: A post-Foucauldian analysis of mental illness and involuntary civil confinement. *Social Justice, 26*(1), 177–207.

Bateson, G. (1972). *Steps to an ecology of mind.* New York: Ballantine.

Becker, E. (1963). *The denial of death.* New York: Free Press.

Belden, G., Hyatt, M., & Ackley, D. (1993). *Towards the learning organization.* Toronto: Institute of Cultural Affairs.

Berger, P., & Luckmann. T. (1967). *The social construction of reality.* Garden City, NJ: Anchor.

Black, M. (1962). *Models and metaphors.* Ithaca, NY:Cornell University Press.

Bolman L., & Deal, T. (1991). *Reframing organizations.* San Francisco: Jossey-Bass.

Bondeson, U. V. (1989). *Prisoners in prison societies.* New Brunswick, NJ: Transaction.

Bowles, M. (1990). Recognizing deep structures in organizations. *Organization Studies, 11,* 395–412.

Bowles, M. (1991). The Organizational shadow. *Organizational Studies, 12,* 387–404.

Bradshaw, P., & Newell, S. (1993). Exploring deep structures of power using Jungian dream analysis. Working Paper, Schulich School of Business, Toronto.

Bridges, W. (1994). *Job shift*. Reading, MA: Addison-Wesley.

Briggs, J., & Peat, F. D.(1989). *Turbulent mirror: An Illustrative guide to chaos theory and the science of wholeness*. New York: Harper & Row,

Bright, C. (1996). *The powers that punish: Prison politics in the era of the 'Big House', 1920–1955*. Ann Arbor: University of Michigan Press.

Bureau of Justice Statistics (1994). *Prisoners in 1993*. Washington, DC: U.S. Department of Justice.

Burell, G. (1984). Sex and organizational analysis. *Organizational Studies, 5*, 97–118.

Burell, G. (1992). The Organization of pleasure. In: M. Alvesson & H. Willmott (Eds), *Critical management studies*. London: Sage.

Burell, G., & Morgan, G. (1979). *Sociological paradigms and organizational analysis*. London: Heinemann Educational Books.

Butz, M. R. (1992a). The fractal nature of the development of the self. *Psychological Reports, 71*, 1043–1063.

Butz, M. R. (1992b). Chaos: An omen of transcendence in the psychotherapeutic process. *Psychological Reports, 71*, 827–243.

Butz, M. R. (1997). *Chaos and complexity: Implications for psychological theory and practice*. Washington, DC: Taylor & Francis.

Camp, G. M., & Camp, C. G. (1985). *Prison gangs: Their extent, nature, and impact on prisons*. Washington, DC: U.S. Government Printing Office.

Clegg, S. (1989). *Frameworks of power*. London and Newbury Park, CA: Sage.

Clemmer, D. (1958). *The Prison community*. New York: Rinehart & Co.

Cornell, D. (1991). *Beyond accommodation: Ethical feminism, deconstruction and the law*. New York: Routledge.

Cox, V., Paulus, P., & McCain, G. (1984). Prison crowding research: The relevance for prison housing standards and a general approach regarding crowding phenomena. *American Psychologist, 39*, 1145–1152.

Denhardt, R. B. (1981). *In the shadow of organization*. Lawrence, KA: Regents Press.

Drucker, P. F. (1993). *Post-capitalist society*. New York: Harper Business.

Duncan, M. G. (1996). *Romantic outlaws, beloved prisons: The unconscious meanings of crime and punishment*. New York: New York University Press.

Durham, A. (1994). *Crisis and reform: Current issues in American punishment*. Boston, MA: Little, Brown & Co.

Durkheim. E. (1934). *The Division of labor in society*. London: McMillan.

Eberle, T. S. (1996). Dislocation policies in western Europe: Past, present and future. *Annals of the American Academy of Political and Social Sciences, 544*, 127–139.

Ekland-Olson, S. (1991). Crowding, social control, and prison violence: Evidence from the post-Ruiz Years in Texas. *Law and Society Review, 20*, 389–421.

Elliott, E., & Kiel, L.D.. (1996). *Chaos theory in the social sciences: Foundations and applications*. (Eds.) Ann Arbor: University of Michigan Press.

Farrington, K. (1992). The Modern prison as total institution? Public perception versus objective reality. *Crime and Delinquency, 38*(1), 6–26.

Fayol. H. (1949). *General and industrial management*. London: Pitman.

Finn, P. (1996). No-frills prisons and jails: A movement in flux. *Federal Probation, 60*, 3.

Fleisher, M. S. (1989). *Warehousing violence*. Newbury Park:, CA: Sage.

Fleisher, M. S. (1996). Management assessment and policy dissemination in federal prisons. *The Prison Journal, 76*(1), 81–91.

Follett, M. P. (1973). In: E. M. Fox and L. Urwick (Eds.), *Dynamic administration: The collected papers of Mary Parker Follett*. London: Pitman.

Ford, C. M., & Gioia, D. A. (Eds) (1995). *Creative action in organizations*. Thousand Oaks, Ca: Sage.

Foucault, M. (1972). *The archaeology of knowledge*. A. Sheridan (Trans.). New York: Pantheon.

Foucault, M. (1977). *Discipline and punish*. New York: Penguin Books.

Freud, S. (1953). *The complete psychological works of Sigmund Freud*. London: Hogarth.

Fromm, E. (1961). *Marx's concept of man*. New York: Ungar.

Fromm, E. (1971). *The crisis of psychoanalysis*. New York: Cape.

Frost, P. J., Mitchell, V. F., & Nord, W.R. (Eds) (1995) *Managerial reality*. New York: Harper Collins.

Frost, P. J., & Egri, C. (1991). The political process of innovation. In: L Cummings & B. Staw (Eds), *Research in organizational behavior*, Greenwich, CT: JAI Press.

Galbraith, J. K. (1967). *The new industrial state*. London: Hamish Hamilton.

Garfinkel, H. (1967). *Studies in ethnomethodology*. Englewood Cliffs, NJ: Prentice-Hall.

Garland, D. (1990). *Punishment and modern society*. Oxford: Clarendon Press.

Gergen, K. (1985). The social constructionist movement in modern psychology. *American Psychologist, 40*, 266–275.

Gergen, K. J. (1991). *The saturated self*. New York: Basic Books.

Gherardi, S. (1995). *Gender, symbolism and organizational cultures*. Thousand Oaks, CA: Sage.

Gilbreth, F. B. (1911). *Motion study*. New York: Van Nostrand.

Goffman, E. (1961). *Asylums*. New York: Anchor Books.

Goffman, E. (1963). *Stigma: Notes on the management of a spoiled identity*. New York: Simon & Schuster.

Grapendaal, M. (1990). The inmate subculture in Dutch prisons. *British Journal of Criminology, 30*, 3.

Gulick, L., & Urwick. L. (Eds) (1937). *Papers in the science of administration*. New York: Institute of Public Administration, Columbia University.

Habermas, J. (1970a). On systematically distorted communications. *Critical Inquiry, 13*, 205–218.

Habermas, J. (1970b). Towards a theory of communicative competence. *Critical Inquiry, 13*, 360–375.

Habermas, J. (1975). *Legitimation crisis*. Boston: Beacon Press.

Hamm, M. (1990). Legislator attitudes toward polices to reduce prison crowding. *Criminal Justice Policy Review, 3*, 220–235.

Hammer, M., & Champy, J. (1993). *Reengineering the corporation*. New York: Harper.

Harvey, J. (1995). *The abeline paradox*. New York: Fress Press.

Hayles, N. K. (1990). *Chaos bound: Orderly disorder in contemporary literature and science*. Ithaca, NY: Cornell University Press.

Heffernan, E. (1972). *Making it in prison: The square, the cool, and the life*. New York: Wiley-Interscience.

Henderson, J. D.(1990). *Causes, preventive measures, and methods of controlling riots and disturbances in correctional institutions*. Loral, MD: American Correctional Association.

Henry, I. (Ed.) (1991). *Creative management*. London: Sage.

Henry, S., & Milovanovic, D. (1996). *Constitutive criminology: Beyond postmodernism*. London: Sage.

Hirschhorn, L (1988). *The workplace within*. Cambridge, MA: MIT Press.

Horkheimer, M. (1972). *Critical theory: Selected essays*. New York: Herder.

Horkheimer, M. & Adorno, T. (1973) *Dialectic of enlightenment*. London: Allen Lane.
Howe, A. (1994). *Punish and Critique: Toward a feminist analysis of penality*. London: Routledge.
Howells, J., & Wood, M.. (1993). *The globalization of production and technology*. London: Belhaven.
Hunt, G., Riegel, S., Morales, T., & Waldorf, D. (1993). Changes in prison culture: Prison gangs and the case of the 'pepsi generation'. *Social Problems, 40.*
Institute for Economic and Policy Studies (1988). *The economic impact of Corcraft Correctional industries in New York state*. Alexandria, VA: The Policy Center.
Irwin, J. (1977). The changing social structure of men's prisons, In: D. Greenberg (Ed.), *Corrections and Punishment*. Beverly Hills: CA: Sage.
Irwin, J. (1980). *Prisons in turmoil*. Boston: Little, Brown.
Irwin, J., & Austin, J. (1997) *It's about time*. Belmont, CA: Wadsworth.
Irwin, J., & Cressey, D. (1962). Thieves, convicts, and the inmate subculture. *Social Problems, 10*, 142–155.
Jacobs, J. J. (1977). *Statesville*. Chicago University of Chicago Press.
Jacobs, J. J. (1983). Prisons: Prison subculture. In: S. H. Kadish (Ed.), *Encyclopedia of crime and justice* (p. 1224). New York: The Free Press.
Johnson, R. (1987). *Hard time: Understanding and reforming the prison*. Monterey, CA: Brooks/Cole.
Jung, C. G. (1971). *The portable Jung*. New York: Viking.
Katz, D., & Kahn, R. L. (1978). *The social psychology of organizations*. New York: Wiley.
Kauffman, S. A. (1993). *The origins of order: Self organization and selection in organizations*. New York: Oxford University Press.
Kaufman, K. (1988). *Prison officers and their world*. Cambridge, MA: Harvard University Press.
Kickert, W. J. M. (1993). Autopoiesis and the science of (public) administration: Essence, sense and nonsense. *Organizational Studies, 14*, 261–278.
Klofas, J., Stojkovic, S., & Kalinich, D. (1990).*Criminal justice organizations: Administration and management*. Pacific Grove, CA: Brooks/Cole.
Knights, D., & Willmott, H. (1989). (Eds). *Labour process theory*. London: MacMillan.
Kuhn, T. (1970). *The structure of scientific revolution*. Chicago: University of Chicago Press.
Kurshan, N. (Ed.) (1992). *Cages of Steel: The politics of imprisonment in the United States*. Washington, DC: Masonnevre Press.
Lasch, C. (1979). *The culture of narcissism*. New York: Warner Books.
Lester, D., Braswell, M., & Van Voorhis, P. (1992). *Correctional counseling*. 2d ed. Cincinnati, OH: Anderson Publishing.
Lewin, R. (1992). *Complexity: Life at the edge of chaos*. New York: McMillan.
Luhmann, N. (1995). *Social systems*. Stanford, CA: Stanford University Press.
Marx, K. (1975). *Early writings*. Harmondsworth: Penguin.
Marx, K. (1976). *Das Kapital*. Harmondsworth: Penguin.
Marx, K., & Engels, F. (1846). *The German ideology*. London: Lawrence & Wishart.
Manning, P. (1979). Metaphors in the field: Varieties of organizational discourse. *Administrative Science Quarterly, 24*, 660–671.
Manning, P. (1988). *Symbolic communication: Signifying calls and the police response*. Cambridge, MA: MIT Press.
Manning, P. (1992). *Organizational communication*. New York: Aldine.
Marcuse, H. (1955). *Eros and civilization*. Boston: Beacon Press.

Marquart, J. W. (1983). *Cooptation of the kept: Maintaining control in a southern penitentiary.* Ann Arbor Mich: University Microfilms.

Mason, R. O., & Mitroff, I. (1981). *Challenging strategic planning assumptions.* New York: John Wiley.

Maturana, H., & Varela, F. (1980). *Autopoiesis and cognition: The realization of the living.* London: Reidl.

McCarthy, B. J. (1988). Responding to the prison crowding crisis: The restructuring of a prison system. *Criminal Justice Policy Review, 2*, 3–20.

McCorkle, R., Miethe, T., & Drass, K. (1995). The roots of prison violence: A test of the deprivation, management, and 'not-so-total' institution models. *Crime and Delinquency, 41*(3), 317–331.

McNeil, K. (1978). Understanding organizational power: Building on the Weberian legacy. *Administrative Science Quarterly, 23*, 65–90.

Michels, R. (1949). *Political parties.* New York: Fress Press.

Miller, J. G. (1978). *Living systems.* New York: McGraw-Hill.

Milovanovic, D. (1992). *Postmodern law and (dis)order: Psychoanalytic semiotics, chaos and juridic exegeses.* Liverpool: Deborah Charles.

Milovanovic, D. (1997). (Ed.). *Chaos, criminology, and social justice.* New York: Praeger.

Milovanovic, D., & Henry, S. (1991). Constitutive penology. *Social Justice, 18*, 204–224.

Mitroff, I., & Linstone, R. H. (1993). *The unbounded mind.* New York: Oxford University Press.

Mooney, J. C., & Reiley, A. P. (1931). *Onward industry.* New York: Harper & Row.

More, H. W. (Ed.) (1977) *Criminal justice management.* St. Paul, MN: West.

Morgan, G. (1980). Paradigms, metaphors and puzzle solving in organizational analysis. *Administrative Science Quarterly, 25*, 605–622

Morgan, G. (1981). The schismatic metaphor and its implications for organizational analysis. *Organization Studies, 2*, 23–44.

Morgan, G. (1983). More on metaphor: Why we cannot control tropes in administrative science. *Administrative Science Quarterly, 28*, 601–607.

Morgan, G. (1989). *Creative organization theory.* Newbury Park, CA: Sage.

Morgan, G. (1993). *Imaginization: The art of creative management.* Newbury Park, CA: Sage.

Morgan, G. (1996). Is there anything more to say about metaphor. In: D. Grant and C. Oswick (Eds), *Metaphor and organizations.* London: Sage.

Morgan, G. (1997). *Images of organization.* 2nd ed. Thousand Oaks: Sage.

Morgan, G., & Zohar, A. (1995). Achieving quantum change: Incrementally!!!. Working Paper. Toronto: Schulich School of Business.

Murton, T. (1976). *The dilemma of prison reform.* New York: Praeger.

Mouzelis, N. (1979). *Organization and bureaucracy.* 2nd ed. London: Routledge & Kegan Paul.

Ott, J. S. (1989). *The organizational cultural perspective.* Pacific Grove, CA: Brooks/Cole.

Owen, B. A. (1988). *The reproduction of social control: A study of prison workers at San Quentin.* New York: Praeger

Pauchant, T. C., & Athos, I. A.. (1992). *Transforming the crisis prone organization.* San Francisco: Jossey Bass.

Pedler, M. J., Burgoyne, J. A., and Boydell, T. H. (1991). *The learning company.* London: McGraw-Hill.

Pettigrew, A. M. (1973). *The politics of organizational decision making* London: Tavistock.

Pfeffer, J. (1981). *Power in organizations.* Marshfield, MA: Pitman.

Plato. (1941). *The Republic.* Oxford: Clarendon.

Prigogine, I., & Stengers, I. (1984). *Order out of chaos.* New York: Bantam Books.

Richards, S. C. (1995). *The structure of prison release*. New York: McGraw-Hill Custom Press.

Richards, S. C. (1998). Critical and radical perspectives on community punishment: Lessons from the darkness. In: J. I. Ross (Ed.), *Cutting the edge: Current perspectives in radical/critical criminology and criminal justice* (pp. 122–144). New York: Praeger.

Rifkin, J. (1995). *The end of work*. New York: Putnam.

Ritzer, G. (1996). *The McDonaldization of society*. Newbury Park, CA: Sage.

Salaman, G. (1978). Towards a sociology of organizational structure. *Sociological Quarterly, 26*, 519–544.

Scarbrough, H., & Corbett, J. M. (1992). *Technology and organization*. London and New York: Routledge.

Scharf, P. (1977). The just community. *New Society, 21*, 104–105.

Schon, D. A. (1963). *Invention and the evolution of ideas*. London: Tavistok.

Schroeder, A. (1976). *Shaking it rough*. Garden City, NJ: Doubleday.

Schutz, A. (1967). *Collected papers 1: The problem of social reality*. The Hague: Martinus Nijhoff.

Schwartz, H. S. (1990). *Narcissistic processes and corporate decay*. New York: NYU Press.

Scott, W. R., and Christensen, S. (Eds.) (1995). *The institutional construction of organizations*. Thousand Oaks, CA: Sage.

Sellin, T. (1976). *Slavery and the penal system*. New York: Elsevier.

Senge, P. (1990). *The fifth discipline*. New York: Doubleday.

Shostak, A. B. (Ed.) (1996). Impacts of changing employment. *Annals of the American Academy of Political and Social Sciences, 544* (Special Issue).

Shover, N., & Einstadter, W. (1988). *Analyzing American corrections*. Belmont, CA: Wadsworth.

Silberman, M. (1995). *A world of violence: Corrections in America*. Belmont, CA: Wadsworth.

Silverman, I. J., & Vega, M. (1996). *Corrections: A comprehensive view*. St Paul, MN: West.

Snarr, R. W., & Wolford, B. I. (1985). *Introduction to corrections*. Dubuque, IO: William C. Brown.

Steadman, H. J., McCarty, D. W., & Morrissey, J. P. (1989). *The mentally ill in jail: Planning for essential services*. New York: Guilford Press.

Steadman, H. J., M. A. McGreevy, Morrissey, J. P., Callahan, L. A., Robbins, P. C., & Cirincione, C. (1993). *Before and after Hinckley: Evaluating insanity defense reform*. New York: Guilford Press.

Stewart, I. (1989). *Does God play dice?*. New York: Basil Blackwell.

Swigert, V. L., & Farrell, R. A. (1976). *Murder, inequality, and the law*. Lexington, MA: D.C. Health.

Sward, K. (1948). *The legend of Henry Ford*. New York: Rinehart.

Sykes, G. (1958). *The society of captives: A study of a maximum security prison*. Princeton, NJ: Princeton University Press.

Sykes, G., & Messinger, S. (1960). The inmate social system. In: R. Cloward (Ed.), *Theoretical Studies in the Social Organization of the Prison*. New York: Social Science Research Council.

Taylor, F. W. (1911). *Principles of scientific management*. New York: Harper & Row.

Territo, L, Halsted, J., & Bromley, M. L. (1998). *Crime and justice in America: A humanistic perspective*. Boston, MA: Butterworth-Heinemann

Thomas, C. W. (1970). Toward a more inclusive model of the inmate culture. *Criminology, 8*, 251–262.

Thomas, C. W., & Foster, S. C. (1972). Prisonization and the inmate contraculture. *Social Problems, 20*, 229–239.

Thomas, C. W., & Peterson, D. (1977). *Prison organization and inmate subcultures.* Indianapolis: Bobbs-Merrill.

Thomas, J., & O'Maochatha, A.. (1989). Reassessing the critical metaphor: An optimistic revisionist view. *Justice Quarterly, 6*, 143–172.

Toch, H. (1998). *Corrections: A humanistic perspective.* New York: Harrow and Heston.

Toch, H., Adams, K., & Grant, D. J. (1989). *Coping: Maladaptation in prisons* New Brunswick, NJ: Transaction.

Touraine, A. (1977). *The self production of society.* Chicago: University of Chicago Press.

Touraine, A. (1988). *The return of the actor.* Minneapolis: University of Minnesota Press.

Useem, B., Camp, G. C., Camp, G. M., & Dugan, R. (1995). *Resolution of prison riots.* National Institute of Justice Research Brief.

Vanek, J. (1975). *Self management.* Harmondsworth: Penguin.

Weber, M. (1946). In: H. Gerth, and C. W. Mills (Eds), *From Max Weber.* New York: Oxford University Press.

Weber, M. (1947). *The theory of social and economic organization.* London: Oxford University Press.

Weil, K. E. (1979). *The social psychology of organizations.* Thousand Oaks, CA: Sage.

Weick, K. E. (1995). *Sensemaking in organizations.* Thousand Oaks, CA: Sage.

Weinstein, D., & Weinstein, A. (1993). Postmodernizing (macro)sociology. *Sociological Inquiry, 63*(2), 224–237

Wilson, E. O. (1975). *Sociobiology: The new synthesis.* New York: Belknap.

Wittenberg, P. (1996a). Language and communication in prison. *Federal Probation, 60*, 4.

Wittenberg, P. (1996b). Power, influence, and the development of correctional policy. *Federal Probation, 60*, 2.

Wooden, W. S., & Parker, J. (1982). *Men behind bars: Sexual exploitation in prison.* New York: Plenum Press.

Young, T. R. (1992). Chaos theory and human agency: Humanist sociology in a postmodern age. *Humanity and Society, 16*(4), 441–460.

Young, T. R., & Arrigo, B. A. (1998). *Chaos and crime: From criminal justice to social justice.* Albany, NY: SUNY Press.

Zedlewski, E. M. (1987). *Making confinement decisions.* Washington, DC: U.S. Department of Justice.

Zimmer, L. (1989). Solving women's employment problems in corrections: Shifting the burden to administrators. *Women and Criminal Justice, 1*, 55–79.

Zohar, A. (1995). *Paradoxical dimensions of organization.* Working Paper, Toronto: Schulich School of Business.

Zupan, L. (1992). Men guarding women: An analysis of the employment of male correction officers in prisons for women. *Journal of Criminal Justice, 20*, 297–309.

Zweig, C., & Abrams, J. (1991). (Eds), *Meeting the shadow.* New York: Tarcher.

CASES CITED

Castello v. Wainwright, 397 F.Supp. 20 (M. D.FLA 1975).

Jones v. United States, 463 U.S. 354 (1983).

IMPEACHMENT WORK IN THE MENENDEZ BROTHERS' MURDER TRIAL: THE INTERACTIONAL ACHIEVEMENT OF FACTICITY, CREDIBILITY AND ACCOUNTABILITY

Stacy Burns

ABSTRACT

This chapter delineates a specific domain of recurrent legal work, lawyers' impeachment work on cross-examination at trial. Impeachment techniques are invoked as part of the claims-making contests between an interrogator and adverse witness which characterize much cross-examination. The research describes certain lawyers' impeachment practices in a detailed way which makes visible their endogenous organization and contingently produced features. The study also finds that witnesses are not without resources and sometimes resist the impugning implications of the questioning. The chapter shows that impeachment of an adverse witness at trial is an incessantly local and contingent achievement.

INTRODUCTION

This study focuses on a specific and recurrent domain of lawyers' work: the impeachment of adverse witnesses on cross-examination at trial.[1] Attempting to

Sociology of Crime, Law and Deviance, Volume 2, pages 233–256.

ISBN: 0-7623-0680-7

impeach a witness on cross-examination is in many ways just the opposite of everyday civil interactions in which persons usually enable one another to 'save face' (Goffman, 1959, p. 10). It concerns the invocation and use by cross-examining lawyers of an array of 'front-destructive' actions oriented to discrediting the managed public presentation of the adverse witness (Ibid., p. 115). Impeachment is a critical trial practice which can be an effective and sometimes dramatic device for discrediting a witness, especially by showing that s/he has been purposefully lying. But impeachment does not just happen; it must be made to happen in skillful ways to sway the decisions of jurors. This paper investigates some of the ways attorneys do this work (with greater or lesser skill and effectiveness) and how these witnesses respond in the face of adversarial questioning and attempted impeachment.

In their daily practice, lawyers engage in many kinds of professional tasks conducted in a wide range of settings and under varying formal and practical constraints. Impeachment work during cross-examination at trial is readily distinguishable from other types of professional legal activities, for example, drafting a pleading or 'lawyer's letter', writing an appellate brief or conducting research in the law library. Although professional skills are used to accomplish these projects, such legal work does not center on real-time interactional techniques and practices. Impeachment work also differs from other routine kinds of legal tasks which are accomplished interactionally, such as negotiating with opposing counsel, advising a client or arguing a motion to a judge or an appellate panel.

Perhaps impeachment work at trial most closely resembles lawyers' work in examining adverse witnesses at oral depositions. The deposition of a witness is a method of pre-trial 'discovery' conducted by lawyers outside the presence of a jury which helps to ascertain the evidence and commit the witness to a definite account of the relevant facts and events in the case. Like trial, a deposition is a formal proceeding in which the witness answers questions under oath and penalty of perjury. Also like trial, deposition testimony is recorded by a court reporter whose transcript becomes part of the permanent record in the case and is often later used at trial as the material for attempting impeachment during cross-examination. Most importantly, like cross-examination at trial, a deposition involves conducting an interrogation of an adverse witness which is highly improvisational in nature, as James and Hazard point out:

> [T]he greatest advantage that an oral deposition has over other methods of discovery is the flexibility it affords in probing a witness by requiring immediate, on-the-spot answers to oral questions (without the chance to rehearse phraseology and content which written

> interrogatories provided in advance would afford) and by allowing the examiner to frame questions as he goes along, in the light of the answers as they are given
>
> (James & Hazard, 1977, p. 181)

Impeachment work at trial usually involves courtroom interrogation by a plaintiff or defense attorney of an adverse witness who has already been questioned on direct examination by the opposing counsel and who has testified harmfully to the cross-examiner's side.[2,3] Efforts to impeach a witness are undertaken before and for the jury (and judge) as the overhearing audience (Heritage, 1985). Impeachment work by trial lawyers is oriented to demonstrating for the jury (and judge) which witnesses and accounts should be believed (or discredited) and, accordingly, how responsibility for the actions and events at issue should be allocated (see Emerson & Messenger, 1977).

The Menendez brothers' murder trial is a criminal trial. What is at issue in such a criminal case is the adjudication of criminal responsibility through the produced achievement and public recognition of either a 'crime' (warranting the state's right to impose punishment on the convicted), or an acquittal (requiring the state to formally relinquish charges against the absolved). Trial lawyers employ a full range of practices to effectuate a win. The tactics they use are oriented to maximizing their client's advantage and minimizing the effectiveness of damaging testimony and other harmful evidence. Central among the professional skills of trial lawyers are the practices by which they attempt to impeach adverse witnesses.

Impeachment of witnesses is necessarily accomplished in real-time and in material detail through sequences of questions and answers which the examining lawyer asks and the witness answers. However, there have been surprisingly few close studies of actual interaction in the courtroom. Even fewer studies provide finely detailed descriptions of such processes, apart from a limited number of recent empirical studies (e.g. Atkinson & Drew, 1979; Atkinson, 1992; Drew, 1985, 1990, 1992; Conley & O'Barr, 1990; O'Barr, 1982). Thus, just what comprises impeachment work in the courtroom remains largely unexplored analytic territory.

DATA AND METHODS

This research is based on videotape records of naturally occurring courtroom activities and interactions and involves close examination of cross-examination in its real-time audio-visual detail. The analysis is also informed by a larger qualitative study of witness examination in the first Menendez brothers' murder trial, drawing upon comprehensive observations of the entire trial and detailed analysis of ten extended cross-examination interchanges.[4]

The analysis relies upon prior studies in ethnomethodology and conversational analysis (Garfinkel, 1967, 1996; Garfinkel & Sacks, 1970; Sacks, Schegloff & Jefferson, 1974; Lynch, Livingston & Garfinkel, 1983; Moerman & Sacks, 1988; Drew & Heritage, 1992). Ethnomethodology provides a distinctive approach to the study of law and legal settings. It examines the practical daily tasks and routine work activities of lawyers for their produced regularities and 'local orderliness' (Garfinkel, 1988, 1996). In ethnomethodological studies are to be found empirical specifications of "social order as the order of ordinary affairs and studies of context as local environments of practical action" (Macbeth, 1991, p. 282 and see Garfinkel, 1988, 1996, p. 11).

Ethnomethodological studies of work are best pursued by following Garfinkel's urgings "to find, collect, specify and make instructably observable the local endogenous production . . . of immortal . . . society's most ordinary organizational things . . . and to provide for them . . . in their ongoingly, procedurally enacted coherence of substantive ordered phenomenal detail" (Garfinkel, 1996, pp. 6–7). This paper centers on two actual cross-examinations in the Menendez Brothers' murder trial and detailed consideration of the core legal skills and professional practices used by the attorneys in conducting them. The episodes were selected based upon their heuristic value for exploring both the nature of cross examination interchange and the processes of impeaching a witness through such exchanges.

Ethnomethodology shares with the research approach of conversational analysis a focus on the constitutive practices and "competencies that underlie intelligible, socially organized interaction" (Greatbatch & Dingwall, 1994, p. 87). Conversational analysis is a data-driven approach which utilizes audio and audio/video records of actual conversations and detailed transcripts of those recordings (see Appendix) to examine the regularities of coordinated social action as these are achieved in ordinary conversation (Sacks, Schegloff & Jefferson, 1974).

The present study is based in large part on videotape recordings of the first Menendez brothers' murder trial and the analysis of cross-examination interchanges in that trial. Yet, reliance on such records may be questioned by analysts who have argued that the presence of cameras in the courtroom fundamentally alters ordinary trial dynamics (e.g. lawyers making speeches and otherwise "playing for the media"). Indeed, the retrial, following two hung juries in the first trial was not televised by order of the court primarily for this reason. The high profile nature of such trials has prompted commentators to question their representativeness as exemplars of routine criminal justice. In view of the gavel-to-gavel coverage, media interest and sensational nature of the facts in the first Menendez brothers' murder trial, a word of caution seems in order. However, these concerns have little to do with the actual use to be made of the videotaped testimony as data in this paper. Even when the data studied arise out of a sensational trial, if looked at through "mundane-colored glasses," primordial features of legal work are there to be found (see Schegloff, 1988–9, pp. 216–217). This paper seeks to examine the " 'special' event" of the televised first Menendez brothers' murder trial with mundane-colored glasses and "turn a topically transient occurrence into a source of longer lasting analytic resources" (Ibid. p. 218).

BACKGROUND FACTS AND APPLICABLE LAW

The first Menendez brothers' murder trial was held in Van Nuys, California from July through mid-December 15, 1993. The Menendez brothers were charged with first degree murder in connection with the August, 1989 shotgun killings of their parents in the family's posh Beverly Hills home (Lyle was 21 years old and Eric was 18 years old at the time of the killings).

Two separate juries were impanelled in the first trial. Lyle Menendez' jury was designated the 'gold' jury and was comprised of five men and seven women. The 'blue' jury for Eric Menendez had six men and six women. Some of the evidence was presented to both juries simultaneously and other evidence was presented before just one jury or the other. The first trial lasted six months (including jury deliberations) and concluded when both juries deadlocked. The brothers were later retried in the same Van Nuys courtroom by a single jury, found guilty of first degree murder and sentenced to life imprisonment.

The two cross-examination exchanges presented here have been lifted from the ongoing flow of emergent events in the first Menendez trial and comprise just a few brief minutes of that lengthy trial. Nonetheless, these moments serve as pieces in the competing factual accounts which the adverse witness and lawyer are striving through their interactional practices to build and sustain.

Prior to the first trial, the Menendez brothers had confessed to committing parricide, but contended that the homicides were justified because they were acting in self defense at the time of the killings. In particular, both defendants claimed that they had been physically and emotionally abused by their parents and also sexually abused by their father (Lyle from about six years old to seven and Eric from about six to eighteen, at the time of his father's death). The defendants further contended that in the few days before the killings, they were in a 'panic' after Lyle Menendez learned that his 18 year old younger brother Eric was still being sexually abused by their father. This was part of the defense claim at trial that the defendants subjectively believed that their use of deadly force was necessary in 'self-defense'.[5]

Witnesses may be impeached by various forms of 'intrinsic' and 'extrinsic' evidence. Impeachment by intrinsic evidence usually makes reference to the witness' own testimony, for example by showing that the testimony is inherently unbelievable; that the witness has repeatedly exhibited a faulty memory or an inability to recall (see Burns, 1996); that the witness has some defect in the capacity to observe, comprehend or recount the matters attested to (James & Hazard, 1977, p. 247 and see Heritage, 1984, pp. 214–215); or that the witness' demeanor or conduct on the stand indicates that they are testifying falsely.[6]

By contrast, impeachment by extrinsic evidence commonly involves the use of some external factual evidence or record (although there is almost no limit to the form which the extrinsic evidence invoked for such purposes may take). Extrinsic evidence used in impeachment may include photographs or videotapes (e.g. a secretly obtained video showing a plaintiff in a civil action to be less severely injured than s/he contends); contrary sworn deposition or trial testimony; a prior inconsistent written statement not given under oath (e.g. a medical record or police report); or other documentary or testimonial proof of extrinsic facts which renders the present testimony highly improbable or even impossible (e.g. evidence that the lighting in the area was so poor that the witness could not possibly have identified the accused). Impeachment by extrinsic evidence is an especially effective and powerful form of impeachment which, in the extreme case, can reveal a lie and establish perjury.[7,8]

Each of the two episodes considered in this study exhibit methods by which attorneys attempt to impeach a hostile witness primarily through the use of extrinsic evidence. The impeachment procedure in the first episode may be characterized as a two-step process. First, cross-examining counsel verifies the details of the witness' version of the facts and events (thus pinning the witness to a specific and definite account). Then the lawyer confronts the witness with an extrinsic fact which on its face contradicts the witness' testimony. This two-step impeachment procedure is an attempt by the attorney to lead the witness out onto a limb which is then cut off so that the witness' version will come crashing down.

In the second episode, the attorney tries to impeach the witness with prior inconsistent statements and testimony given under oath. The impeachment device in this episode relies upon the general presumption (of lay persons and psychologists of memory) that memories fade over time (see e.g. Luh, 1922). Thus, statements made by a witness closer in time to the actual events in question are presumed more likely to be accurate than the witness' current testimony. As happens in the second episode, this impeachment is commonly initiated by asking a question such as: "Isn't it true that your memory of these incidents was better at the time you gave your original statement than it is today because that statement was made closer in time to the actual events?" The preference built into such a question is for the witness to respond affirmatively: "Yes, my memory was better closer in time to the events." If the witness disagrees, s/he is subject to being challenged to explain why s/he now contends (contrary to common sense) that h/er present memory is better and should be believed over h/er prior statement.

Before presenting the data episodes, some final brief comments about the brothers' testimony are in order. Lyle Menendez had previously testified that several days before the killings, he learned of his younger brother's continuing abuse and confronted their father Jose Menendez about it. Lyle further testified that he threatened the father that if the sexual molestation of Eric did not stop, he would go public and reveal the family's 'dirty secret' to the authorities (the incestuous homosexual abuse). Both brothers also testified that following the confrontation between Lyle and Jose Menendez, their parents became 'unsettlingly distant', such that they came to believe that their father and mother planned to kill them rather than risk revelation of the molestation and the brothers believed that their lives were in imminent danger. In the moments just before the killings, the defendants testified that their parents went into the family den, closed the double doors and 'secreted themselves'. (In fact, the Menendez parents were merely watching television and eating ice cream sundaes). The brothers stated that believing their parents were about to kill them, they armed themselves with 12-gauge shotguns, entered the den and shot both parents to death.[9]

ANALYSIS

FIRST EPISODE

"Mr. Menendez, Did You Know Big 5 Stopped Carrying Handguns in March of 1986?"

Background

This episode involves the testimony previously given by defendant Eric Menendez on direct examination about the events which led up to the purchase of the shotguns they used in the killings. In particular, Eric testified that several days before the August 1989 killings, he and his brother were in a state of uncontrollable fear for their lives and went to the local Big 5 store in near-by Santa Monica to buy handguns in order to protect themselves. However, the brothers testified that after selecting two handguns, they were told by a store clerk that there was a fifteen day waiting period for the purchase of handguns in California and that they could not leave the store with the guns that day. Lyle and Eric further testified that when they left the Big 5 store they remained in a panic and drove aimlessly onto the southbound San Diego freeway, simply ending up in San Diego where they bought the shotguns and ammunition used to kill their parents. The brothers claimed that the subsequent cash purchase of the shotguns in San Diego, 130 miles from Beverly Hills and Santa Monica, was a matter of mere happenstance and was not evidence of premeditation in the killings.

However, from the prosecution's perspective, the undisputed facts surrounding the San Diego gun purchase strongly support an alternative and contrary interpretation, i.e. that the killings were *not* the result of the brothers' legitimate fear of deadly attack. Instead, the prosecutors viewed the fact that the Menendez brothers purchased the shotguns far away in San Diego, used a false name, false identification and a false address and paid cash for their purchase (when they had numerous credit cards and usually charged everything) as indicating that the killings were premeditated. (The prosecution also emphasized in their closing argument additional evidence of criminal intent presented at trial, including that the defendants had attempted to set up an alibi, picked up every shell casing after the killings, disposed of their bloody clothing and the shotguns and faked a 911 emergency call seeking help following the killings.)

Episode One:

[P is the prosecutor; D is the defense counsel and W is the witness defendant Eric Menendez. The transcript is slightly edited].

P: Now. (4.0) You're telling the truth about everything in this case aren't you?
W: I'm telling the truth to the best that I can.
P: Ok. (1.0) And even though you've lied in the past – you're telling the truth now, aren't you?
W: Yes I am.
P: Did you truly go to the Santa Monica Big 5 store on the morning of August eighteenth to buy these handguns?
W: *Def*initely [without a doubt I did.
P: [And you actually – You and Lyle did?
W: Yes.

P: And you and Lyle actually looked in this case and you selected these guns and you were told you couldn't leave the store *with* those handguns because there's a fifteen day wait.
W: Yes.
(8.0)
P: Mr. Menendez, (1.0) did you know that Big 5 stopped carrying handguns in March of 1986?
(3.0)
W: No I don't know that.
P: Okay. Now af[ter
W: [Ah, Mr. Kiriama. There *were* guns there and we *did* look at them. And he did say that we could not carry them anymore.
P: Now. (1.0) After you went to this Big 5 store in Santa Monica, what happened?

Findings

I have given this episode the name, "Did you know Big 5 stopped carrying handguns in March of 1986," to encapsulate the moment when Eric Menendez is confronted by the prosecutor with an extrinsic fact which on its face appears to contradict the witness' account of events. Eric and Lyle claimed that as part of their desperate actions to protect themselves from their parent's deadly threat, they had first tried to purchase handguns from the local Big 5 store. However, during the course of the cross-examination the prosecutor suggests that this claim is a lie because of 'the fact' that Big 5 had actually stopped selling handguns over three years before the defendants claimed they looked at handguns there. This extrinsic fact has the potential to completely annihilate defendants' version of the gun purchase, since of course they could not have first looked at handguns at the Santa Monica Big 5 store if the store had not stocked such guns for over three years.

In confronting the witness with such a potentially inconsistent extrinsic fact, the cross-examiner poses an apparent puzzle for the witness to solve, that of potentially "disjunctive experiences of the same world" (Pollner, 1975, p. 417. Also see Heritage, 1984, p. 213). Thus, without a credible explanation forthcoming from Eric in real-time, the confrontation exposes Eric to being shown to be a liar and, more significantly, to a jury verdict convicting the defendants of murdering their parents.

This cross-examination segment is part of a longer exchange which began several minutes earlier between the cross-examiner (prosecutor, Lester Kiriama) and the witness. We enter the episode at the point where the prosecutor has just elicited from Eric Menendez a detailed account of the brothers' attempted purchase of handguns at Big 5 and is summarizing the details of the defendants' account:

> Mr. Menendez, do you recall your brother stating in his testimony that you had basically selected the guns, you were ready to purchase the guns and then you found out that you couldn't leave the store with the guns because you had the 15 day waiting period.

The foundation for the impeachment in the episode begins to be laid when the questioner asks the witness to affirm that he has been telling the truth, a common sequential feature in setting up a witness for impeachment. This is done by the prosecutor asking a series of three questions, which by their progressive precision, increasingly bind the witness more and more closely to the facts of his account. The first question in the sequence inquires broadly about the witness' general truth-telling, "Now you're telling the truth about everything in this case, aren't you?" The witness' response to the question is a rather cautious and indefinite, "I'm telling the truth to the best that I can." The prosecutor's next question asks the witness more specifically, "and even though you've lied in the past, you're telling the truth now, aren't you?" The question itself editorializes about the witness' credibility and the witness responds with a more certain answer, "Yes I am." The prosecutor then asks the most factually-specific question, "Did you truly go to the Santa Monica Big 5 store on the morning of August eighteenth [1989] to buy these handguns?" The witness' response (fitted to the heightened precision of the question) upgrades his degree of definiteness about the facts he has attested to, stating: "*Def*initely, without a doubt I did."

Through the series of questions, the prosecutor re-commits the witness to his specific version of the facts about the attempted handgun purchase. (Eric agrees that he and Lyle went to Big 5 on a given date, looked at handguns, selected the guns they wanted to purchase, but were then told by the clerk that there was a 15-day wait to buy the guns). The witness is thereby set up for the impending impeachment.

An eight second silence follows which sets off what is to happen next. The prosecutor then asks the crucial impeachment question, "Mr. Menendez, did you know Big 5 stopped carrying handguns in March of 1986?" The witness' version of the facts is thereby 'juxtaposed' (Pomerantz, 1988–89) with the purportedly contradictory fact that Big 5 stopped selling handguns three years before Eric and his brother claimed to have looked at them there.

The 'fact' that Big 5 stopped carrying handguns in 1986 is embedded in and presupposed by the question. If true, of course, such a fact strongly suggests that Eric (and Lyle) are lying about looking for handguns at the local Big 5 prior to purchasing shotguns far away in San Diego. This would substantially undermine the genuineness of their self-defense claim. The presumed fact is incorporated into the question by structuring the inquiry at the level of the witness' knowledge of the fact and not at the factual level of whether it indeed *is* a fact that Big 5 stopped selling handguns in March of 1986. In so phrasing the query, counsel attempts to constrain the witness to accept the presupposed fact as true. Such a lawyers' practice is the flip-side of phrasing questions which assert de-ontologizing claims that take a stance of disbelief toward the witness and/or h/er testimony and create the opportunity for the jury to infer that the matter testified about is not a fact (e.g. "Now tell me about this time where you claim to have had this conversation with Eric Menendez . . ." (Burns, 1996)).

The lawyer has set up the interchange prospectively for the jury to hear so that, if the witness stumbles in response to the confrontational question, it won't be missed by the jurors. There is about a three second silence before the witness verbally responds and on the video record Eric can be seen to hesitate during this silence and then begin to slowly shake his head. This may be some demeanor evidence of lying to be weighed by the jury.

Eric's verbal response, like the confrontational question, is structured at the level of what the witness knows. However, the tense of the question and the answer do not match. Eric does not respond with a past tense answer to the past tense question (such as: Q: "Did you know . . ." A: "No, I didn't know that"). Rather, he replies, "No I *don't* know that." This response is arguably half-way between answering the question on the basis of the witness' knowledge, (i.e. at the level the question was posed) and answering it at the level of the factual premise contained in the question. By building the presumed fact into the question and inquiring into the witness' knowledge of this fact, the lawyer makes it harder for an inexperienced witness to contest the fact or, without a question pending, to deny that it is a fact.

After his initial response, Eric retakes the floor by overlapping the prosecutor's next question to amplify his answer. Eric's further response directly takes issue with the factual presupposition of the question. The witness states, with hearable emphasis, "Mr. Kiriama. There *were* guns there and we *did* look at them and he *did* say that we could not carry them anymore." (There is some emphatic eyebrow work by the witness visible on the video record. He raises his eyebrows which remain high for the entire statement and seem to peak at the audible points of emphasis ["there *were* guns there and we *did* look at them and he *did* say that we could not carry them anymore"]).

It is curious that in Eric's amended response, he says, "he *did* say we could not carry them anymore." The amended response, including the selection of the phrase "could not *carry* them anymore," retains the structure and part of the phraseology of the original confrontational question: "Mr. Menendez, did you know Big 5 stopped carrying handguns in March of 1986?" Like the question, Eric's response refers to 'Mr. Kiriama' by surname and uses the word 'carry'. The response appears to smear or blur the difference between testimony that Big 5 'stopped carrying' handguns and Eric's claim that he was not allowed to 'carry' the handguns out of the store.

During the impeachment effort, witness Eric Menendez never admits to lying and does not dramatically break down on the stand. The prosecutor in this case arguably could have exaggerated and made more for the jury of the critical impeachment moment, for example, by asking the witness a series of follow-up questions highlighting the inconsistency (e.g. "Well, Mr. Menendez, wouldn't you agree that you could not in fact have looked at handguns at the Santa Monica Big 5 when you claim that you did if the store had stopped stocking handguns three years earlier?" or, "Do you have some reason to believe that Big 5's stock records may be inaccurate?").[10]

In a sense, the episode ends in an interesting sort of stand-off. Rather than drawing out a line of questioning to earmark the contradiction, counsel drops the matter, leaving Eric's final response unaddressed to pursue a new topic of inquiry, asking the witness, "Now . . . after you went to this Big 5 store in Santa Monica, what happened?" At this point in the testimony, the jury is left with two competing yet irreconcilable versions of the facts at issue and without a conclusive basis for deciding between them (see Pollner, 1975). However, the next day Eric testifies on re-direct examination that he must have been mistaken about it being a Big 5 store where he and Lyle first looked at handguns, and the prosecution later calls a Big 5 employee to the stand who testifies that business records show that the store stopped stocking handguns in March of 1986. The prosecutor eventually argues to the jury in closing argument that Eric was impeached on this point, but in the end, it is up to the jury to decide whether the witness' lack of credibility has been exhibited on cross and the witness has been successfully impeached.

SECOND EPISODE

"Would You Say Your Memory Today, Four Years Later, Is Better Than Your Memory Was in July of 1990?"

Background

This episode involves witness Judalon Smith, who is somewhat of a 'turncoat' witness from the perspective of the prosecution. Before the brothers' arrest,

Eric Menendez had confessed to his psychotherapist, Dr. Jerome Oziel, that he and his brother had murdered their parents. For his own protection, Dr. Oziel asked Judalon Smith to be secretly present in the waiting area of his office when the Menendez brothers were scheduled to meet him there. (Lyle found out about Eric's confession and wanted to discuss this with Dr. Oziel). If necessary, Dr. Oziel wanted a witness to the fact that his life was in danger so he could breach the psycho-therapist-patient privilege and notify authorities that the brothers posed a threat to his life. (There is an exception in California to the legal obligation of a psychotherapist to preserve the confidentiality of doctor-patient communications if the patient poses an imminent threat to the safety of another).

Previously in a separate civil court proceeding, Judalon Smith testified under oath that while in an adjacent office, she overheard Lyle Menendez boast to Dr. Oziel that he and his brother "had committed the perfect crime." She also testified that she heard Lyle threaten Dr. Oziel. This testimony about a 'perfect crime' statement by Lyle is evidence which would weaken defendants' contention that they killed in genuine self defense. Additionally, if Lyle had threatened Dr. Oziel, this conduct is arguably evidence of consciousness of guilt amounting to an admission.

During her testimony in the present trial, witness Smith turns hostile to the prosecution and contradicts both helpful points of her prior testimony. She now denies that she heard Lyle threaten Dr. Oziel or say that he and his brother had executed 'the perfect crime'. In an effort to assert that witness Smith's own previous testimony was true, the prosecutor (Pamela Bozanich) in the following interchange attempts to show that the witness' current testimony contradicts her prior testimony, which should be believed over her present testimony because it was given closer in time to the events in question.

It is also significant to the prosecution that Judalon Smith's changed testimony contradicts the testimony of key prosecution witness Dr. Oziel, who has already told the Menendez juries about Eric's confession, the 'perfect crime' statement and the threat by Lyle. Ms. Smith was first called by the defense on direct examination to discredit Dr. Oziel and cast doubt on his testimony. In fact, she may be out to get Dr. Oziel, her former therapist with whom she had previously had an affair (according to the admission of both).[11]

Episode Two:

[P is the prosecutor; D is the defense counsel, W is the witness Judalon Smith and J is the judge. The transcript is slightly edited].

P: Now . . . you testified in . . . Santa Monica on July the thirtieth and July the thirty first of 1990. Correct?
(2.5)
W: Yes.
P: And when you testified did you have to raise your arm and swear an oath to tell the truth?
W: Yes.
P: And when you testified . . . in July of 1990, did you understand that you were testifying under penalty of perjury?
W: Yes.
P: And when you testified, did you tell the truth?
W: Yes I told the truth as I *believed* it and *knew* it at the time.
P: Would you say your *memory today* (.) four years later is *better*? than your memory was in July of 1990?
W: In some (.) aspects ye-s because I now have my *own* memory (.) not the memory Oziel planted in me.
P: Okay. When you testified in front of Judge Albrecht . . . did you tell him (.) that you were testifying from a memory which had been implanted in you by someone else?
W: At the time of that testimony um (.) I (.) was not in therapy yet. I had not been diagnosed. I was actually in deni-al. I believed I was perfectly fine, that nothing *was* wrong with me. It was my appearance in that courtroom and my inability to know what went on, the disassociation that happened (.) *frightened* me. And as a result of that I then agreed to go into therapy.
P: When you refer to dissociation, what are you talking about?
W: Um (1.5) *Appar*ently, what I've a learned (.) about that incident was um (4.0) when (1.0) *I*: (.) came into what *I* perceived as a hostile environment (.) and something that was (.) um part of the (1.0) initial trauma (.) that I had suffered (.) which was (1.0) co:ming and being in a room with the Menendez *bro*thers (1.0) um I did something which (.) is called disassociating where I (.) somehow *sep*arated myself (.) and (2.0) I was there and I answered questions from some *el*ement of my *con*sciousness but (.) um
[P looks over to jury]
(2.0)
P: But you weren't really there? [P looks over to jury again]
W: I'm not a psychologist((laughing)) so I – I'm not re-ally the *per*son to ask to explain to you what (.) ha:ppened [P rubs her cheek] I'm (1.0) ya know I'm just (.) the victim.
P: Ms. Smith, when you testified then in front of Judge Albrecht would it be

your testimony then that you were in a dissociative *state* (.) is that correct? (1.5) And you didn't really know what you were doing?//would that be fair to say?
D: [((inaudible objection))
W: *Corr*ect.
J: Overruled.
P: And when you filled out the affidavit, (.) the four drafts you made of the affidavit, were you dis'sociating during those *four* drafts?
W: I'm not *con*scious (.) that I was doing that but I was *suf*fering (.) at that time. I (.) still had had *no* treatment for (.) um the things that I was later diagnosed with which (.) are the same kind of traumas and disorders and syndromes that people who are prisoners of war (.) ah (.5) are diagnosed with and come back with.
P: Were you disassociating when you talked to Diane Sawyer about what you had heard (.) in Dr. Oziel's office?
(3.0)
W: I'm not aware or qualified ((laughing)) to say that.
P: Now Ms. Smith,
(2.0)
W: All I can say is that at that time I had not been diagnosed or gone into treatment (1.0) I was operating on the basis that I believed I was perfectly fine (1.0) and normal.

Findings

In this episode, the prosecutor attempts to impeach the witness based upon several prior inconsistent statements she made, including several under oath. The witness presents an explanation for her prior inconsistent statements which turns upon a 'folk' expert psychological/psychiatric diagnosis and opinion regarding her mental condition at the time she gave those statements.[12] When the prosecutor attempts to probe this explanation, the witness asserts that she is not a psychologist. The prosecutor is thus faced with a witness who attempts to articulate and rely upon the substance of a purported 'expert' opinion about her prior mental state, while at the same time avoid accountability for explaining that opinion or responding to questions concerning it.

Reminiscent of the first episode, a series of foundational questions is asked by the prosecutor about whether the witness told the truth in her prior statements: the prosecutor confirms with the witness that she gave prior testimony before a court of law; that her testimony was given under oath; and that she understood that she was testifying under penalty of perjury. The prosecutor then asks the witness, "And when you testified [in the prior court proceeding], did you tell the truth?" Notably, unlike Eric Menendez' firm confirmation of truth-telling in the first episode, here witness Smith gives a somewhat unusual response. The witness testifies that she told the truth as she "*believed* it and *knew* it at the time." This qualification by the witness raises the difference between 'truth' and 'belief' and is obviously quite self-serving. In common discourse, truth is something that one knows for sure, while belief is something which is accepted as true without certain proof. In saying, "I told the truth as I believed it or knew it at the time," the witness tries to manage the contradiction between her prior and present testimony by 'perspective smearing', i.e. invoking a difference between the truth as she knew it at the time of her earlier testimony versus the truth as she now knows it to be. The response revises and relativizes the witness' earlier inconsistent testimony to be merely what she then 'believed' to be true and suggests that now the witness has learned that her previous testimony was not in fact true.

The prosecutor, seeking to establish the prior testimony as true, next asks the crucial question, 'Would you say your *memory today*, four years later is *better*? than your memory was in July of 1990?' This is a lawyer's gambit which is routinely deployed in instances where a witness recants or contradicts an earlier statement. The device turns on the presumption that a witness cannot reasonably claim to have a better memory years after the incident in question than s/he did at a time nearer to the subject events, and builds into the question the presumption that the witness' previous testimony is more accurate and credible than her present testimony.

Instead of admitting that her current memory cannot be better than it was four years earlier, witness Judalon Smith claims that "in some aspects yes because I now have my own memory, not the memory Oziel planted in me." This novel response that her memory may indeed be better today poses an obstacle for the cross-examiner because it avoids acknowledging that memories fade over time and in fact suggests that fading memory is not the issue. If the witness had responded that her memory was better previously, the attorney could have proceeded to the next step in the impeachment effort and juxtaposed the substance of the prior and present testimony to suggest that the witness' present testimony is less credible. However, because of the witness' specific response, the questioner is forced to address the witness' assertion that she told the truth before and she is telling the truth now, notwithstanding that the two versions materially contradict one another and could be the basis for a perjury charge.[13]

The examining attorney must now deal with a witness who contends that her previous testimony lacks credibility because it was based upon a 'planted memory', not her own memory. Counsel attempts to probe the issue by asking, "when you testified earlier in front of Judge Albrecht did you tell him you were testifying from a memory which had been implanted in you?" This suggests (tongue in cheek) that perhaps it was unreasonable for the witness to proceed to testify if she did not have a memory of her own.

The witness does not directly answer the question, but begins an extended explanation: "At the time of the testimony," she was unaware of her brainwashed condition through no fault of her own. The cross-examiner then pursues the 'disassociation' excuse, asking "When you refer to disassociation, what are you talking about?" The cross-examiner's question is open-ended (and arguably ill-advised) because it invites the witness to offer a narrative. By giving the witness an opportunity to expound upon her explanation and the meaning of disassociation, the prosecutor in effect solicits testimony from the witness that amounts to an impermissible expert opinion from a lay witness.[14]

In an elaborate response, the witness claims that it is reasonable to believe that her prior inconsistent testimony was honest (albeit erroneous) because that testimony was given while she was 'disassociating' and emanated from a memory that had been implanted in "some element of [her] consciousness" by psychologist Dr. Oziel. During the witness' response, the prosecutor can be seen on the video record to look skeptically over to the jury in an effort to claim, with them, that the response is bizarre and should not be believed. The prosecutor continues her stance of disbelief toward the testimony by next asking, "But you weren't really there?" Counsel again looks collusively to the jury. The witness laughs at the question and avoids answering it by formulating herself as "just the victim" (Loeske, 1993) who is "not a psychologist" and is not qualified to answer it.

The prosecutor next asks a series of questions which strain the credibility of the witness. The questions concern whether she was 'disassociating' at the specific times she previously gave different statements and testimony about the Menendez case which were favorable to the prosecution. She asks the witness, were you in a 'dissociative state' when you testified before Albrecht; were you 'dis'sociating' when you filled out four drafts of your affidavit; and were you "disassociating when you talked to Diane Sawyer?" In asking these questions, the interrogator emphasizes for the jury the many occasions on which the witness was purportedly 'disassociating' and also implies that the witness appeared to have been acting with full consciousness at these times and thus that her explanation for the inconsistent testimony is implausible and the previous testimony is true. However, these questions may also harm the prosecution, because the witness takes each as an opportunity to further buttress her claim that the inconsistencies in her testimony can be reasonably explained away by her having been brainwashed by Dr. Oziel, or because the unusual explanation may mean to the jury that none of her testimony, past or present, can be believed.

The prosecutor in this interchange is faced with a witness who engages in 'identity smearing'. At times, the witness seeks to justify having given prior inconsistent testimony by providing what amounts to an expert diagnosis and opinion. However, when the cross-examiner attempts to probe this explanation, the witness asserts that she is 'not qualified' to give expert testimony. By infusing her lay testimony with expert terminology, opinions and diagnoses, the witness tries to smear her identity between that of a lay person and an expert witness. S/he maneuvers to provide a credible excuse for her prior inconsistent testimony and at the same time avoids answering questions about that opinion. This poses difficult work for the examiner in trying to make the witness responsible for prior inconsistent testimony.

IMPLICATIONS AND CONCLUSION

This has been a study of lawyers' work in attempting to impeach adverse witnesses through detailed real-time questioning in a criminal murder trial. Such work is often determinative for juries in deciding the fate of defendants. As legal scholar John Cole (1988) observes, during criminal trial and adjudication, the defendant simultaneously "exists in the space of being guilty and not guilty and awaits creation as one or the other" (p. 919). Criminal trial activities have as their product the formal public display and recognition of the essential character of the accused and h/er conduct with reference to the law and legitimate moral order.

Impeachment work by trial lawyers on cross-examination is variously directed to persuasively establishing, through credible and legally consequential evidence and accounts, the pertinent facts and events that support h/er client's case and undermine the opponent's. As such, impeachment work involves claims-making contests in which the interrogating lawyer and the adverse witness compete to articulate, substantiate and sustain (or defeat) a particular version of the facts, evidence and witness' credibility in the case in the face of a competing alternative account (see Emerson & Messenger (1977)).

During cross examination, the adverse witness and cross examiner vie to dissolve what, in Pollner's terms, might otherwise become a 'reality disjuncture':

> [A] yet-to-be completed ironicization [of experience] in that a choice is yet to be made as to which of the competing versions of the world will prevail as definitive to that world . . .

> [F]or the ironic function to move to completion, a choice must be made as to which determination will be adhered to and which will be disregarded' (Pollner, 1975, p. 414).

Thus, impeachment work serves to establish as a foundational matter what will be the prevailing context which will be treated as 'real' to the exclusion of the contrary view and within which the law, facts, evidence and credibility of witnesses in the case will be accorded meaning.

A witness' testimony may appear to be erroneous or contradictory for various reasons and there may or may not be reasonable explanations for this. The witness may indeed be trying to deliberately "create impressions opposite to the real facts" and may be intentionally evasive or non-responsive, manipulate the question to maneuver a non-damaging response around it, or offer an excuse or explanation instead of directly answering (Morrill, 1979, p. 54). However on other occasions, the witness may in fact be attempting to testify truthfully, but for one reason or another genuinely fail to recollect or make an honest mistake (Ibid.).

For the lawyer faced with damaging testimony from an adverse witness, the task on cross-examination is to impeach the witness and/or h/er account. If the attorney is unable to show the witness to be an outright liar, s/he tries to point to (or create) errors, implausibilities, contradictions, inconsistencies, etc. in the witness' testimony. Savvy cross-examiners regularly embed damaging presumptions and inferences into their questions, or build questions which exhibit the attorney's disbelief that the testimony is true (Burns, 1996). Thus, it is sometimes possible for skillful counsel to show the witness to be intentionally evasive or deliberately fabricating even when s/he is actually trying to testify honestly and sincerely.

But, impeachment is an incessantly local, contingent (and defeasible) enterprise and achievement. From this viewpoint, the interactions between attorneys and adverse witnesses are highly contingent and dynamic two-way productions, in which both sides thrust and parry in an effort to score points. We have seen that notwithstanding institutionalized asymmetries in professional knowledge, interactional power and participation rights between lawyers and witnesses, witnesses are not without resources and are sometimes able to resist or 'smear' the impugning implications of the cross-examiner's questions.

The impeachment practices analyzed here are especially fitted to the trial courtroom and to other kinds of legal settings in which person properties are routinely constituted (Emerson, 1969; Holstein, 1993). More comprehensively, however, the study of impeachment practices directs attention to the intricate ways in which persons contingently, collaboratively and defeasibly take positions on "what really happened," and what is a "lie," and thereby locate one another within orders of intelligibility, credibility, morality and consequence (Garfinkel, 1967; Sacks, 1979; Macbeth, 1996. Also see Button, 1992).

NOTES

1. The author gratefully acknowledges comments by Robert Emerson, John Heritage, Mark Peyrot and Doug Macbeth. I thank Harold Garfinkel for inviting me to present an earlier version of this paper at the Annual Meeting of the Pacific Sociological Association in San Diego, California, 1994.

2. The term 'impeachment' as used by lawyers is an irremediable member's gloss. Even definitionally among lawyers, there is no uniform meaning of impeachment. Most broadly, it has been defined to mean "the adducing of proof that a witness is unworthy of belief" (Black's Law Dictionary, 4th ed.). There are certain standard bases for impeaching the credibility of a witness in court, such as by challenging the witness' ability to perceive, remember or recount the matter about which s/he testifies or by establishing the existence of contradictory evidence or bias, interest or other motive to testify untruthfully (see, e.g. *California Evidence Code*, section 780). Furthermore, a jury is usually instructed by the judge before deliberations that a witness who is willfully false in one portion of his testimony may be disbelieved in the whole of his/her testimony (see, e.g. BAJI 2.22, *California Jury Instructions* (Civil), 7th ed.). Impeachment work occurs in both civil and criminal trials and is highly consequential to both. In both types of trial, impeachment serves to implicate and make visible central matters of evidence, credibility, facticity, truth and accountability in any given case, (See Garfinkel, 1967; Emerson, 1969).

3. Direct examination is aptly characterized as a 'friendly' exchange between the questioner and the witness. Direct examination is the first examination of a witness on the merits by the party on whose behalf s/he is called (Black's Law Dictionary, 4th ed.). Its purpose is to elicit facts relevant to establishing the elements of the parties' claims or defenses in the litigation. In general, the lawyer during direct examination creates the opportunity for the witness to provide a favorable version of events, which version is typically edited and selectively slanted to highlight or exaggerate helpful facts, evidence and accounts, while at the same time minimizing, qualifying, explaining away or omitting harmful facts and accounts. The accounts proffered on direct have commonly been prepared and rehearsed in advance between counsel and the witness. Thus, rather than consisting of the hostile exchange which is characteristic of cross-examination, direct examination is more akin to a collaborative 'co-telling' (See G. Jefferson, 1978).

4. The ten data episodes were collected from broadcast coverage of the trial provided by a local cable station and by Courtroom Television Network ('Court TV'). Court TV's gavel-to-gavel coverage has for the first time made readily available for fine grained interactional analysis the real time micro-processes and materially organized tasks and skills which comprise the work of trial examination and advocacy. The episodes were transcribed using a simplified version of the Jefferson notational system (Atkinson & Heritage, 1984, pp. ix–xi).

5. If established, 'perfect self defense' serves as a legal justification for a crime. When the actor acts in perfect self-defense, s/he is privileged to engage in the conduct, even though s/he performed the physical part of the crime of murder and although s/he did so intentionally. By contrast, 'imperfect self defense' serves to mitigate a charge of murder to voluntary manslaughter. Following a hung jury for both defendants at the first trial, the judge at the second Menendez trial ruled that an objective standard of self defense governed the case, rejecting the claim of the defendants that their subjective belief that they were about to be attacked should be controlling.

6. 'Demeanor' relates to the witness' look and personal appearance and includes such matters as h/er expressions, gestures, tone of voice, the hesitancy or readiness with which answers are given and 'air of candor' (Black's Law Dictionary, 4th ed., p. 517). The trier of fact hears the trial testimony, observes the demeanor of the witness on the stand and makes factual determinations based in large part on the assessment of their credibility.

7. This is not to say that complete annihilation of a witness on cross-examination is not possible by intrinsic means (see Wellman, 1936, p. 64; Morrill, 1979, p. 71 for examples).

8. Perjury is the criminal offense of making a false statement under oath where such statement is material to the matter of inquiry and known by the witness to be false.

9. The evidence also showed that the parents had never verbally or physically threatened the lives of either defendant. No medical records of either defendant were introduced into evidence to show that prior to the killings either brother had been hospitalized for any abuse-related injury.

10. Compare the impeachment work of the highly talented defense counsel in the episode analyzed in Burns, 1996, pp. 25–28.

11. Judalon Smith eventually sued Dr. Oziel in Santa Monica Superior Court for assault and battery, the civil equivalent of criminal rape. Ms. Smith also contended that she was a criminal rape victim, but the prosecutor's office refused to bring rape charges against Dr. Oziel, who was to be the prosecution's star witness in the Menendez trial. Therefore, witness Smith is unfriendly to the prosecution and apparently in retaliation, she recants her earlier testimony which was favorable to the prosecution.

12. To testify as an expert, a witness must be qualified as an expert and must have special knowledge, skill, experience, training or education about the subject of his/her testimony (See California Evidence Code, section 720 and see Heritage, 1984, p. 215). Even assuming that the witness was a qualified psychological or psychiatric expert, she would not be permitted to render a self-diagnosis which on its face could not be considered to be impartial or objective.

13. Pollner, 1974, pp. 51–53, points out that despite disagreements about what actually happened among experiencers and observers of an event, what results is not an agreement that multiple worlds exist, but rather the use of several member's procedures to reconcile the competing versions of their ordinary affairs.

14. When this improper expert opinion is sought by the prosecutor, defense counsel notably does not object or move to strike the answer. That no objection to the question or motion to strike the response is made by the defense when such objection and motion would almost certainly be sustained by the court perhaps indicates that the defense views the solicited testimony as helpful.

REFERENCES

Atkinson, J. M. (1992). Displaying Neutrality: Formal Aspects of Informal Court Proceedings. In: P. Drew & J. Heritage (Ed.), *Talk at Work* (pp. 199–211). Cambridge: Cambridge University Press.

Atkinson, J. M., & Drew, P. (1979). *Order in Court*. London: Macmillan.

Atkinson, J. M., & Heritage J. (Eds) (1984). *Structures of Social Action: Studies in Conversational Analysis*. Cambridge: Cambridge University Press.

Burns, S. (1996). Lawyers' Work in the Menendez Brothers' Murder Trial. *Issues in Applied Linguistics, 7*(1), 19–32.

Button, G. (1992). Answers As Interactional Products: Two Sequential Practices Used in Job Interviews. In: P. Drew & J. Heritage (Eds), *Talk at Work* (pp. 212–231). Cambridge: Cambridge University Press.

Cole, J. (1988). Thoughts From the Land of And. *39 Mercer Law Review, 907*.

Conley, J., & O'Barr, W. (1990). *Rules Versus Relationships: The Ethnography of Legal Discourse*. Chicago: University of Chicago Press.

Drew, P. (1985). Analyzing the Use of Language in Courtroom Interaction. In: T. van Dijk (Ed.), *Handbook of Discourse Analysis, vol. III: Discourse and Dialogue* (pp. 133–148). London: Academic Press.

Drew, P. (1990). Strategies in the Contest Between Lawyer and Witness in Cross-Examination. In: J. N. Levi & A. G. Walker (Eds), *Language in the Judicial Process* (pp. 39–64). New York: Plenum.

Drew, P. (1992). Contested Evidence in Courtroom Cross-Examination: The Case of a Trial for Rape. In: P. Drew & J. Heritage (Eds), *Talk at Work* (pp. 470–520). Cambridge: Cambridge University Press.

Drew, P., & Heritage, J. (Eds) (1992). *Talk at Work*. Cambridge: Cambridge University Press.

Emerson, R. (1969). *Judging Delinquents*. Chicago: Aldine.

Emerson, R. (1992). Public Bureaucracies. In: *Law, Politics and Society, Vol. 12* (pp. 3–29). Stamford, CT.: JAI Press.

Emerson, R., & Messinger, S. (1977). The Micro-Politics of Trouble. *Social Problems, 25*, 121–35.

Garfinkel, H. (1967). *Studies in Ethnomethodology*. Englewood Cliffs, NJ: Prentice-Hall.

Garfinkel, H. (1988). Evidence for Locally Produced, Naturally Accountable Phenomena of Order, Etc.: An Announcement of Studies, *Sociological Theory, 6*, 103–109.

Garfinkel, H. (1996). Ethnomethodology's Program. *Social Psychology Quarterly, 59*(1), 5–21.

Garfinkel, H., & Sacks, H. (1970). On Formal Structures of Practical Actions. In: J. C. McKinney & E. A. Tiryakian (Eds), *Theoretical Sociology* (pp. 338–366). New York: Appleton Century Crofts.

Goffman, E. (1959). *The Presentation of Self in Everyday Life*. New York: Doubleday.

Greatbatch, D., & Dingwall, R. (1994). The Interaction Construction of Interventions by Divorce Mediators. In: J. P. Folger & T. Jones (Eds), *New Directions in Mediation* (pp. 84–109). Thousand Oaks, CA: Sage Publications.

Heritage, J. (1984). *Garfinkel and Ethnomethodology*. Cambridge: Polity.

Heritage, J. (1985). Analyzing News Interviews: Aspects of the Production of Talk for an 'Overhearing' Audience. In: T. van Dijk (Ed.), *Handbook of Discourse Analysis, Vol. III: Discourse and Dialogue* (pp. 95–119). London: Academic Press.

Holstein, J. (1993). Court-Ordered Insanity. New York: Aldine de Gruyter.

James, F., & Hazard, G. (1977). *Civil Procedure*, 2nd ed. Boston: Little, Brown and Company.

Jefferson, G. (1978). Sequential Aspects of Storytelling in Conversation. In: J. N. Schenkein (Ed.), *Studies in the Organization of Conversational Interaction* (pp. 219–248). New York: Academic Press.

Loeske, D. (1993). Constructing Conditions, People, Morality and Emotions: Expanding the Agenda of Constructionism. In: G. Miller & J. Holstein (Eds), *Constructionist Controversies* (pp. 207–216). New York: Aldine De Gruyter.

Luh, C. W. (1922). The Conditions of Retention. *Psychol. Monograph.*, *31*(142), 27.

Lynch, M., Livingston, E., & Garfinkel, H. (1983). Temporal Order in Laboratory Work. In: K. Knorr-Cetina & M. Mulkay (Eds), *Science Observed* (pp. 205–238). London: Sage.

Macbeth, D. (1991). Teacher Authority as Practical Action. *Linguistics and Education*, *3*, 281–313.

Macbeth, D. (1996). The Discovery of Situated Worlds: Analytic Commitments, or Moral Orders? *Human Studies*, *19*, 267–87.

Moerman, M., & Sacks, H. (1988). On Understanding in the Analysis of Natural Conversation. In: M. Moerman (Ed.), *Talking Culture* (pp. 180–186). Philadelphia: University of Pennsylvania Press.

Morrill, A. (1979). *Trial Diplomacy*. Chicago: Court Practice Institute, Inc.

O'Barr, W. (1982). *Linguistic Evidence: Language, Power and Strategy in the Courtroom*. New York: Academic Press.

Pollner, M. (1974). Mundane Reasoning. *Philosophy of the Social Sciences*, *4*, 35–54.

Pollner, M. (1975). 'The Very Coinage of Your Brain': The Anatomy of Reality Disjunctures. *Philosophy of the Social Sciences*, *5*, 411–430.

Pomerantz, A. (1988–89). Constructing Skepticism: Four Devices to Engender the Audience's Skepticism. *Research on Language and Social Interaction*, *22*, 293–313.

Sacks, H. (1979). Hotrodder: A Revolutionary Category. In: G. Psathas (Ed.), *Everyday Language* (pp. 7–14). New York: Irvington Press.

Sacks, H. (1992). In: G. Jefferson (Ed.), *Lectures on Conversation (1964–1972)*. Oxford: Blackwell.

Sacks, H., Schegloff, E., & Jefferson, G. (1974). A Simplest Systematics for the Organization of Turn Taking in Conversation. *Language*, *50*, 696–735.

Schegloff, E. (1988–9). From Interview to Confrontation: Observations on the Bush-Rather Encounter. *Research on Language and Social Interaction*, *22*, 215–240.

Wellman, F. (1936). *The Art of Cross-Examination*. New York: MacMillan.

APPENDIX: DATA TRANSCRIPT NOTATIONS

The notations in the transcripts use the conventions developed by Gail Jefferson (in Atkinson and Heritage, 1984 at pp. ix-xvi). The principal symbols used are:

( )	= Numbers in parentheses indicate the occurrence and duration of pauses in tenths of a second
[	= Marks the point at which overlapping talk begins

]	= Marks the point at which overlapping talk ends
=	= Notes the end of one utterance and the start of the next with no gap or overlap
-	= Indicates the point at which a word is cut off
(( ))	= Double parentheses indicate the transcriber's descriptions
*do*ing	= Indicates some form of emphasis by means of pitch
MONTH	= Capital letters are used to indicate that an utterance or part of an utterance is produced with louder amplitude than the surrounding talk
slee::p	= Colons indicate a stretch of the immediately prior sound
.,?	= punctuations note falling, continuing and rising intonation, respectively

CONTRIBUTIONS TO THEORY DEVELOPMENT

DEFICIENCIES IN THE SOCIOLOGY OF SEX WORK

Ronald Weitzer

ABSTRACT

It has been noted that "there are few areas in the social sciences which have generated as much anecdotal evidence, opinion, general musing, moral outrage, and theorizing" as prostitution (Earls & David 1989: 7). These are not the only deficiencies, however, in the literature on prostitution and other sex work. This article critically examines this body of work and suggests remedial measures to help overcome some serious problems in the field and to fill some major gaps in the literature.

OVERGENERALIZATION

'Sex work' may be defined as sexual services or performances in exchange for material compensation.[1] Examples include prostitution, pornography, stripping, and telephone sex. The 'sex industry' refers to the owners, managers, workers, and organizations involved in the sale of sex. All too often, however, the terms 'sex work', 'sex industry', 'prostitution', and 'pornography' are used by scholars in a sweeping fashion, masking important differences within each category. Overgeneralization is especially common in analytical writings (e.g. Barry, 1979; Jeffreys, 1997; McElroy, 1995), but it also can be found in some empirical studies (e.g. Farley & Barkan, 1998; Hoigard & Finstad, 1992), where authors move from the subgroup on which their research is based to larger claims about 'prostitutes' or 'sex workers'. When it comes to

Sociology of Crime, Law and Deviance, Volume 2, pages 259–279.

ISBN: 0-7623-0680-7

prostitution, the most serious blunder is that of equating all prostitution with street prostitution, ignoring entirely the indoor side of the trade. In the United States and many other countries, only a minority of prostitutes work the streets, yet they have received the lion's share of attention.[2] From the limited literature available, it appears that there is significant variation both between and within the categories of street and indoor prostitution, in at least five respects.

(1) *Stratification.* The world of prostitution is stratified. Street prostitutes occupy the lowest order and receive the strongest dose of stigma; upscale workers, such as call girls, are somewhat less reviled. Since their work is largely invisible to the public, the latter experience little direct censure. Bryan (1966: 450) writes that 'the call girl rarely experiences moral condemnation through interpersonal relations'. There is also a hierarchy *within* each level. Street prostitution is structured by race, income, age, appearance, drug use, and locale – all of which shape workers' daily experiences. In many cities, for example, black, white, and Latino prostitutes work in different areas, and whites earn more than persons of color (Cohen, 1980; Porter & Bonilla, 2000). Drug-addicted workers also differ strikingly from non-addicts in their willingness to engaged in unsafe sexual practices and accept low prices. Among indoor workers, status increases from massage parlor workers to brothel workers to escorts and independent call girls (Heyl, 1979). Workers themselves often draw distinctions between their work and that of others in the industry, distinctions that usually include some disparagement of other types of workers (Sheehy, 1973). The world of male prostitution is similarly stratified, though there appears to be more mobility between the ranks than is true for female prostitution (Luckenbill, 1986; West, 1993).

(2) *Control over Working Conditions.* Prostitutes vary in their access to resources for protection, their freedom to refuse clients, their dependence on managers and other third parties, and the job alternatives available should they decide to leave sex work (Chapkis, 2000; Davidson, 1998; Heyl, 1979). Control over these conditions is generally lowest at the bottom of the hierarchy – among streetwalkers – but there are some respects in which mid-level workers (in massage parlors and brothels) have less control over working conditions than independent street prostitutes, who do not work for pimps.

(3) *Work Experiences.* Workers differ in whether and how often they experience victimization and exploitation: Assault, rape, and robbery are occupational hazards for streetwalkers, but relatively rare among offstreet workers who have not been recruited by force or fraud (Lowman & Fraser,

1995; McElroy, 1995; Perkins & Bennett, 1985: 239, 300; Perkins, 1991: 290; Prince, 1986: 495, 512). Risk of exposure to sexually transmitted diseases varies between street and indoor workers. HIV infection rates are highest among street prostitutes who inject drugs, but rare among call girls (Seidlin, 1988) and among women working in legal brothels in places such as Nevada, Australia, and Holland (Pyett, 1996). Not one of Nevada's legal brothel workers has tested HIV-positive since testing was mandated in 1985 (*Las Vegas Review Journal*, 1998). Other variations include whether the work is even minimally rewarding, other than financially, and the nature of the sexual encounter itself – ranging from fleeting transactions to encounters that sometimes resemble dating experiences (with conversation, flirtation, gifts, hugging, kissing). Call girls are much more likely than streetwalkers to engage in these activities as well as to receive massages and oral sex from clients (Lever & Dolnick, 2000). Gender appears to make a difference. The limited research on male and transgender prostitutes suggests that, in general, they derive more sexual satisfaction from their work than female prostitutes (Weinberg, Shaver & Williams, 1999; West, 1993). In a remarkable study of Brazil, transgenders reported a high degree of enjoyment of their sexual relations with clients; moreover, prostitution was the only sphere of life where they were able to build self-esteem (Kulick, 1998).

(4) *Adjustment to their Work*, Prostitution does not have a uniform effect on workers' self-images and psyches. One study found that streetwalkers exhibited significant psychological problems, while call girls, brothel workers, and massage parlor workers were 'capable of handling themselves well, manifesting good emotional controls, being well aware of conventionality, and doing well in the occupation of their choice' (Exner et al., 1977: 483). Research on 75 streetwalkers and 75 call girls in California and 150 women working in Nevada's legal brothels found that 97% of the call girls reported an *increase* in self-esteem after they began working in prostitution, compared with 50% of the brothel workers and only 8% of the streetwalkers (Prince, 1986: 454). Call girls expressed positive views of their work; brothel workers were 'generally satisfied' with their work; but street prostitutes evaluated their work more negatively (Prince, 1986: 497). Other studies of call girls, masseuses, escorts, strippers, and telephone sex workers reported that the workers felt the job had at least some positive effect on their lives, took some pride in their work, or believed that they were providing a valuable service (Bryan, 1966; Chapkis, 1997; Farley & Davis, 1978; Flowers, 1998; Foltz, 1979; Lever & Dolnick, 2000; Rich & Guidroz, 2000; Salutin, 1971; Thompson & Harred, 1992; Verlarde & Warlick, 1973; West, 1993). While they dislike certain aspects of

the work, they are also more likely than street prostitutes to describe positive aspects of their work.

(5) *Impact on the Community.* Street and off-street prostitution have very different effects on the surrounding community; street prostitution is visible and often disruptive of the peace, whereas indoor prostitution has little, if any, negative impact on the environment and, if discreet, there is normally little public awareness of it (Reynolds, 1986). Street prostitution is associated with a host of problems, including disorderly conduct, sex acts in public places, the discarding of condoms and syringes in public areas (public health hazards), harassment by customers of non-prostitute women, harmful effects on local children, and loss of business to local merchants (Weitzer, 2000). Such adverse effects on local communities, real or perceived, explains why contemporary antiprostitution campaigns are largely directed at street prostitution rather than the indoor trade. In cities throughout the country, residents living near prostitution strolls have increasingly mobilized to drive prostitution off their streets.

We have less information on other kinds of sex work, but the available data point to similar variations and nuances in how the work is structured, in workers' status and resources, and in work experiences. Sex work involves varying degrees of exploitation, victimization, and agency. For individual actors, the work may be distasteful in some respects but rewarding in others, may involve coercion at some times and consent at other times, and may be experienced as both somewhat 'exploitative' and somewhat 'empowering' (Bell, Sloan & Strickling, 1998; Chapkis, 1997; McLeod, 1982; Sanchez, 1997). Roberta Perkins writes that 'for prostitutes, sex work is a mixture of pleasant and unpleasant experiences' (Perkins, 1991: 292). Victimization is utterly real, especially among street prostitutes and those who have been trafficked into sex work against their will; these women clearly need more police protection and assistance from service providers. But other sex workers are much less vulnerable to violence, exercise more control over their work, and derive at least some psychological or physical rewards from what they do. None of this is meant to discount similarities across types of sex work, but scholars need to examine both similarities *and* differences and to investigate *all* varieties of sex work. Toward this end, we need a more careful examination of the ways in which sex workers themselves experience and describe their work – negatively, positively, or indifferently. Unfortunately, much academic writing neglects workers' own conceptions and instead superimposes upon them one or another essentialist paradigm.

ESSENTIALISM

Overgeneralization is most apparent among those who take an essentialist position on sex work. It is remarkable how much of the writing on sex work comes from ideologues holding staunch views about the 'essential' nature of this kind of work. One camp claims that 'sales of sexual labor are wrong because they are inherently alienating or damaging to human happiness' (Satz, 1995: 64). Analysts who accept this view maintain that sex work requires absolute *condemnation*. Exemplifying this approach are the radical-feminist works of Andrea Dworkin (1981, 1997), Catherine MacKinnon (1987, 1989), Kathleen Barry (1979, 1995), and Sheila Jeffreys (1997) – who reduce commercial sex to male violence against women and 'sexual slavery' whether or not it involves outright coercion or manipulation. These writers favor the term 'prostituted women' over 'prostitutes' because it underscores their lack of agency and directs attention to the males who use them.

Strident condemnation of sex work appears in some empirical studies as well. In their book on street prostitution in Norway, Cecilie Hoigard and Liv Finstad insist that prostitution is an 'abomination' and a form of 'brutal oppression' (Hoigard & Finstad, 1992: 76, 183). These gratuitous evaluations are based more on the authors' personal opinions than on their findings, and such judgments only detract from their otherwise important research results. Another example is a book on pornography by Gail Dines, Robert Jensen, and Ann Russo, which is filled with unsubstantiated, categorical assertions about the role of porn in subordinating women and brutalizing male viewers. Even though the authors (all professors) insist that the book is a scholarly work, they take pains to dismiss empirical research ("instead of being paralyzed by the limitations of social science") and rely instead on their own convictions and some anecdotal testimonials from 'victims' of pornography (Jensen 1997: 5). Gayle Rubin (1984: 301) calls this perspective a 'demonology' that selects the 'worst available examples' of sex work and presents them as representative. The radical feminist literature on sex work, she argues, is filled with 'sloppy definitions, unsupported assertions, and outlandish claims' (Rubin 1993:36).

The opposite type of essentialist bias is *romanticization*. Shannon Bell (1995: 16) describes her anthology, *Whore Carnival*, as 'a recognition and commendation of the sexual and political power and knowledge of prostitutes', which sounds rather celebratory. Others offer glowing assessments, if not outright praise. Wendy McElroy (1995: 148) flatly states, 'Pornography benefits women, both personally and politically', a claim that receives only limited support in her book. Similarly, Nadine Strossen (1995: 166) argues that 'pornography provides information about women's bodies and techniques for facilitating female sexual pleasure, which is otherwise sadly lacking in our society'. This claim needs qualification in light of the fact that most pornography caters to *male* sexual pleasure, with enlightenment about women's sexuality secondary at best. Other appreciative analyses of sex work include the works of Bell (1987), Carmen & Moody (1985), Delacoste & Alexander (1987), Elias et al. (1998), Kipnis (1996), Nagle (1997), Soble (1986, 1996), and Williams (1989).

The polarities of condemnation and glorification are reflected in a larger feminist debate. Radical feminist prohibitionists denounce all sex work as the ultimate expression of women's oppression by men, while the other, liberal side (including both feminist and non-feminist analysts) insists that consensual commercial sex services are legitimate and valuable occupations that differ little from other jobs and may even be liberating for the workers, insofar as they are freed from the drudgery of low-paying, dead-end jobs (Chancer, 1993; Freeman, 1990; Jolin, 1994; Overall, 1992). These are diametrically opposed paradigms, turning on different images of the workers involved: quintessential *victims* of male domination vs. *workers* who exercise control over their lives.

This protracted 'sex war' has generated more heat than light, more sniping than understanding of sex work. None of the recent contributions to this debate has broken new ground; each simply recapitulates arguments made countless times before. The problem is not limited to excessive heat, however; it also involves gross *distortion* of the subject matter. Grand claims about the fundamental character of sex work (1) clash with the reality of variation documented above, (2) superimpose a framework that obscures instead of illuminates social structures, processes, and outcomes, and (3) neglect entirely male and transgender work. The latter lacuna presents a serious and unanswered challenge to these analysts. Both camps concentrate on female workers and male consumers; if their claims are not meant to apply to male and transgender workers, sex work cannot be said to have any 'intrinsic' properties. If their claims (e.g. oppression and objectification) are meant to apply to males and transgenders, then this needs to be spelled out theoretically and documented empirically. The argument that a functionally equivalent power relationship characterizes female, male, and transgender sex workers' interactions with their clients – with all three types of workers subordinate to other men – seems plausible, but the empirical literature on male and transgender sex work is of little help in resolving this problem. The little we do know suggests a more complex picture than simple equivalence between male, transgender, and female workers. To identify core similarities and key differences, we desperately need systematic, comparative studies of male, female, and transgender workers (e.g. Ronai & Cross, 1998; Weinberg, Shaver & Williams,

1999). A further wrinkle in essentialist assumptions is presented by *female* customers of male and female workers. Although they constitute a small fraction of the market, female customers present the same challenge to grand claims about sex work as do male and transgender workers. Does the customers' gender influence the character of the encounter? Is gender domination or any other kind of oppression present or absent in exchanges between female customers and male and female sex workers? These questions have yet to be tackled empirically or theoretically.

A third approach, somewhat less essentialist than the others, is *normalization*. First, there is the argument that sex work is no different from other work. Eileen McLeod (1982: 28), for example, argues that prostitution is much like other 'women's work' and that both female prostitutes and other women 'barter sex for goods', albeit less conspicuously in the case of non-prostitutes. This argument is overdrawn. In terms of workers' own experiences of sexual labor and associated working conditions, there are some important qualitative differences between sex work and other work. Precisely because it involves *sexual* contact or performance it is rather different – again, in terms of workers' own experiences – from other jobs, even those that require physical contact such as massage and physical therapy. Sex work is work, but *not* just like any other kind of work.

A second type of normalization seeks to humanize the workers. This is laudable – up to a point. It may indeed be the case that most porn actors are 'like all the rest of us' in their everyday lives, as Wendy McElroy (1995: 148) suggests; that many or most prostitutes are 'ordinary women' as Roberta Perkins (1991: 348) says; or that they are 'ordinary and likable' people, as Arlene Carmen & Howard Moody (1985: 88) claim. But this is *not* how they are generally perceived by the public. At the societal level, sex work is still regarded as highly *stigmatized* work, and most of those involved in the buying and selling of sex feel compelled to remain in the closet. Fully 72% of the American public say that pornography 'degrades women' by portraying them as 'sex objects' (*Time*, 1986), and 62% think that pornographic materials lead to a 'breakdown of morals' (Davis & Smith, 1994). When asked their opinion of the idea of 'men spending an evening with a prostitute', 61% of Americans considered this morally wrong (*Time*, 1977). Most Americans favor either more curbs or stricter enforcement of existing laws against prostitution, pornography, stripping, and commercial telephone sex (see Table 1). In this context of societal stigma, it is especially valuable for researchers to document, rather than simply assert, the ways in which these workers lead 'normal' lives.

Sound analysis neither demonizes nor celebrates sex work. Evaluations are based instead on concrete evidence, not the author's preconceptions. It is

Table 1. Public Attitudes Toward Sex Work

	% agreeing
Prostitution should be illegal[1]	70
Need stricter laws to control pornographic books and movies[2]	77
Media should publish names and photos of men convicted of soliciting prostitutes[3]	50
Female strippers at clubs should be illegal[4]	46
Male strippers at clubs should be illegal[5]	45
Phone numbers offering sex talk should be illegal[6]	76
Massage parlors and porn shops that might be centers for casual sex should be closed[7]	70

[1] Gallup Poll, May 28–29, 1996, N = 1,019
[2] NBC News/*Wall Street Journal* Poll, June 10–14, 1994, N = 1,502
[3] *Newsweek* Poll, January 26–27, 1995, N = 753
[4] Gallup Poll, 1991 (*Gallup Poll Monthly*, no. 313, October)
[5] Gallup Poll, 1991 (*Gallup Poll Monthly*, no. 313, October)
[6] Gallup Poll, 1991 (*Gallup Poll Monthly*, no. 313, October)
[7] *USA Today* Poll, August 4–8, 1988, N = 1,283 registered voters

always a challenge for observers to suspend value judgments on highly charged topics, but it is imperative that this subject be examined as objectively as possible.

THEORETICAL POVERTY

Theoretical advancement in the field has been conspicuously lacking over the past two decades. Much of the empirical literature is totally descriptive (e.g. Armstrong, 1978; Weiner, 1996). Important research findings are therefore presented in a way that fails to maximize our understanding of sex work. Another body of literature can be criticized for being overly 'theoretical': marked by lofty, grand theorizing disconnected from empirical data and instead driven by a particular political agenda – exemplified in the perspectives of condemnation and celebration reviewed above.

What is needed instead are 'middle-range' theories that use empirical findings to explain specific aspects of sex work: e.g. motivations for entry, socialization processes, stratification patterns, reasons why men buy sex, forces shaping popular perceptions and official policies, and so forth. Only a few empirical studies are theoretically driven in the strict sense: labelling and control theories have been used to explain why individuals become sex workers

(Davis, 1971; Gray, 1973; Rosenbaum, 1975), differential association to describe socialization into sex work and the learning of techniques to neutralize stigma and preserve self-esteem (Dressel & Peterson, 1982; Foltz, 1979; Heyl, 1977; Salamon, 1989; Thompson & Harred, 1992), status-defense theory to explain the rise of antiporn crusades (Zurcher & Kirkpatrick, 1976), and constructionist and resource-mobilization theories to analyze the prostitutes' rights movement (Jenness, 1993; Weitzer, 1991). Most writers, however, have neither built upon this tradition of theoretical work nor tested other approaches. Most feminist analyses, again, debate the core properties of sex work rather than test propositions with empirical data. Theoretically-driven research is sorely needed.

METHODOLOGICAL FLAWS

Many studies of sex workers can be faulted on methodological grounds. Some authors fail to describe how and where they contacted research subjects. Others fail to include comparison groups (non-prostitutes matched on background characteristics, e.g. age, social class), without which it is impossible to know if the findings reported for a prostitute sample differ significantly from those of non-prostitutes. Those few studies that do include appropriate control groups yield mixed results. Some find significant differences between prostitutes and controls on family characteristics and childhood experience of victimization, while others find no significant differences (e.g. Earls & David, 1989; Nadon, Koverola & Schludermann, 1998).

An even more serious problem is the widespread reliance on unrepresentative, convenience samples. While random sampling of most populations of sex workers and customers is impossible, too often the findings and conclusions drawn from non-random samples are not properly qualified as non-generalizable. Victimization studies – which examine individuals' experiences of assault, rape, and robbery – are a case in point. Street prostitutes appear to experience high rates of violence in the course of their work, but the samples used in such studies typically consist of women who had contacted service agencies, were approached on the street, or were interviewed in jail (Davis, 2000; Farley & Barkan, 1998; James & Meyerding, 1977; McKeganey & Barnard, 1996; Silbert & Pines, 1982). The high victimization rates reported in these studies are vulnerable to self-selection bias; That is, the most desperate segment of the population or those persons who are most frequently or seriously victimized may be most likely to approach service providers or agree to interviews. Generalizing from prostitutes in custody to the population of prostitutes is also problematic, just as with other types of incarcerated offenders. Yet the implications of this sampling bias typically are neglected in the published reports; moreover, the victimization rates reported are often reproduced in the secondary literature and in newspaper reports without disclosing the sampling technique and its limitations.

A third flaw is the small samples used in many studies. It is rare to find a sample size over 200 (exceptions include Lever & Dolnick, 2000; Pomeroy, 1965; Weiner, 1996); some are in the 30–50 range; the majority are much smaller. Numerous articles and entire books have been based on a handful of interviews: Between 14 and 17 individuals were interviewed by Gray (1973), Jackman, O'Toole & Geis (1963), Miller (1995), Romenesko & Miller (1989), and Verlarde (1975) – to mention but a few. Goldstein (1983) relied on six madams in drawing conclusions about occupational mobility in the world of prostitution, and Knox (1998) interviewed only six workers in his study of male prostitution. Qualitative studies are valuable in their own right, for the richness of the data presented, and it is acknowledged that barriers to access limit the number of respondents that can be drawn from any highly stigmatized population. But the question remains: Do the small samples adequately tap the full range of experiences within the population? If not, the conclusions drawn only partially reflect social reality.

NEGLECT OF MEN

Male sexual behavior is less socially constrained than female sexual activity. Women are labelled as 'good girls' or 'bad girls', whereas males are subject to less dichotomized evaluation; promiscuous females are branded as 'sluts', whereas male promiscuity is a badge of honor.

This double standard also colors popular images of the individuals involved in commercial sex transactions. When we think of pornography and other sex work, we tend to think of female actors, despite the fact that the sex trade involves at least two parties (at least one of whom is usually male), and despite the fact that customers are much more numerous than the workers who service them. In the wider culture, female sex workers are quintessential deviant women, whereas the customers are seen as essentially normal men. There is, of course, some stigmatization of customers: there is a perception that men who 'have to pay for it' have some personal deficiency or perversity, and such men risk censure if others discover their involvement in this seedy activity (Campbell, 1998: 156). But this stigma is much less than what the prostitute experiences. One may be a bit surprised to learn that a male friend has visited a prostitute, but shocked to learn that a female friend has sold sex. The very terminology used – whore, hooker, harlot, slut – is heavily laden with opprobrium. By contrast, customers are referred to as johns, tricks, clients, and patrons – fairly tame labels.

A gender disparity is also evident in the research literature, which is predominantly concerned with female sex workers rather than male workers, managers, or customers. The only arena where men have been studied routinely is in experimental research on viewers of pornography (Donnerstein, Linz & Penrod, 1987). Little research has been done on the men who sustain prostitution – the customers – who are difficult but not impossible to access. Many customers are middle aged, middle class, and married, but we are only beginning to understand their motivations, attitudes and behavior (Atchison, Fraser & Lowman, 1998; Campbell, 1998; Hoigard & Finstad, 1992; Holzman & Pines, 1982; Jordan, 1997; McKeganey & Barnard, 1996; Monto, 2000; Winick, 1962). Similarly, studies of male sex workers comprise only a fraction of the literature (Browne & Minichiello, 1996; Dressel & Peterson, 1982; Luckenbill, 1986; Pittman, 1971; Salamon, 1989; van der Poel, 1992; West, 1993). These studies point to some important differences in the ways male and female sex workers experience their work (e.g. in the amount of stigma, violence, control over the encounter, and sexual gratification involved), but much more research is needed to corroborate these differences. The voluminous body of research on pornography offers almost nothing on gay porn (Duncan, 1989). This is an important lacuna because there seem to be important differences between gay and straight pornography, specifically in terms of its significance and reception in gay and straight cultures. Porn carries substantial stigma in the straight world, whereas it appears to hold a fair amount of esteem within the gay community, where porn actors are revered (Stychin, 1992; Thomas, 2000).

Pimps are individuals (usually male) who are actively involved in promoting the prostitution of others and benefiting materially from that association. Almost never studied directly, the little we know about pimps comes mainly from prostitutes, not from the pimps themselves (Milner & Milner, 1972; James, 1973). We need to know more about the dynamics of recruitment, socialization, surveillance, exploitation, coercion, and trafficking. We also need to avoid treating pimping as a monolithic enterprise; pimping arrangements vary in terms of emotional, economic, coercive, and sexual relationships (Davidson, 1998; Hoigard & Finstad, 1992). Research is scarce on both street-level pimps and the managers and owners of indoor establishments (Goldstein, 1983; Hausbeck & Brents, 2000; Heyl, 1977).

Law enforcement has mirrored the larger cultural bias favoring men. Traditionally, the act of patronizing a prostitute was not a crime in the United States and most other countries. Among the standard arguments for the law's exclusive focus on prostitutes were that they profit from the transaction, are repeat offenders, and present a public nuisance. The fact that johns also benefit from the purchase of sexual services and also contribute to a disorderly street scene was ignored. An underlying reason for differential legal treatment was the status gulf between male patrons and 'women of ill repute'. Prostitutes were outcasts, but their clients were seen as valuable members of society, even if they occasionally dabbled in unsavory sexual liaisons. Even some leading scholars have offered arguments in support of this double standard. Punishing the men who patronize prostitutes, claim Benjamin & Masters (1965: 385–386), would cause 'substantial loss to society' because these men are 'an important source of support for family, business, church, state, and other institutions'. The idea of penalizing johns is thus a 'senseless and harmful' proposition. Men's roles and social standing are held up as grounds for excusing their illicit sexual escapades.

Institutionalized bias is sometimes explicitly condoned by criminal justice officials. The renowned Model Penal Code and its official commentary stipulate that persons who patronize a prostitute should be treated more leniently than persons who sell sexual favors – because of the prevalence of extramarital sex, a 'popular understanding' favoring the clients, and the notion that the law should target the sellers of vice (American Law Institute 1980: 468). Each point is arguably a dubious rationale for gender bias, as some courts have pointed out. Some police officers continue to view prostitution through a gendered prism. In New York City in 1984, the head of the police department's Public Morals Division, Captain Jerome Piazza, defended the policy of not arresting johns on the grounds that 'You can ruin a lot of marriages by making a "john" collar' (*New York Times*, 1984). A Phoenix vice detective expressed similar compassion for arrested customers: 'It's really sad. The men are between thirty and sixty, white, with good jobs and families. When their cases go to court, it almost inevitably means a divorce. I hate to see it'.[3] Similar sympathies were not extended to the prostitutes.

Since the 1960s the act of soliciting a prostitute has been criminalized by every state in the country, though many state laws continue to punish patronizing less severely than prostitution (Posner & Silbaugh, 1996: 156), and law enforcement still falls most heavily on the prostitute. Arrests of johns in most American cities are either sporadic or non-existent. In very few jurisdictions are prostitutes and their patrons apprehended in equal proportion.

The disparity is even more striking when we take numbers into account. Customers greatly outnumber prostitutes, yet only about 10% of prostitution-related arrests in the United States are male customers; hence, the proportions of each population represented in the arrests reveals even more extensive gender discrimination.

While it is very difficult for police to build cases against pimps – because prostitutes are so reluctant to testify against them – the same cannot be said for prostitutes' customers. In the United States, the police have sufficient legal powers to arrest the johns, and it cannot be claimed that these arrests are more difficult than arrests of prostitutes. Female police officers have posed as streetwalkers and made arrests with ease. Even male officers dressed as women can make these arrests – an occasional practice in some cities.

Bias is also evident in post-arrest outcomes (Bernat, 1985; Kandel, 1992; Lowman, 1990). In most cities first-time arrested customers are routinely offered pretrial diversion rather than prosecution. In 1993 in Seattle, for example, 69% of the prostitutes charged with solicitation were convicted, whereas only 9% of the customers were convicted, largely because most were offered diversion (Seattle Women's Commission 1995). Customers who are prosecuted and convicted typically receive lower fines or find themselves less likely to be jailed than prostitutes (Atchison, Fraser & Lowman, 1998; Lowman, 1990). Only recently have the authorities in some cities begun to arrest customers in substantial numbers, but this is exceptional. State penal codes treat patronizing as a misdemeanor, but a significant segment of the population favors stiffer punishment in the form of public shaming of johns. A recent poll found that half of the American public endorsed the idea of punishing men convicted of soliciting prostitutes by placing their names and pictures in the news media (see Table 1).

Much more research is needed on the treatment of prostitution by the courts and police (Benson & Matthews, 2000; Bernat, 1985; Kandel, 1992; Lowman, 1990; Pearl, 1987). The literature has been preoccupied with the deviant actors to the neglect of the social control apparatus. Major questions remain to be investigated: How do police vice squad officers view their work, prostitutes, and johns? Are male prostitutes treated similarly to female prostitutes? Are alternative sanctions (e.g. community service) better suited than fines and jail to encouraging prostitutes to leave the trade (Weitzer, 1999)? What happens when a jurisdiction shifts its law enforcement emphasis from prostitutes to johns? Do arrests of the customers help to reduce street prostitution? What about the new 'johns schools' (in San Francisco, Las Vegas, and several other cities) where arrested customers participate in a day-long series of lectures on prostitution in lieu of an arrest record (Monto, 2000; Weitzer, 2000)? Does this experience affect the men's attitudes toward prostitution and toward women, as the lectures intend? Does it deter them from buying sex in the future?

ORGANIZATIONS, INTEREST GROUPS, AND THE STATE

Most studies examine *individual actors*, and this usually means street prostitutes or male subjects in pornography experiments. Individual-level analyses illuminate only part of the picture, however. We need studies of specific sectors of the sex industry – escort agencies, brothels, massage parlors, telephone sex agencies, pornography firms – focusing on how they recruit, train, and manage workers, on social relations within the agency, on their responses to external controls and threats, and on other organizational dynamics. Massage parlors were studied in the 1970s, but no recent investigations have been conducted. Organizational studies of other types of sex work are scarce. Reichert & Frey (1985) studied hotel prostitution in Las Vegas; Nevada's brothel industry has been examined by Hausbeck & Brents (2000); and telephone sex agencies have been studied by Flowers (1998) and Rich & Guidroz (2000) – but this is just the tip of the iceberg.

Nor is there much research on organized intervention in sex work on the part of interest groups and service providers. Lobbying groups have received fairly little coverage. Prostitutes' advocacy groups have been studied in the United States (Boles & Elifson, 1998; Jenness, 1993; Weitzer, 1991) and some other countries (Kempadoo & Doezema, 1998; McLeod, 1981; Sullivan, 1997), but studies have yet to be done on groups such as the Exotic Dancers Alliance and the Adult Film Association of America, which represents 200 producers and distributors of X-rated films and videos. Antipornography campaigns have been studied (Duggan & Hunter, 1995; Zurcher & Kirkpatrick, 1976), but in-depth studies of antiprostitution organizations are non-existent (except Davis, 2000).

State intervention, other than law enforcement, has rarely been investigated. Blue-ribbon commissions on pornography have been studied in the United States, Canada, and Britain (Hawkins & Zimring, 1988; Kanter, 1985; Simpson, 1983; Vance, 1986), but beyond this little is known about the forces shaping changes in contemporary public policy and enforcement practices (exceptions include Downs, 1989; Larsen, 1992; Roby, 1969; Sullivan, 1997; Weitzer, 1999, 2000).

CONCLUSION

The literature on sex work is deficient in several respects. Too many analysts make generalizations that are demonstrably unwarranted. The corrective here is simple: limit research conclusions to the subgroup studied, and instead of reifying 'sex work' and 'prostitution', disaggregate and analyze specific types of sex work. Essentialist claims are problematic since they are grounded largely, if not entirely, on analysts' value judgments or political agendas rather than research findings. Any review of the empirical literature would present a challenge to grandiose assumptions about the 'fundamental nature' of sex work. Theoretical underdevelopment bedevils the field. Most writings are either completely atheoretical or grand-theoretical; missing are middle-range propositions based on empirical findings. Methodological flaws come in various flavors, and I have questioned the conclusions drawn from studies utilizing small and unrepresentative samples and lacking control groups. It could be argued that it is better to have no data than misleading data based on questionable research designs, particularly when the study is picked up by the media and the findings reported without proper qualifications. Distortion is evident, for example, in recent articles in *Life* magazine (Foglino, 1998) and the *New York Times* (Nieves, 1999), which reported victimization rates of prostitutes in the United States by citing findings from Farley & Barkan's (1998) unrepresentative sample of street prostitutes in one city: San Francisco. Finally, the literature is marked by a preoccupation with female actors to the neglect of male sex workers, customers, managers, and profiteers, and by a neglect of the organizations involved in promoting, fighting, or regulating sex work.

Huge gaps remain in the scholarly literature. We need more research on telephone sex work, off-street prostitutes of all stripes, the porn industry, gay and lesbian pornography, legal prostitution systems, the dynamics of law enforcement, and the forces influencing law and public policy. We know precious little about contemporary brothels, transgender prostitutes, and call girls and escorts. We need much more research on the men involved at all levels – customers, workers, managers, producers, owners. Fresh research is called for even on those topics that have already attracted substantial attention. For example, the literature on massage parlors is confined to the 1970s. Have these establishments changed since that time? We have a sizeable body of research on street prostitution, but even here we need to know more about the micro-level dynamics of this trade (which requires careful ethnographic observation) and about the stratification of street work and its effects on the workers. This deviant world does not offer easy access to the outsider, which helps to account for the scarcity of research in some key areas. But gaining access is best viewed as a challenge rather than a iron barrier.

ACKNOWLEDGMENT

I am grateful to Kathleen Guidroz, Martin Monto, and Lois West for comments on this article.

NOTES

1. The term 'sex work' is rejected by analysts who take the view that 'work' is a misnomer for something that is fundamentally about 'oppression', but my generic use of the term implies no such evaluation. Later in the paper I suggest that essentialist claims about the inherent nature of sex work are simplistic and misleading.

2. Though it is impossible to arrive at precise figures on the number of workers in this hidden domain, a rough estimate is that street prostitution accounts for 10–20% of all prostitution in the United States (Alexander, 1987). A recent study of London estimated that 12% of the city's prostitutes worked on the street (Matthews, 1997).

3. Interview by Julie Pearl; transcript on file with author and at Hastings Law Journal.

REFERENCES

Alexander, P. (1987). Prostitutes Are Being Scapegoated for Heterosexual AIDS. In: F. Delacoste & P. Alexander (Eds), *Sex Work: Writings by Women in the Sex Industry*. Pittsburgh: Cleis.

American Law Institute (1980). *Model Penal Code and Commentaries, Part II, Sections 240.0 to 251.4*. Philadelphia: American Law Institute.

Armstrong, E. (1978). Massage Parlors and their Customers. *Archives of Sexual Behavior, 7*, 117–125.

Atchison, C., Fraser, L., & Lowman, J. (1998). Men Who Buy Sex. In: J. Elias, V. Bullough, V. Elias, & G. Brewer (Eds), *Prostitution*. Amherst, NY: Prometheus.

Barry, K. (1979). *Female Sexual Slavery*. Englewood Cliffs, NJ: Prentice Hall.

Barry, K. (1995). *The Prostitution of Sexuality*. New York: NYU Press.

Bell, H., Sloan, L., & Strickling, C. (1998). Exploiter or Exploited: Topless Dancers Reflect on Their Experiences. *Affilia, 13*, 352–368.

Bell, L. (Ed.) (1987). *Good Girls/Bad Girls: Feminists and Sex Trade Workers Face to Face*. Seattle: Seal Press.

Bell, S. (1995). *Whore Carnival*. New York: Autonomedia.

Benjamin, H., & Masters, R. E. L. (1965). *Prostitution and Sexual Morality*. London: Souvenir Press.

Benson, C., & Matthews, R. (2000). Police and Prostitution: Vice Squads in Britain. In: R. Weitzer (Ed.), *Sex for Sale: Prostitution, Pornography, and the Sex Industry*. New York: Routledge.

Bernat, F. (1985). New York State's Prostitution Statute: Case Study of the Discriminatory Application of a Gender Neutral Law. In: C. Schweber & C. Feinman (Eds), *Criminal Justice Politics and Women*. New York: Haworth.

Boles, J., & Elifson, K. (1998). Out of CASH: The Rise and Demise of a Male Prostitutes' Rights Organization. In: J. Elias, V. Bullough, V. Elias, & G. Brewer (Eds), *Prostitution*. Amherst, NY: Prometheus.

Browne, J., & Minichiello, V. (1996). Research Directions in Male Sex Work. *Journal of Homosexuality, 31*, 29–56.

Bryan, J. (1966). Occupational Ideologies and Individual Attitudes of Call Girls. *Social Problems, 13*, 441–450.

Campbell, R. (1998). Invisible Men. In: J. Elias, V. Bullough, V. Elias, & G. Brewer (Eds), *Prostitution*. Amherst, NY: Prometheus.

Carmen, A., & Moody, H. (1985). *Working Women: The Subterranean World of Street Prostitution*. New York: Harper and Row.

Chancer, L. (1993). Prostitution, Feminist Theory, and Ambivalence. *Social Text, 37*, 143–171.

Chapkis, W. (1997). *Live Sex Acts: Women Performing Erotic Labor*. New York: Routledge.

Chapkis, W. (2000). Power and Control in the Commercial Sex Trade. In: R. Weitzer (Ed.), *Sex for Sale: Prostitution, Pornography, and the Sex Industry*. New York: Routledge.

Cohen, B. (1980). *Deviant Street Networks: Prostitution in New York City*. Lexington, MA: Lexington.

Davidson, J. O'Connell (1998). *Power, Prostitution, and Freedom*. Ann Arbor: University of Michigan Press.

Davis, N. (1971). The Prostitute: Developing a Deviant Identity. In: J. Henslin (Ed.), *Studies in the Sociology of Sex*. New York: Appleton-Century-Crofts.

Davis, N. (2000). From Victims to Survivors: Working with Recovering Street Prostitutes. In: R. Weitzer (Ed.), *Sex for Sale: Prostitution, Pornography, and the Sex Industry*. New York: Routledge.

Davis, J., & Smith, T. (1994). *General Social Survey: Cumulative Codebook*. Chicago: NORC.

Delacoste, F., & Alexander, P. (Eds) (1987). *Sex Work: Writings by Women in the Sex Industry*. Pittsburgh: Cleis.

Donnerstein, E., Linz, D., & Penrod, S. (1987). *The Question of Pornography: Research Findings and Policy Implications*. New York: Free Press.

Downs, D. (1989). *The New Politics of Pornography*. Chicago: University of Chicago Press.

Dressel, P., & Peterson, D. (1982). Becoming a Male Stripper. *Work and Occupations, 9*, 387–406.

Duggan, L., & Hunter, N. (Eds) (1995). *Sex Wars*. New York: Routledge.

Duncan, D. (1989). Trends in Gay Pornographic Magazines: 1960 through 1984. *Social Science Research, 73*, 95–98.

Dworkin, A. (1981). *Pornography: Men Possessing Women*. New York: Putnam.

Dworkin, A. (1997). *Life and Death*. New York: Free Press.

Earls, C. M., & David, H. (1989). Male and Female Prostitution: A Review. *Annals of Sex Research, 2*, 5–28.

Elias, J., Bullough, V., Elias, V., & Brewer, G. (Eds) (1998). *Prostitution*. Amherst, NY: Prometheus.

Exner, J., Wylie, J., Leura, A., & Parrill, T. (1977). Some Psychological Characteristics of Prostitutes. *Journal of Personality Assessment, 41*, 474–485.

Farley, F., & Davis, S. (1978). Masseuses, Men, and Massage Parlors. *Journal of Sex and Marital Therapy, 4*, 219–225.

Farley, M., & Barkan, H. (1998). Prostitution, Violence, and Posttraumatic Stress Disorder. *Women and Health, 27*, 37–49.

Flowers, A. (1998). *The Fantasy Factory: An Insider's View of the Phone Sex Industry.* Philadelphia: University of Pennsylvania Press.

Foglino, A. (1998). Quitting the Streets. *Life*, November, 96–100.

Foltz, T. (1979). Escort Services: An Emerging Middle Class Sex-for-Money Scene. *California Sociologist, 2*, 105–133.

Freeman, J. (1990). The Feminist Debate Over Prostitution Reform. *Berkeley Women's Law Journal, 5*, 75–109.

Goldstein, P. (1983). Occupational Mobility in the World of Prostitution: Becoming a Madam. *Deviant Behavior, 4*, 267–279.

Gray, D. (1973). Turning Out: A Study of Teenage Prostitution. *Urban Life and Culture, 1*, 401–425.

Hausbeck, K., & Brents, B. (2000). Inside Nevada's Brothel Industry. In: R. Weitzer (Ed.), *Sex for Sale: Prostitution, Pornography, and the Sex Industry.* New York: Routledge.

Hawkins, G., & Zimring, F. (1988). *Pornography in Free Society.* New York: Cambridge University Press.

Heyl, B. (1977). The Madam as Teacher: The Training of House Prostitutes. *Social Problems, 24*, 545–555.

Heyl, B. (1979). Prostitution: An Extreme Case of Sex Stratification. In: F. Adler and R. Simon (Eds), *The Criminology of Deviant Women.* Boston: Houghton Mifflin.

Hoigard, C., & Finstad, L. (1992). *Backstreets: Prostitution, Money, and Love.* University Park: Pennsylvania State University Press.

Holzman, H., & Pines, S. (1982). Buying Sex: The Phenomenology of Being a John. *Deviant Behavior, 4*, 89–116.

Jackman, N., O'Toole, R., & Geis, G. (1963). The Self-Image of the Prostitute. *Sociological Quarterly, 4*, 150–161.

James, J. (1973). Prostitute-Pimp Relationships. *Medical Aspects of Human Sexuality, 7*, 147–160.

James, J., & Meyerding, J. (1977). Early Sexual Experience and Prostitution. *American Journal of Psychiatry, 134*, 1381–1385.

Jeffreys, S. (1997). *The Idea of Prostitution.* North Melbourne, Australia: Spinifex.

Jenness, V. (1993). *Making it Work.* New York: Aldine de Gruyter.

Jensen, R. (1997). Introduction in Gail Dines, Robert Jensen, and Ann Russo. *Pornography.* New York: Routledge.

Jolin, A. (1994). On the Backs of Working Prostitutes: Feminist Theory and Prostitution Policy. *Crime and Delinquency, 40*, 69–83.

Jordan, J. (1997). User Buys: Why Men Buy Sex. *Australian and New Zealand Journal of Criminology, 30*, 55–71.

Kandel, M. (1992). Whores in Court. *Journal of Law and Feminism, 4*, 329–352.

Kanter, M. (1985). Prohibit or Regulate? *Osgoode Hall Law Journal, 23*, 171–194.

Kempadoo, K., & Doezema, J. (Eds) (1998). *Global Sex Workers: Rights, Resistance, and Redefinition.* New York: Routledge.

Kipnis, L. (1996). *Bound and Gagged: Pornography and the Politics of Fantasy in America.* New York: Grove.

Knox, M. (1998). Negotiations and Relationships among Male Sex Workers and Clients in Liverpool. In: J. Elias, V. Bullough, V. Elias, & G.Brewer (Eds), *Prostitution.* Amherst, NY: Prometheus.

Kulick, D. (1998). *Travesti: Sex, Gender, and Culture among Brazilian Transgendered Prostitutes.* Chicago: University of Chicago Press.

Larsen, E. N. (1992). The Politics of Prostitution Control: Interest Group Politics in Four Canadian Cities. *International Journal of Urban and Regional Research, 16*, 169–189.

Las Vegas Review Journal (1998). March 7.

Lever, J., & Dolnick, D. (2000). Clients and Call Girls: Seeking Sex and Intimacy. In: R. Weitzer (Ed.), *Sex for Sale: Prostitution, Pornography, and the Sex Industry*. New York: Routledge.

Lowman, J. (1990). Notions of Formal Equality Before the Law: The Experience of Street Prostitutes and Their Customers. *Journal of Human Justice, 1*, 55–76.

Lowman, J., & Fraser, L. (1995). *Violence Against Persons who Prostitute: The Experience in British Columbia*. Ottawa, Canada: Department of Justice.

Luckenbill, D. (1986). Deviant Career Mobility: The Case of Male Prostitutes. *Social Problems, 31*, 283–296.

McElroy, W. (1995). *XXX: A Woman's Right to Pornography*. New York: St. Martin's.

McKeganey, N., & Barnard, M. (1996). *Sex Work on the Streets: Prostitutes and Their Clients*. Buckingham: Open University Press.

MacKinnon, C. (1987). *Feminism Unmodified*. Cambridge: Harvard University Press.

MacKinnon, C. (1989). *Toward a Feminist Theory of the State*. Cambridge: Harvard University Press.

McLeod, E. (1981). Man-made Laws for Men? The Street Prostitutes' Campaign against Control. In: B. Hutter & G. Williams (Eds), *Controlling Women: The Normal and the Deviant*. London: Croom Helm.

McLeod, E. (1982). *Working Women: Prostitution Now*. London: Croom Helm.

Matthews, R. (1997). *Prostitution in London: An Audit*. Middlesex, UK: Middlesex University.

Miller, J. (1995). Gender and Power on the Streets: Street Prostitution in the Era of Crack Cocaine. *Journal of Contemporary Ethnography, 23*, 427–452.

Milner, C., & Milner, R. (1972). *Black Players*. Boston: Little Brown.

Monto, M. (2000). Why Men Seek out Prostitutes. In: R. Weitzer (Ed.), *Sex for Sale: Prostitution, Pornography, and the Sex Industry*. New York: Routledge.

Nadon, S,, Koverola, C., & Schludermann, E. (1998). Antecedents to Prostitution: Childhood Victimization. *Journal of Interpersonal Violence, 13*, 206–221.

Nagle, J. (Ed.) (1997). *Whores and Other Feminists*. New York: Routledge.

New York Times (1984). November 14.

Nieves, E. (1999). For Patrons of Prostitutes, Remedial Instruction. *New York Times*, March 18.

Overall, C. (1992). What's Wrong with Prostitution? *Signs, 17*, 705–724.

Pearl, J. (1987). The Highest Paying Customer: America's Cities and the Costs of Prostitution Control. *Hastings Law Journal, 38*, 769–800.

Perkins, R. (1991). *Working Girls*. Canberra: Australian Institute of Criminology.

Perkins, R., & Bennett, G. (1985). *Being a Prostitute*. London: George Allen and Unwin.

Pittman, D. (1971). The Male House of Prostitution. *Transaction, 8*, 21–27.

Pomeroy, W. (1965). Some Aspects of Prostitution. *Journal of Sex Research, 1*, 177–187.

Porter, J., & Bonilla, L. (2000). Drug Use, HIV, and the Ecology of Street Prostitution. In: R. Weitzer (Ed.), *Sex for Sale: Prostitution, Pornography, and the Sex Industry*. New York. Routledge.

Posner, R. A., & Silbaugh, K. B. (1996). *A Guide to America's Sex Laws*. Chicago: University of Chicago Press.

Prince, D. A. (1986). *A Psychological Study of Prostitutes in California and Nevada*. Doctoral dissertation, United States International University, San Diego, California.

Pyett, P. (1996). Risk Practices for HIV Infection and other STDs Amongst Female Prostitutes Working in Legalized Brothels. *AIDS Care*, *8*, 85–94.

Reichert, L., & Frey, J. (1985). The Organization of Bell Desk Prostitution. *Sociology and Social Research*, *69*, 516–526.

Reynolds, H. (1986). *The Economics of Prostitution*. Springfield: Charles Thomas.

Rich, G. J., & Guidroz, K. (2000). Smart Girls Who Like Sex: Telephone Sex Workers. In: R. Weitzer (Ed.), *Sex for Sale: Prostitution, Pornography, and the Sex Industry*. New York: Routledge.

Roby, P. (1969). Politics and the Criminal Law: Revision of the New York State Penal Law on Prostitution. *Social Problems*, *17*, 63–109.

Romenesko, K., & Miller, E. (1989). The Second Step in Double Jeopardy: Appropriating the Labor of Female Street Hustlers. *Crime and Delinquency*, *35*, 109–135.

Ronai, C., & Cross, R. (1998). Dancing with Identity: Narrative Resistance Strategies of Male and Female Strippers. *Deviant Behavior*, *19*, 99–119.

Rosenbaum, K. (1975). Female Deviance and the Female Sex Role. *British Journal of Sociology*, *26*, 169–185.

Rubin, G. (1984). Thinking Sex: Notes for a Radical Theory of the Politics of Sexuality. In: C. Vance (Ed.), *Pleasure and Danger*. Boston: Routledge and Kegan Paul.

Rubin, G. (1993). Misguided, Dangerous, and Wrong: An Analysis of Antipornography Politics. In: A. Assiter and A. Carol (Eds), *Bad Girls and Dirty Pictures*. London: Pluto.

Salamon, E. (1989). The Homosexual Escort Agency. *British Journal of Sociology*, *40*, 1–21.

Salutin, M. (1971). Stripper Morality. *Transaction*, *8*, 12–22.

Sanchez, L. (1997). Boundaries of Legitimacy: Sex, Violence, Citizenship, and Community in a Local Sexual Economy. *Law and Social Inquiry*, *22*, 543–580.

Satz, D. (1995). Markets in Women's Sexual Labor. *Ethics*, *106*, 63–85.

Seattle Women's Commission (1995). *Project to Address the Legal, Political, and Service Barriers Facing Women in the Sex Industry*. Report to the Major and City Council, Seattle.

Seidlin, M. (1988). Prevalence of HIV Infection in New York Call Girls. *Journal of AIDS*, *1*, 150–154.

Sheehy, G. (1973). *Hustling*. New York: Delacoste.

Silbert, M., & Pines, A. (1982). Victimization of Street Prostitutes. *Victimology*, *7*, 122–133.

Simpson, A. W. B. (1983). *Pornography and Politics: A Look Back at the Williams Committee*. London: Waterloo.

Soble, A. (1986). *Pornography: Marxism, Feminism, and the Future of Sexuality*. New Haven: Yale University Press.

Soble, A. (1996). *Sexual Investigations*. New York: New York University Press.

Strossen, N. (1995). *Defending Pornography*. New York: Anchor.

Sullivan, B. (1997). *The Politics of Sex: Prostitution and Pornography in Australia Since 1945*. Cambridge: Cambridge University Press.

Stychin, C. F. (1992). Exploring the Limits: Feminism and the Legal Regulation of Gay Male Pornography. *Vermont Law Review*, *16*, 857–900.

Thomas, J. A. (2000). Gay Male Video Pornography: Past, Present, and Future. In: R. Weitzer (Ed.), *Sex for Sale: Prostitution, Pornography, and the Sex Industry*. New York: Routledge.

Thompson, W. E., & Harred, J. L. (1992). Topless Dancers: Managing Stigma in a Deviant Occupation. *Deviant Behavior*, *13*, 291–231.

Time (1977). Poll conducted July 26–31, 1977. N = 1,044 registered voters.

Time (1986). Poll conducted July 7–9, 1986, N = 1.017.

van der Poel, S. (1992). Professional Male Prostitution: A Neglected Phenomenon. *Crime, Law, and Social Change*, *18*, 259–275.

Vance, C. (1986). The Meese Commission on the Road. *The Nation*, August 2–9: 76–82.

Verlarde, A. (1975). Becoming Prostituted. *British Journal of Criminology*, *15*, 251–263.

Verlarde, A., & Warlick, M. (1973). Massage Parlors: The Sensuality Business. *Society*, *2*(November), 63–74.

Weinberg, M., Shaver, F., & Williams, C. (1999). Gendered Sex Work in the San Francisco Tenderloin. *Archives of Sexual Behavior*, *28*, 503–521.

Weiner, A. (1996). Understanding the Social Needs of Streetwalking Prostitutes. *Social Work*, *41*, 97–105.

Weitzer, R. (1991). Prostitutes' Rights in the United States: The Failure of a Movement. *Sociological Quarterly*, *32*, 23–41.

Weitzer, R. (1999). Prostitution Control in America: Rethinking Public Policy. *Crime, Law, and Social Change*, *32*, 83–102.

Weitzer, R. (2000). The Politics of Prostitution in America. In: R. Weitzer (Ed.), *Sex for Sale: Prostitution, Pornography, and the Sex Industry*. New York: Routledge.

West, D. J. (1993). *Male Prostitution*. Binghamton, NY: Haworth.

Williams, L. (1989). *Hard Core: Power, Pleasure, and the 'Frenzy of the Visible'*. Berkeley: University of California Press.

Winick, C. (1962). Prostitutes' Clients' Perceptions of the Prostitutes and of Themselves. *International Journal of Social Psychiatry*, *8*, 289–297.

Zurcher, L., & Kirkpatrick, R. G. (1976). *Citizens for Decency: Antipornography Crusades as Status Defense*. Austin: University of Texas Press.

CRIME IN HIGH PLACES: A CRIMINOLOGICAL PERSPECTIVE ON THE CLINTON CASE

David O. Friedrichs

ABSTRACT

Crime in high places – including alleged presidential crime – has been largely neglected by criminologists. A newly emerging integrative criminology, as articulated by Gregg Barak, provides one point of departure for exploring crime in high places. In the present article some dimensions of a traditional criminological approach are identified and discussed. The case of President Clinton is explored as a means of demonstrating both the value of and the limitations of a criminological analysis of crime in high places. It also reveals some of the benefits and costs of exposing such crime. A deeper understanding of crime in high places requires an integrated criminology drawing upon traditional criminological concerns as well as insights from many different disciplines.

INTRODUCTION

The relative neglect of crime in high places by criminologists is surely one of the most enduring paradoxes of the discipline.[1] Some dimensions of this paradox are addressed here.[2] The case of President Clinton – a case of various alleged crimes – serves as a focus for illustrating some of the key issues

Sociology of Crime, Law and Deviance, Volume 2, pages 281–300.

ISBN: 0-7623-0680-7

involved. Insofar as the president may be said to occupy 'the highest place' in American society his case has both unique elements and elements in common with other cases of crime in high places.

Why is crime in high places – and presidential crime in particular – so neglected by criminologists? Some criminologists have attended to presidential crime – e.g. William Chambliss (1988), David Simon (1996) and Frank Hagan (1997) – although much of what is written about such crime (or alleged crime) takes the form of exposing and discussing specific cases and circumstances.[3] I have in earlier work attempted to identify the reasons for the general neglect of crime in high places, but briefly stated this neglect can be attributed to: ideological resistance; professional socialization; lack of payoff in career terms; absence of funding; methodological barriers; and the ultimate complexity of such crime (Friedrichs, 1995, 1996a). Much of the alleged wrong-doing of the president and others in high places may be viewed as something other than crime. In this same vein, one response to the allegations of presidential wrong-doing is to view such wrong-doing as a proper concern for journalists, political scientists, and historians, but not for criminologists. Possibly this low level of attention assumes that criminological analysis has little if anything to add to an understanding of such wrong-doing that cannot be gained from other disciplinary vantage points. Nevertheless it is intrinsically remarkable if criminology has a bottomless well of analysis applicable to the delinquency of inner city youths and relatively little to contribute to the analysis of crimes of the most powerful adults in our society. Of course we have by now a substantial literature on white collar crime, and some work on state crime or governmental crime, but a far more modest literature on crime at the very top (e.g. see Friedrichs, 1996b, 1998). Yet the direct and residual costs of crime in high places are immense.

INTEGRATING CRIMINOLOGY AND DISINTEGRATED CRIMINOLOGY

Can crime in high places be understood, or explained, by criminological theory? Virtually all broad surveys of criminological theory begin with an acknowledgment of the almost bewildering number of different theories. The testing of at least many of these theories has been a major criminological exercise through much of the 20th century. Toward century's end an intensification of efforts to integrate different theories has taken place (e.g. Arrigo 1995; Bernard & Snipes, 1996; Messner, Krohn & Liska, 1989; Vaughan, 1997).[4] Gregg Barak (1998), in *Integrating Criminologies*, provides us with arguably the most thorough effort at integration to date. I take the following to be some of the principal themes set forth by Barak: First, criminology needs to more fully integrate the widest possible range of disciplines (including such emerging disciplines as feminist studies, media studies, and ethnic studies) into criminological analysis; second, macro and micro levels of analysis need to be addressed simultaneously; third, social control and crime must be understood in their interrelated aspect; fourth, objective-positivist social science needs to be reconciled with, or treated as complementary to, subjective-romantic hermeneutics; fifth, a modernist and a post-modernist criminology need to be synthesized; and sixth, breadth of knowledge should be privileged over depth. Barak's effort at integration is itself unusual in its breadth, since more typically the integrative effort in criminology is limited to integration of already-existing theories within the discipline (e.g. see Bernard & Snipes, 1996).[5] As we move into the future the emphasis on integration is likely to intensify. But I also agree with Barak's (1998, p. 211) observation that "the current state of theoretical development, data measurement, and methodological sophistication is still too primitive, in most cases, to be able to discern the relative impact or weight of converging variables of crime and crime control." At the present stage, and perhaps indefinitely, the move toward integration must be interpretive. However, studies undertaken in a positivist framework can certainly contribute to our understanding of crime in high places, within limits. Finally, I adopt here the premise that one cannot even begin to hope that one might understand crime in high places other than by the application of an integrated, interdisciplinary approach.[6]

A mainstream, positivist view holds that theoretical and methodological sophistication should enable us to produce highly complex, multi-layered theoretical models of crime in high places, and these theories should lend themselves to empirical testing. The critical (especially postmodern) view holds that the notion of demonstrating empirically valid causal schemes is an illusion; rather, we should expose the manner in which explanatory schemes are constituted.[7] I lean toward the latter view, and find the concept of 'intertextuality' – or the notion that our human environment is "so complex and interactive as to defy disentanglement" (Rosenau, 1992, pp. 32–33) – especially persuasive. However, I also happen to believe that it should be part of the reintegrative project to seek ways in which such different approaches to our understanding of crime complement each other. I am not persuaded, then, that despite obvious fundamental differences on underlying assumptions, we must choose one approach entirely to the exclusion of the other. In Barak's (1998) terms a 'hyperintegrative' criminology brings together what can be learned from a modernist and a postmodernist approach. In the section that follows I will attend to some of the traditionally criminological dimensions of an attempt to understand crime in high places. A guiding premise here, however, is that the distinctly criminological dimension only offers us a partial explanation of an endlessly complex phenomenon that also requires input and insights from a broad range of disciplinary perspectives, on many different levels.

UNDERSTANDING CRIME IN HIGH PLACES: THEORETICAL INITIATIVES FROM CRIMINOLOGY

In the context of traditional explanations for criminality, crime, and criminalization we confront a number of paradoxes or conundrums when it comes to the crimes of those in high places. Discussion of criminological theory – or explaining crime – does not always clearly differentiate between explaining criminality, or the propensity of individuals and entities to commit crime; explaining crime, or the occurrence of an event involving the violation of law; and explaining criminalization, or the process of certain activities being defined as criminal.[8] Crime generally – including crime in high places – can only be properly understood by addressing these several levels of explanation. In the case of crime in high places, however, emphasis on one or the other of these levels of explanation tends to be especially intense. In part this sharp difference of emphasis obviously reflects partisan biases. The challenge for a criminological analysis is to assess evidence supporting the different levels of explanation as dispassionately as possible. Of course any such analysis is likely to be distorted to an indeterminate degree by some forms of bias. Certainly much of the literature on criminality has emphasized either biogenetic factors, psychological influences or subcultural values associated principally with conventional offenders and economically disadvantaged individuals. The literature on crime traditionally has focused on structures of opportunity and situational factors within a lower class setting, although we now have a substantial parallel literature on white collar/corporate crime. And on criminalization perhaps the largest proportion of the existing literature has addressed the criminalization of victimless crime or public order crime, as well as the justice system processing of conventional offenders. Further development of all three levels of understanding, as applicable to crime in high places, is surely needed.

Criminality in High Places

Criminality has traditionally been associated with 'differentness' – including the notion of biogenetic differentness – or exposure to a process of socialization promoting attitudes and patterns of behavior at odds with the law.[9] If criminality is associated disproportionately with lower class individuals some theorists hold that criminality may be more pronounced among elites, or at least as pronounced among elites as among the underprivileged. For example, John Hagan (1989) and Alex Thio (1988) have developed 'structural' or 'power' theories of crime, which claim that criminality is more pronounced among the powerful and privileged than among the powerless and underprivileged. In this interpretation, the advantaged have stronger deviant motivations, enjoy greater deviant opportunities, and are subject to weaker controls. The claim about stronger deviant motivations is based on the contention that the powerful are potently conditioned to aspire to material success, and accordingly they experience relative deprivation much more strongly than do the underprivileged or the powerless. Charles Tittle (1995) has formulated a control balance approach to crime that views criminality, or non-conformity, as a function of control exerted over others and control exerted by others. In a U-shaped model deviance is most pronounced among those with the most, and the least, control in either sense. These theoretical initiatives provide a context for some of the observations that follow.

We have no reliable evidence (or reason to believe) that biogenetic factors play a role with elite offenders, although the internalization of a form of pathological greed or egotism has been attributed to such offenders. But from a conventional vantage point it seems perplexing that people who typically seem to have opportunities for various degrees of legitimate (licit) success available to them, and have so much at risk in law-violating, engage in criminal conduct. Perhaps those at the extreme ends of the social order – that is, those at the top and those at the bottom – are disproportionately risk-takers, insofar as it is difficult to be truly successful without taking significant risks, while those at the bottom may feel they have little to lose by taking risks. Law-breaking in at least some circumstances is one form of risk-taking that may be viewed as providing opportunities not available by complying with the letter of the law. Risk-taking, on this view, is correlated with socioeconomic or political status in a curvilinear pattern.

One reason why many ordinary members of society refrain from law-breaking is that they are unwilling to endure the consequences of investigation, adjudication, and punishment. Those who reach the higher levels in society may be somewhat more resilient than most people, and because of their resilience they make decisions and choices less constrained by the possibility of stressful and painful outcomes. They may also be especially likely to assume that they enjoy a form of immunity from such prosecution, that they are "citizens above suspicion." Any such assumption, of course, has been severely tested in the recent era.

In one controversial interpretation elite criminal conduct may have a self-destructive dimension, in the sense that the potential costs so dramatically outweigh any potential gains. The cases of Richard Nixon and the Watergate crimes, and Michael Milken and the securities law crimes, come to mind (Woodward & Bernstein, 1976; Stewart, 1991). In such cases one can hardly dismiss the role of psychological factors, such as a sense of not being deserving of success and on at least a subconscious level wanting to destroy a circumstance that is on that basis inherently stressful. In an alternative interpretation, however, Nixon and Milken were long successful in their devious or illegal activities and believed they could continue to get away with such activities indefinitely.

Crime in high places has been differentiated from at least some forms of conventional crime on the grounds that it is especially likely to be planned, goal-oriented, and rational. Much such crime does seem to have these attributes. But not all crime in high places is rational, or well thought out. Those in high places often have quite extraordinary demands on their time. Arguably a factor in a certain proportion of such crime is the absence of clearly thinking through the consequences of the law-breaking. In any case, it is surely an error to regard crime in high places as uniformly a product of fully deliberated and carefully planned choices.

If we attempt to explain the involvement – direct or indirect – of presidents in illegal activities in terms of the criminality of presidents formidable obstacles are encountered. On the one hand the lives of presidents are more fully documented than those of almost any other members of our society. On the other hand it is difficult to obtain a truly objective, comprehensive account of the character of a particular president. Surely Thomas Jefferson and Abraham Lincoln have been among the most venerated of American Presidents. Yet in their own lifetime, and in the numerous biographies written of them, quite different assessments of their character have been produced. In the more recent era Presidents Kennedy and Johnson – and now Clinton – have been subjected to contrary assessments of their character. Arguably those who reach such high offices are also disproportionately likely to have complex characters with different (and sometimes contradictory) tendencies, insofar as such complexities and contrary tendencies may contribute to their success in dealing with very different types of constituencies, personalities, situations, and the like. Are dissemblers (or those with a flexible conception of truth-telling) and individuals with an instrumentalist orientation toward other people more likely than those without such attributes to reach the presidency? Much historical evidence would seem to suggest that for presidential candidates and presidents highly principled integrity and blunt truth-telling has not necessarily been rewarded. It would be quite interesting if we were able to construct reliable indices of moral character of high-level politicians and look for positive or negative correlations with electoral success as well as success in office.[10] The methodological difficulties of doing so within a positivist framework seem especially formidable to me; however, much has surely been revealed and can be further examined with an interpretive framework.

In his seminal work on white collar crime E. H. Sutherland (1949, pp. 257, 265) discounted the significance of personality or psychological factors in explaining this form of crime. For the most part the subsequent literature on white collar crime has followed Sutherland's lead on this. On balance Sutherland was surely correct that other types of factors are far more influential. On the other hand, one can go too far in the other direction if one entirely discounts any role for personality and character. Without deifying Lincoln and demonizing Nixon it is possible to identify both personality and character issues that contributed to significant differences in their ultimate reputations.

Crime in High Places

For a crime to occur some form of opportunity, and a set of conditions, must exist.[11] The facilitating conditions include various elements of the environment. Those in high places – including presidents – operate in a singularly complex environment, culturally, organizationally, and structurally. A form of "structural schizophrenia' might be said to exist, insofar as this environment includes many contradictory demands. Altogether, the president must respond to contradictory forces, pressures, influences, and constituencies. The environment is also singularly dynamic, with shifting 'weights' in terms of agenda priorities. The president is simultaneously operating in a political environment, an administrative environment, a symbolic environment, a historical environment, and so forth. The political environment, in particular, can be described as inherently criminogenic, insofar as it produces powerful pressures to make corrupt deals (in at least some sense of the term) in return for campaign financing or some other form of political benefit. In other environments within which the president is performing pressures often operate in quite a different direction. In criminological terms the challenge is to identify just how the pressures arising in these overlapping environments interact, and whether criminogenic pressures can be shown to have increased over time within the context of such interaction.

Is the choice of activities such as illegal campaign funding practices best understood as a function of institutionalization and habituation: that is, are these illegal practices so thoroughly embedded in the operation of the political system that engaging in them is quite routine and occurs without significant deliberation; or do such practices represent a rational, deliberate choice involving a careful assessment of risks and rewards involved in complying with law, or failing to do so? Indeed, both dimensions surely operate, but the challenge here is to determine which dimension is dominant, and how these dimensions interact.

Altogether those in high places are privy to, or part of, a structure of opportunities for illegal conduct that has some unique dimensions, and incorporates some singular pressures. This structure of opportunity seems to render a certain level of crime quite inevitable; the extent to which legitimate alternatives are available in the special environment within which people in high places operate has to be more fully explored.

Criminalization of Activities in High Places

The process of criminalizing at least certain activities of those in high places is undertaken by various entities, with a range of objectives. These entities include legislators, judges, prosecutors, regulators, policing agencies, and the media. The motives involved range from a disinterested, principled commitment to ethical governance to outright partisan goals oriented toward achieving political victory. In at least some interpretations the dynamic character of the relationship between the different entities has changed. For example, the media tends to play a more central and influential role. The bottomless appetite for scandal in an environment involving a vastly expanded and ferociously competitive media, then, is one element in the criminalization equation. Gail Collins (1998) attributes the insatiable appetite for gossip to the erosion of privacy in conjunction with a desire for intimacy. Although many of the scandals highlighted by the media do not necessarily involve violations of law the media attention to scandal in high places surely contributes to the criminalization environment. Law-makers overall have more incentives to create new categories of crime pertaining to those in high places than repealing (or striking down) existing categories. Morgan & Reynolds (1997) claim that in the post-Watergate era the mere appearance of ethical impropriety has become increasingly subject to rules; they claim that this form of expanded criminalization has only benefited certain parties with vested interests (e.g. journalists, interest groups and the like) while promoting broader levels of distrust without benefits to the general public. Furthermore, the Independent

Counsel differs from conventional prosecutors in important ways.[12] Once appointed to investigate a particular high-level target, and having expended significant funds toward that end, the Independent Counsel may be regarded as under significant pressure to find crime, in part by stretching existing classifications of crime to fit new circumstances.

Altogether, we have to understand more fully how the various entities identified above interact in the process of criminalization of activities of those in high places.

THE CASE OF PRESIDENT CLINTON

Surely no case of alleged criminal misconduct – with the possible exception of the O. J. Simpson case – was subjected to more discussion and analysis during the 1990s than that involving President Clinton, (and, more marginally, First Lady Hillary Clinton). In addition to the sex scandals and related perjury or obstruction of justice charges (especially the Paula Jones and Monica Lewinsky matters), these allegations have included violations of law relating to the Whitewater land deal, missing billing records, withheld documents, effectively bribing or manipulating potentially damaging witnesses, partisan use of FBI files, improper activities relating to the firing of Travel Office personnel, and legally questionable activity relating to political fund raising (Paul, 1999, p. 15).[13] Of course avowed enemies of President Clinton have alleged still more charges of serious wrong-doing, including complicity in drug smuggling and the deaths of various associates or potentially hostile witnesses.

The Clinton scandals have been largely – almost wholly – disregarded by criminologists.[14] The Clinton case brings into sharp relief many of the conundrums and paradoxes involved in crime in high places, and the application of a criminological analysis of such crime. Of course as a crime in 'the highest place' it is also a very special case.

The Clinton Scandals as Criminality

A persistent response to the Clinton scandals is this. How could such an exceptionally bright and capable man, with so much going for him, get himself into so much trouble?[15] Although there were on-going differences of opinion on the specific guilt or innocence of the President with regard to particular allegations, the combined force of the Starr Report and his own admissions compel any disinterested observer to acknowledge that he has lied on various levels (and, most would say, committed perjury) and seems to have exercised poor judgment in a number of circumstances. Some themes in the large literature and endless dialogue on the President's personality and character include the following: driving ambition from an early age; a willingness to dissemble (or, some would insist, lie outright) in any circumstance where telling the truth might be politically inadvisable (or catastrophic); willingness to form bonds with people and use them for his own purposes, but discard them with no second thought if they become a liability, or no longer serve his purposes; an ability to 'compartmentalize', to be able to focus on policy issues in the midst of the overwhelming pressure of allegations of misconduct; and extraordinary resilience, or the ability to bounce back from setbacks.[16] At least some of these attributes are linked with his somewhat unstable or dysfunctional family background and circumstances. His sexual involvement with a 21-year old White House intern, Monica Lewinsky, struck many observers as so incomprehensibly foolhardy and self-destructive that they reasoned that some form of deeply engrained or pathological compulsive behavior pattern was involved. On the crucial matter of dissembling, or lying, three basic possibilites could be identified: President Clinton told the truth in objective terms (although, of course, he himself has conceded not doing so, or at least misleading people, in connection with the Lewinsky matter); President Clinton willfully lied, in objective terms; and President Clinton internalized a postmodern perspective disputing the very idea of objective truth, and regards truth as a construct, devoid of fixed meaning![17]

On the matter of personality and character traits a number of observations can be made. First, such traits are hardly unique to this president: in particular, he shares at least some of these traits with Franklin Roosevelt and John Kennedy. Second, many others members of society – perhaps especially those who enjoy some form of conventional success – share such traits as well. Third, and obviously, having such traits does not necessarily correlate with engaging in criminal conduct. Fourth, other presidents who have been accused of some form of misconduct – most notably, President Nixon – have had some dramatically different personality attributes. Nevertheless, if we accept the list of personality or character attributes of President Clinton as relatively accurate it is also not difficult to see how they might contribute to a disposition to make ill-advised, unethical, or outrightly illegal choices in the service of blind ambition, to have the confidence that one will either get away with such choices or survive any fallout from them, and the willingness to let others in one's circle take the fall on one's behalf. If criminality is in some respects a construct, and the application of the notion of criminality to President Clinton's conduct is open to challenge, such attributes nevertheless seem clearly implicated in the Clinton scandals.

The Clinton Scandals as Crime

Those in high places surely encounter opportunities – sometimes, quite extraordinary – for engaging in unethical or illegal activity that the ordinary citizen certainly does not encounter. Within some political milieus opportunities for wrong-doing, institutionalized practices linked with such wrong-doing, and cultural values promoting such wrong-doing are especially pronounced. In the case of Clinton, in one reading, the political milieu of his home state, Arkansas, was especially conducive to crime and morally questionable activities. A number of Clinton's close associates in this milieu – with Webster Hubbell as one conspicuous example – were at some point found guilty of significant violations of law.

If Clinton was complicit in the violation of campaign funding laws the question arises: Was he simply sucked into a circumstance of institutionalized practices that happen to be illegal, or did he play an active role in the creation of opportunities for obtaining campaign funds in return for some form of political consideration? Political realities render it quite difficult to envision realistic reforms that would truly diminish (to say nothing of eliminate) the structure of opportunity for wrong-doing in conjunction with campaign funding.

In this context as well the structural forces in the larger society and culture that produce an environment fostering the forms of crime being addressed here must be considered. The criminological issue here is whether the conditions promoting such crime have expanded during the period of Clinton's presidency. Any such expansion would have to be understood in terms of a multiplicity of political, economic, social, and cultural forces.

The Clinton Scandals as Criminalization

In one interpretation the Clinton crimes are best understood in terms of criminalization. First, especially in the post-Watergate era, we have the criminalization of some forms of political behavior or activity not previously treated as criminal. Second, we have the actions of specific prosecutors (in this case Independent Counsel Kenneth Starr) in their attempt to criminalize President Clinton's actions. These specific attempts at criminalization are widely criticized – not without some justification – as a reflection of partisan politics. In addition to the special prosecutors office other branches of the government – notably the federal judiciary (including the U.S. Supreme Court) and the Congress play a role in the criminalization endeavor through rulings on behalf of or against the prosecutorial initiatives and through the use of

Congressional hearings and an impeachment trial. Then, given the extraordinarily high profile nature of the allegations, we have the role of the media and the general public. In both realms we have a contest between those endorsing and supporting the criminalization effort and those criticizing and opposing this effort.[18] The ultimate failure of the President's adversaries to obtain a conviction in the Senate impeachment trial was clearly a reflection of both the weakness of their case on legal grounds and the absence of broad public and political support for such a conviction.

Some Implications of the Clinton Scandals for Crime in High Places in the Postmodern 21st Century

The so-called Clinton crimes – if indeed crimes they were – can only be understood as an outcome of a broad complex of factors, operating on various different levels. I believe criminology can make a contribution to the understanding of the Clinton scandals by delineating some of the different criminological dimensions involved, indicating some of their interconnections, and placing the events in question in a comparative context: that is, an assessment of how the alleged crimes differ from, and what they have in common with, other forms of crime. The Clinton scandals collectively can only be understood in terms of a complex interaction of factors on many different levels, and include personality/character/judgment factors, structures of opportunity and political or cultural milieu factors, and myriad criminalization factors. The preceding analysis has been restricted principally to traditionally criminological dimensions. But an underlying premise of this paper is that a thorough understanding of the Clinton scandals is only possible through the introduction of many different disciplinary perspectives (including history, political science, psychology, and communication studies) insofar as these scandals involve an endlessly complex interaction of many different factors on many different levels. Accordingly, the Clinton scandals necessarily call for an integrative criminological approach.

CORE ELEMENTS OF AN INTEGRATIVE MODEL FOR UNDERSTANDING CRIME IN HIGH PLACES

An integrative model for understanding crime in high places should begin with a consideration of philosophical assumptions and presuppositions. It must then address the sociohistorical context within which such crime occurs. Next, it attends to the specific political and economic environment during the period under consideration. The socio-cultural context is quite central to the analysis, with some special attention to the critical role of mass communications. Organizational dimensions – of both the setting of crime and entities prosecuting crime – represent another important level of analysis. Then, community and peer group norms, controls and influences, along with the more immediate situational factors surrounding the event or events in question, must be addressed. Social statuses such as class, race and gender must be taken into account. Finally, individualistic (psychological, organic or biogenetic) factors must be explored. Concepts, insights and findings of the widest possible range of disciplines can be drawn upon insofar as they are relevant to advancing an understanding of the particular form of crime under consideration. On all levels of the analysis specific concepts, insights and findings emanating from criminology are an important part of the interpretive scheme. Ideally the links and interconnections between the macro-level and micro-level dimensions of this model can be identified.

Several core themes are highlighted in this model. First, the maldistribution of power, the changing context of power relations, and the dynamics of personal and interpersonal power; second, the constitutive processes whereby certain activities come to be defined as crime; and third, the circumstances under which deviance comes to be normalized within certain milieus.

A truly integrated theoretical framework for understanding crime in high places is an ambitious and large-scale enterprise. In this article, then, I have largely restricted myself to a provisional exploration of the specifically criminological dimensions of such an enterprise.

SOME UNDEREXPLORED CONSEQUENCES OF CRIME IN HIGH PLACES

Crime in high places often has manifest and quite easily identifiable costs, which may take the form of lives lost or public funds lost due to immoral and illegal actions on the part of elites. But such crime also tends to have especially large latent costs. It surely compromises the ability of elites to assume moral leadership in response to various forms of social harm, and other forms of crime (from conventional to white collar crime), and imposes some practical constraints as well. It deflects attention, resources, and energy from an effective response to international and domestic social problems. Somewhat less directly, and arguably not so easily measured, elite crime can surely be hypothesized to generate cynicism and distrust, with multiple significant consequences. How does a highly visible role model who is guilty of crimes or misconduct impact on the young in terms of rationalizations for their own misconduct, or, on the other side, their aspirations regarding public service?

These types of costs of crime in high places, although not easily measured, may be extraordinarily significant. The identifiable and potential costs of wrong-doing on the part of President Clinton, then, are extraordinarily broad, in terms of what happens in foreign policy, responses to international terrorism, and global markets; domestic policy as it relates to the poor, the elderly, children, and other constituencies; long-term effects in terms of impact on the socialization of the young, and so on.[19] The crisis of confidence in a particular leader can contribute to a broader crisis of legitimacy for the political system itself. If one adopts the premises of chaos theory, then, the ultimate ramifications of the Clinton Scandals are literally global, all-encompassing, and never-ending.

SOME UNINTENDED CONSEQUENCES OF FOCUSING ON CRIME IN HIGH PLACES

A focus on conventional crime has as its manifest intention advancing a fuller understanding of such crime, ideally generating effective responses to its prevention and control. Such attention has also contributed to intensifying intergroup suspicion, hostility, and conflicts – as certain social groups come to be blamed for contributing disproportionately to the conventional crime rate – and the production or reinforcement of stereotypes typically harmful to minority group members.

The exposure of crime in high places is necessary on one hand, but also has some counterproductive dimensions. First, it generates distrust and cynicism in relation to elites and the governance of the state. A certain level of distrust and cynicism is very much warranted, and has the constructive effect of reducing the chances of people being deceived and manipulated by elites. On the other side, however, at a certain point levels of distrust and cynicism can become counterproductive, leading to apathy and retreatism from public affairs and issues.

Second, it can promote paranoia regarding state activities, contributing in some cases to conventional forms of terrorist activity. Presumably all (or virtually all) of those among public interest advocacy groups, journalists, or criminologists who expose crime in high places view an incident of domestic terrorism such as the Oklahoma City bombing with utter revulsion and horror. Nevertheless, and in a vein somewhat parallel to the unwitting complicity of the news media in schoolyard shootings through extensive coverage of such events, the exposers of crime in high places may also have some level of unwitting complicity in extreme forms of anti-state activity. It is very difficult to envision an easy (or acceptable) solution to this problem in either of these cases, but it remains necessary to be conscious of the problem.

Third, full and relentless exposure of crime in high places can inevitably interfere with attention to and activity relative to achieving constructive policies. High level offenders tend to differ from low level offenders by virtue of the fact that a significant proportion (many would argue, by far the greatest proportion) of their time is devoted to constructive – often essential – activities, which is generally not the case with conventional offenders. If exposure of Watergate, Iran/contra arms, Whitewater and other such high level alleged violations of law are necessary and ideally productive in terms of deterrence and reform, they also undeniably deflect attention from more directly productive political and policy-making activity.

Fourth, the relentless exposure of crime in high places may deter at least some highly competent individuals from pursuing higher office, not necessarily because they are corrupt and fear exposure but because they are unwilling to endure (and impose on their family) the relentless (sometimes overblown) examination of all their activities, past and present. Reference was made earlier to the argument that the mere appearance of impropriety today becomes the focus of inquiry in a manner that is highly counterproductive (Morgan & Reynolds, 1997). If this is so even politicians or public officials committed to honest, constructive public service may fear that they will be subjected to a demeaning (and costly) inquiry.

Fifth, and following on this point, the intensification of exposure of crime in high places has had the effect, however unintended, of expanding the scope of what is considered fair game for journalistic inquiry, with virtually no aspect of personal and professional conduct immune from such inquiry. The Clinton sex scandal, for example, has inspired more intense exposure of sex-related indiscretions of other politicians.

CONCLUDING OBSERVATIONS

Throughout most of the 20th century conventional crime was the principal focus of criminological attention. We witnessed significant attention to white collar crime, and crime in high places, only in the final decades of the century. With the beginning of a new century we still have remarkably little in the way of a criminology of high (and the highest) places. I have tried to suggest here some elements of such a criminology. The Clinton Scandals were invoked as one way of illustrating the challenges, and conundrums, involved in the construction of a criminology of high places. A traditional criminological framework provides one point of departure in this venture. Criminology can make an important contribution to the vast torrent of analysis and commentary on the Clinton Scandals through systematic comparison of alleged criminality, crime, and criminalization in this special circumstance with criminality, crime, and criminalization in the many other circumstances investigated by criminologists. Ultimately a rich and sophisticated understanding of crime in high places surely requires an interdisciplinary, integrated criminological approach promoted by Greg Barak (1998), drawing upon the perspectives and insights of a diversity of disciplines.

AUTHOR'S NOTE AND ACKNOWLEDGMENTS

This article was first presented as a paper at the Annual Meeting of the American Society of Criminology, Washington, D. C., November, 1998. Helpful comments by the editor of this volume and anonymous reviewers are gratefully acknowledged. I also appreciate the support of a University of Scranton Faculty Research grant.

NOTES

1. As a randomly chosen example, the May, 1998, issue of *Criminology* includes five articles on various aspects of juvenile delinquency, as well as articles on controlling drugs, police subculture, burglary, and homicide. Such a focus or emphasis is generally replicated in other criminological forums, including conference panels.

2. By 'high place' I mean those in major leadership positions. Although any line of demarcation between high (and not-so-high, to say nothing of middling and low) places has an element of arbitrariness, in the context of the national government this designation would obviously include the President, Vice President, chief of staff, members of the cabinet, top commanders of the military branches, Supreme Court justices, senators and congressmen.

3. E.g., see Miller, 1992. Chambliss attributes corruption to structural contradictions of capitalism, and (for government officials) contradictions that lead to the institutionalization of criminal networks (i.e. pressures to use whatever means are necessary to achieve goals, while legitimating law). For Simon and Hagan elite deviance (including that in presidendial administrations) is a function of structural conditions, cultural values, and processes of rationalization. Jack Douglas (1974), the prominent sociology of deviance scholar, and criminologist Milton Mankoff (1974), produced analyses of the Watergate Affair that highlighted the flexible character of the concept of 'legality', shifts in the distribution of power, and structural attributes of a capitalist society.

4. Some earlier initiatives stressed the significance of a multitude of factors in crime and delinquency, but these initiatives did not tend to take a theoretical form (Sutherland & Cressey, 1966, pp. 61–66).

5. Bernard & Snipes (1996) contend that theories often predict different but not necessarily contrary outcomes; they view integration as an alternative to competitive testing of theories, and falsification. Vaughan's (1997) approach to integration calls for

merging macro- and micro-levels of analysis. In a summary statement Lanier & Henry (1998) have systematically identified the following problems with the integrative enterprise: What is to be integrated (e.g. concepts or propositions); how are propositions logically related; what is the nature of causality; on what level should concepts be integrated; and what should be the scope of integration? It is not possible here to address these complex and challenging issues.

6. Osgood (1998) makes a case for 'stealing' ideas more liberally from other disciplines and applying them to criminological phenomena and issues. Of course the obverse of this is that other disciplines can benefit from exposure to criminology.

7. Although these so-called 'discursive' approaches to criminological theorizing have been roundly criticized it is also conceded that most attempts at rigorous, positivistic criminological theory have not been especially fruitful (Leavitt, 1999).

8. See Lacey (1997) on criminalization itself as providing an integrative conceptual framework for uniting basic concerns of criminology, criminal law, and criminal justice.

9. Criminality itself can be conceived of in essentialist and constitutive terms. In my view it is most accurately treated as a complex and variable mix of essentialist and constitutive elements.

10. Marvin Olasky (1998) claims to find a direct link between good character and good presidential leadership, but others (e.g. Berke, 1998) suggest that the opposite may be true.

11. Miethe & Meier (1995) contend that theories of crime must incorporate theories of victimization. In the case of conventional crime lifestyles of victims are significant; as applied to crime in high places the specific character of public knowledge, attitudes and sentiments plays a role.

12. For various perspectives on the Independent Counsel see The Georgetown Law Journal, Volume 85, No. 6, July, 1998. Especially as a consequence of the Clinton case the Independent Counsel law was widely criticized in 1998 and 1999.

13. Although the political fund-raising abuses may be viewed as far more consequential for American society as a whole – especially in relation to growing corporate dominance of the political system – these abuses generated significantly less public interest than did the sexual misconduct and perjury allegations.

14. As one exception to this statement Robert N. Roberts and criminologist Marion T. Doss, Jr. (1997) have explored the Whitewater matter as exemplifying a long line of historical public integrity wars reflecting ideological campaigns to persuade the American public to accept a particular political vision for American society.

15. And parallel questions have been raised in the much more limited case against Hillary Clinton.

16. Of course various other traits have been attributed to President Clinton – such as his capacity to empathize with other people, his gregariousness, his energy, his hot temper, and so forth – but I here selectively emphasize traits that are at least hypothetically relevant to criminological concerns. For the interpretation of one major student of Clinton's life and career see Maraniss (1995, 1999).

17. The individual in the 'highest place' in terms of wealth – Microsoft president Bill Gates – was also accused of assuming an evasive, lawyerly approach to 'truth' in his testimony in the antitrust case against his company (Cohen 1999).

18. One of the criminological challenges arising out of the Clinton Scandals is this: How does the relentless coverage in the media, and the endless discussion and

commentary in many different forums, impact on public perceptions of crime and criminal justice, orientations toward lawful and unlawful behavior, and initiatives or operations on all levels of the criminal justice system? Parallel questions arose in the wake of Watergate, of course. Although much of the impact initially may occur on a subconscious level, and is certainly difficult to measure in any reliable way, any such measurable impact can hardly be discounted. The postmodernist concept of intertextuality is certainly applicable here.

19. For a general review see Alter (1999). On just one of these consequences, for example, historian Ruth Rosen (1998, p. B8) notes: "Bill Clinton squandered a rare opportunity – made possible by his highly publicized pledge to help poor women after he signed the welfare-reform law in 1996 – to move child care to the top of the national political agenda. And it is the children of welfare mothers whose lives will be most affected by the President's dalliances. For this, there can be no forgiveness."

REFERENCES

Alter, J. (1999). The Fallout. *Newsweek*, 28–33.

Arrigo, B. A. (1995). The Peripheral Core of Law and Criminology: On Postmodern Social Theory and Conceptual Integration. *Justice Quarterly, 12*, 447–472.

Barak, G. (1998). *Integrating Criminologies*. Boston: Allyn & Bacon.

Berke, R. L. (1998). In: Presidents, Virtues Can be Flaws (and Vice Versa). *The New York Times* (September 27): Sect. 4, 1.

Bernard, T. J., & Snipes, J. B. (1996). Theoretical Integration in Criminology. In: M. Tonry (Ed.), *Crime and Justice: A Review of the Research* (pp. 301–348). Chicago: University of Chicago Press.

Chambliss, W. J. (1988). *On the Take: From Petty Crooks to Presidents*. 2nd Edn. Bloomington, IN: Indiana University Press.

Cohen, A. (1999). A Tale of Two Bills. *Time*, 52–53.

Collins, G. (1998). *Scorpion Tongues: Gossip, Celebrity, and American Politics*. New York: William Morrow & Co.

Cox, A. (1998). Ethics, Campaign Finance, and Democracy. *Society*, (March–April), 54–60.

Douglas, J. D. (1974). Watergate: Harbinger of the American Prince. *Theory and Society, 1*, 103–110.

Friedrichs, D. O. (1995). State Crime or Governmental Crime: Making Sense of the Conceptual Confusion. In: J. I. Ross (Ed.), *Controlling State Crime: An Introduction* (pp. 53–80). New York: Garland.

Friedrichs, D. O. (1996a). Governmental Crime, Hitler and White Collar Crime: A Problematic Relationship. *Caribbean Journal of Criminology and Social Psychology, 1*, 44–67.

Friedrichs, D. O. (1996b). *Trusted Criminals: White Collar Crime in Contemporary Society*. Belmont, CA: ITP/Wadsworth Co.

Friedrichs, D. O. (Ed.) (1998). *State Crime*. Volumes I & II. Aldershot, UK: Ashgate/Dartmouth.

Garment, S. (1991). *Scandal: The Culture of Mistrust in American Politics*. New York: Doubleday.

Hagan, F. (1997). *Political Crime – Ideology and Criminality*. Boston: Allyn & Bacon.

Hagan, F. E., & Simon, D. R. (1998). From Inslaw to Iraqgate: Elite Deviance (in the Bush Era). *The Justice Professional, 10*, 229–248.

Hagan, J. (1989). *Structural Criminology*. New Brunswick, NJ: Rutgers University Press.

Henry, S., & Milovanovic, D. (1996). *Constitutive Criminology: Beyond Postmodernism*. London: Sage.

Lacey, N. (1997). Criminology, Criminal Law, and Criminalization. In: M. Maguire, R. Morgan, & R. Reiner (Eds), *The Oxford Handbook of Criminology* (pp. 431–450). Oxford, UK: Clarendon Press.

Lanier, M. M., & Henry, S. (1998). *Essential Criminology*. Boulder, CO: Westview.

Leavitt, G. (1999). Criminological Theory as an Art Form: Implications for Criminal Justice Policy. *Crime & Delinquency, 45*, 389–399.

Lull, J., & Hinerman, S. (1997). *Media Scandals*. New York: Columbia UniversityPress.

Mankoff, M. (1974). Watergate and Sociological Theory. *Theory and Society, 1*, 103–110.

Maraniss, D. (1995). *First in His Class – A Biograpy of Bill Clinton*. New York: Simon & Schuster.

Maraniss, D. (1998). *The Clinton Enigma: A Four-and-a-Half Minute Speech Reveals this President's Entire Life*. New York: Simon & Schuster.

Messner, S. F. Krohn, M. D., & Liska, A. (1989). *Theoretical Integration in the Study of Crime and Deviance*. Albany, NY: SUNY Press.

Miethe, T. D., & Meier, R. F. (1995). *Crime and Its Social Context: Toward an Integrated Theory of Offenders, Victims, and Situations*. Albany, NY: SUNY Press.

Miller, N. (1992). *Stealing from America*. New York: Paragon House.

Morgan, P. W., & Reynolds, G. H. (1997). *The Appearance of Impropriety*. New York: The Free Press.

Nye, J, S. Zelikow, P. D., & King, D. C. (1997). *Why People Don't Trust Government*. Cambridge, MA: Harvard University Press.

Olaskey, M. (1998). *The American Leadership Tradition: Moral Vision from Washington to Clinton*. New York: The Free Press.

Osgoode, W. (1998). Interdisciplinary Integration: Building Criminology by Stealing from our Friends. *The Criminologist, 23*, 1, 3–4, 41.

Paul, E. F. (1999). Late Rather than Sooner; Less Rather than More. *Society*, (March/April), 12–17.

Pearson, F., & Weiner, N. (1985). Towards an Integration of Criminological Theories. *Journal of Criminal Law and Criminology, 76*, 16–150.

Roberts, R. N., & Doss, M. T. Jr. (1997). *From Watergate to Whitewater: The Public Integrity War*. Westport, CT: Praeger.

Rosen, R. (1998). Why Feminists are Furious with Bill Clinton – or should be. *Chronicle of Higher Education*, (September 18), B8.

Rosenau, P. M. (1992). *Postmodernism and the Social Sciences*. Princeton: Princeton University Press.

Simon, D. (1996). *Elite Deviance*. 5th ed. Boston: Allyn & Bacon.

Stewart, J. B. (1991). *Den of Thieves*. New York: Simon & Schuster.

Stewart, J. B. (1997). *Blood Sport*. New York: Simon & Schuster.

Sutherland, E. H. (1949). *White Collar Crime*. New York: Holt, Rinehart & Winston.

Sutherland, E. H., &. Cressey, D. R. (1966). *Principles of Criminology*. 7th Edn. Philadelphia: J. P. Lippincott.

Thio, A. (1988). *Deviant Behavior*. 3rd ed. New York: Harper & Row.

Thompson, D. F. (1987). *Political Ethics and Public Office*. Cambridge, MA: Harvard University Press.

Tittle, C. R. (1995). *Control Balance: Toward a General Theory of Deviance*. Boulder, CO. Westview Press.

Vaughan, D. (1997). Situated Action, Theory Elaboration, and Boundary Work: An Agenda for the 21st Century. A Paper Presented at the Annual Meeting of the American Society of Criminology, San Diego, November.

Wills, G. (1996). The Clinton Scandals. *The New York Review of Books* (April 18): 59–67.

Wilson, J. Q., & Herrnstein, R. J. (1985). *Crime and Human Nature*. New York: Touchstone.

Woodward, B., & Bernstein, C. (1976). *The Final Days*. New York: Simon & Schuster.

BEYOND BLACK AND WHITE: ETHNOVIOLENCE BETWEEN OPPRESSED GROUPS

Barbara Perry

ABSTRACT

It is ironic that at the same time that policy makers, scholars and commentators point to the increasing diversity of the United States, they stubbornly persist in collapsing racial and cultural relations into a black-white binary. If we are to make sense of the current state of racial and cultural conflict, it is imperative that we broaden our understanding to recognize the United States for what it is: a multicultural, multiracial and multiethnic community, characterized by multiple and cross-cutting coalitions and cleavages. It is particularly important to acknowledge this in our conversations about hate crime, where minority on minority violence is not unheard of. Two recent conflicts highlight this often overlooked reality: the Crown Heights, NY violence between blacks and Jews in 1991; and the black-Asian-Hispanic conflicts which exploded in the Los Angeles riots of 1992. While obviously signs of the long-standing tensions among and between these similarly marginalized groups, these incidents represent efforts to negotiate identity and place in the United States. These events – and others like them – present the actors with opportunities to do difference through violence.

This paper explores the phenomenon of intercultural violence among and between oppressed groups. This task is made difficult by the lack of

Sociology of Crime, Law and Deviance, Volume 2, pages 301–323.

ISBN: 0-7623-0680-7

literature in this area. Thus what follows is a preliminary analysis, in which I have drawn from the fields of anthropology (e.g. Herdt, 1997; Almaguer, 1995), social work (e.g. Greene), and cultural studies (e.g. West, 1993, 1994; Hooks, 1992, 1994, 1996). Specifically, I argue that minority on minority hate crime can be understood as a resource for 'doing difference'. However, this bias motivated violence is played out within the broader context of white, male, heterosexual supremacy. Consequently, violence between oppressed groups is an attempt to sustain identity, and to attain some semblance of dominance on the part of groups who may otherwise be marginalized. As a test of my proposition, I will examine three illustrative sets of relationships: African American-Asian American; Jewish-African American; and gay men within communities of color.

DOING DIFFERENCE, DOING HATE CRIME

Race, gender, sexuality, and all those dimensions of difference which shape our social world, represent what West & Fenstermaker (1987, 1993) refer to as 'ongoing accomplishments' (see also Messerschmidt, 1993, 1997). Difference, and identity are created through conscious, reflective pursuit and must be established and re-established under varied conditions. In other words, identity-construction is an activity concerned with "managing situated conduct" (West & Zimmerman, 1987: 127), according to socially normative expectations of what constitutes the 'essence' of one's race or gender, for example.

The construction of identity is an interactional accomplishment by which actors perform their 'manliness' or 'womanliness', their 'whiteness' or 'blackness' or 'Asian-ness'. They do so with an eye to how their behavior will be interpreted or evaluated by others. Central to this conceptualization is the notion of 'accountability'. At all times, in all situations, actors are concerned with whether their behavior *will be seen to be* in accordance with approved standards for their assigned identity. Consequently,

> To the extent that members of society know their actions are accountable, they will design their actions in relation to how they might be seen and described by others (West & Fenstermaker, 1987: 25).

Since this enactment is situated within existing relations of power, the conduct will generally repeat and thus support those relations (Winant, 1997, 1998). Conventional culture is consumed with ensuring our awareness of and commitment to traditional notions of gender, sexuality, race, and so on. Movies, advertising, the legal order, even the organization of department stores take for granted the *essential* differences between groups noted earlier. It is in this context that we are continually expected to 'account' for our gendered behavior, for example.

Within the essentialist understanding of identities, there is very little space for ambiguity, or crossing the boundaries between categories of difference. Speaking of gender, specifically, West & Zimmerman (1987: 136) contend that

> A person engaged in virtually any activity may be held accountable for performance of that activity as a *woman* or a *man*, and their incumbency in one or the other sex category can be used to legitimate or discredit their other activities.

In other words, accountability involves the assessment of behavior as either conforming or deviating from culturally normative standards. Whenever we 'do difference' – which is a recurring effort – we leave ourselves open to reward or censure. So it is that we are discouraged from the "attempt to cross the line, to transgress, desert or quit" (Bourdieu, cited in Fine, 1997: 58).

To the extent that individuals or groups 'perform' in a way that corresponds to the 'mythical norm' or in ways that correspond to normative conceptions of one's identity construct, they are held to be doing difference appropriately (Messerschmidt, 1997). In so doing, they uphold the boundaries that separate them from the Other, and ultimately the social relations of power. Conversely, when individuals or groups cross those boundaries, when they fail to perform their identity in normative ways, they are held to be doing difference inappropriately, and thereby leave themselves open to censure. In situations such as this, where subordinate groups attempt to redefine difference they may become vulnerable to attack. With this in mind, Miller's (1995: 57) questions can be answered:

> . . . when does . . . confrontation with difference have negative effects: when does it lead to great difficulty, deterioration, and distortion, and to some of the worst forms of degradation, terror and violence – both for individuals and for groups – that human beings can experience?

The answer to Miller's question: when boundaries are threatened, when subordinate groups seek to redefine their place, when they do difference inappropriately. The threat must be repressed, and the dominance of the hegemonic group reaffirmed. It is in this context that hate crime emerges as a resource for doing difference, and punishing those who do difference inappropriately.

To summarize: when we do difference, when we engage in the process of identity formation, we do so within the confines of structural and institutional norms. In so doing – to the extent that we conform to normative conceptions of identity – we reinforce the structural order. However, not everyone always performs 'appropriately'. Frequently, we construct our gender, or race, or sexuality in ways that in fact challenge or threaten socio-cultural arrangements. We step out of line, cross sacred boundaries, or forget our 'place'. It is in such a context that hate crime often emerges as a means of responding to the threats. The tensions between actors may culminate in violent efforts to reassert the dominance of one, and realign the relative positions of both.

Moreover, it is important to keep in mind that identity is shaped relationally. Both the perpetrator and the victim of hate are continually engaged in the process of constructing their identities:

> . . . it is not only the racist or sexist who constructs difference, but the victim of each or both who seeks to create difference as well. At times, the 'victim' has done so in response to the racism and/or sexism in the society in order to survive, but at other times movements made up of these 'victims' have sought to redefine difference as part of a struggle for power and personhood (Rothenberg, 1992: 48).

Such alternative constructions of difference challenge the carefully molded perceptions of about how the world should be, and what each person's or each group's place is in that world. When confronted with such novelties, one means by which to 'put things right' is through violence. Consequently, hate crime provides a context in which the perpetrator can reassert his/her identity, and at the same time, punish the victim(s) for the individual or collective performance for his/her identity. In other words, hate motivated violence is used to sustain or create the privilege of one group, and to police the boundaries between groups by reminding the Other of his/her 'place'. Perpetrators thus recreate their own masculinity, or blackness, while punishing the victims for their deviant identity performance.

Hate crime, then, is a forceful illustration of what it is to engage in situated conduct. The interactions between groups provide a context in which both compete for the privilege to define difference in ways that either perpetuate or reconfigure hierarchies of social power. Simultaneous and oppositional efforts to do difference set up tensions in which the act of victimization co-constructs the victim and perpetrator. This confrontation is informed by the broader cultural and political arrangements which "allocate rights, privilege and prestige according to biological or social characteristics" (Sheffield, 1995: 438). Perpetrators attempt to reaffirm their collective identity, their access to resources and privilege, while at the same time limiting the opportunities of the victims to express their own needs. The performance of hate violence, then, confirms the 'natural' relations of difference.

CAN WE ALL GET ALONG?

The same diversity that threatens the white majority – and thus underlies hate crime – similarly causes ruptures and discomfort among and between this nation's oppressed groups. These ruptures ultimately revolve around identity and recognition. Yet such struggles for recognition take place on different terms viz. oppressed groups, as opposed to those involving white-minority relations. The white majority excludes or marginalizes subordinate groups. However, within the politics of interethnic conflicts, there emerges an opportunity to acquire or sustain recognition. The driving force here is that

> . . . while the subject desires recognition as human, capable of activity, full of hope and possibility, she receives from the dominant culture only the judgement that she is different, marked and inferior (Young, 1990: 60).

One way to overcome, indeed overturn, this negation is to extrapolate the 'rules of the game' to the context of subordinate ethnic conflict, to mark another as 'different, marked, inferior'. Only in this way can the subordinate establish some semblance of dominance, demanding of other oppressed groups that which is not forthcoming from the hegemonic majority. Punished, repressed and reprimanded for asserting their ethnic identity within view of the majority culture, members of subordinate groups can opt to engage in hate crime as an alternative resource for constructing their identities.

Consequently, minority on minority violence can be viewed within the framework of 'doing difference' (West & Zimmerman, 1987; West & Fenstermaker, 1995; Messerschmidt, 1993) since it too revolves around hierarchical conflict. Interethnic violence among and between subordinate groups "becomes a 'field of possibilities' for transcending class and race discrimination," i.e. a critical resource for doing race, in particular (Messerschmidt, 1993: 103). But it is important to interpret such violence within the master narrative of white, heterosexual, masculine hegemony. That is, minority on minority hate crime is not only about Korean-African American conflict, or African American-Jewish conflict. Rather, it is about how these tensions play out in the context of relations of racial/ethnic/gender subordination. As Ikemoto (1995: 307) contends, "If you experience racism as one marginalized by it, then you may use racism to explain your relationship with other groups and their members." Even in their relationships with one another, members of subordinate groups are "dependent on the will and leftovers of a dominant group" (Ikemoto, 1995: 308). Ultimately, hegemonic constructions of race or gender identity infuse the experiences and interactions of subordinate groups as well.

One might expect the common experiences and marginality of oppressed groups to provide the basis for solidarity rather than division. It is not unreasonable to suppose that they might recognize and exploit their common hardships. Gay men and lesbians share with recent immigrants and native born racial minorities discrimination, harassment, and violent victimization. All are subject to ongoing daily patterns of (mis)treatment which seek to maintain their inferior status.

In practice, what emerges is not always a shared commitment to racial or gender justice, but instead shared antagonisms and hostilities directed toward one another. Their rage at their continued disempowerment is misdirected downwards or sideways toward those who are similarly victimized, rather than upwards toward those who seek to exploit the cleavages. The 1992 Los Angeles riots were a clear illustration of this tendency, representing as they did

> . . . a multi-racial, trans-class, and largely male display of justified social rage. For all its ugly xenophobic resentment, its air of adolescent carnival, and its downright barbaric behavior, it signified the sense of powerlessness in American society (West, 1993: 255).

Ironically, the common 'powerlessness' becomes the basis for conflict rather than community. Even in Los Angeles, the combatants generally attacked one another, rather than posing any direct threat to their white oppressors.

Evidence of these divisions is apparent in opinion polls which illustrate the negative perceptions which oppressed groups hold of one another (Oliver & Johnson, 1984). There are chilling parallels between dominant and subordinate groups' readings of other minority groups. In fact, hegemonic systems of meaning construction are reproduced in the context of interethnic relations. Oliver & Johnson (1984) summarize research findings of interethnic antagonisms manifest in perceptions of power accruing to other minority groups, of unfair employment competition, and negative stereotyping. Similarly, Holmes (1994: B8) reports the findings of a national poll commissioned by the National Conference of Christians and Jews. The results suggest strongly that blacks, Asians and Hispanics generally hold even more negative views of one another than do whites. Forty-six percent of Hispanics and 42% of blacks saw Asians as "unscrupulous, crafty and devious in business." Sixty-eight percent of Asians and 49% of blacks agreed that Hispanics "tend to have bigger families than they are able to support." Thirty-one percent of Asians and 28% of Hispanics believed that blacks "want to live on welfare."

These findings are indicative of the hostilities between groups. Yet they are also indicative of the extent to which dominant world views permeate all members of society. They reinforce the assertion that the society of the United States is grounded in constantly shifting hierarchies of oppression. In other words, they 'revealed the power of racist rhetoric between politically, economically, and culturally disadvantaged groups' (Chun, 1996: 3). Perhaps we should not have expected solidarity after all. Subordinate groups are not immune to the power of hegemonic ideologies. They too are a crucial part of the audience, having listened to, observed and lived within the structures of inequality:

> We must remember that racial minorities having been socialized in a society that sees them as inferior to whites, are equally likely to believe in the inferiority of racial groups other than their own (McClain & Stewart, 1995: 149).

Even immigrants arrive here with pre-packaged ideas of how race and gender operate in the United States, having been 'informed' by American media outlets world-wide. This is especially important in understanding the relationship between African Americans and other minority groups. White supremacy is re-inscribed in the hostility with which other people of color greet blacks in this country:

> In race talk, the move into mainstream America always means buying into the notion of American blacks as the real aliens. Whatever the ethnicity or nationality of the immigrant, his nemesis is understood to be African American . . . Often people of color . . . hold black people responsible for the hostility they encounter from whites. It is as though they see blacks as acting in a manner that makes things hard for everybody else (hooks, 1995: 198–199).

This is what distinguishes minority relations from majority-minority relations: the sense that all Others are in competition for the favors of the white majority; they are in a struggle to assign blame for their relative positions of inferiority. The struggle for economic, political and cultural empowerment becomes a struggle to disempower 'the competition', through violence if necessary.

What follows is an exploratory attempt to make sense of intercultural violence between and among subordinate groups. I have chosen to explore three illustrative sets of relationships in order to focus my discussion. The group and inter-group dynamics explored here cross lines of race, ethnicity, class, gender and sexuality. The salience of each of these components differs, of course, from situation to situation, and from group to group. The cases I have chosen provide insight into very different patterns and relationships which nonetheless share common themes. Consequently, I examine in turn hate motivated violence among and between: African Americans and Asians; Jews and African Americans; and gay men within communities of color.

AFRICAN AMERICAN-ASIAN CONFLICT

The black and Hispanic victimization of Korean shop-owners in the aftermath of the Rodney King verdict illustrates the potential for violence when subordinate groups – especially those with a long history in the United States – are threatened by what is perceived as the empowerment of another oppressed group. Nowhere is the racial animus between blacks and Asians more apparent than in the low-income, predominantly black communities where so many Koreans have established 'mom-and-pop' businesses. As newcomers to the United States, Korean employment opportunities are often blocked by language, educational and, of course, racial barriers. They turn instead to self- and family- employment in the retail trade. In cities like New York, Los Angeles and Atlanta, upwards of 30% of Koreans are small business owners. In this role, they assume the role of commercial 'middle-men' between corporations reluctant to locate in the inner city, and their low-income, non-white clientele. The entrepreneurial stance adopted by Koreans is not welcomed by the African Americans inhabiting these communities. Quite the contrary: from the perspective of African Americans, Koreans are 'foreign' interlopers who have, first of all, forced out black business owners, and secondly, engaged in exploitive practices:

> In the space of a few years, the Korean newcomers have come to own most of the grocery shops, the shops selling wine and spirits, and the delicatessens in the predominantly black areas where they have settled and gone into business. In the process, they have become the target of angry protests and even violence by the black residents (Jo, 1992: 398).

In this light, Koreans are seen to have "jumped the queue" in the struggle for political and economic opportunities tantamount to success. This makes them vulnerable to the opprobrium and resentment of the oppressed communities they serve. This is clear in the response of one black youth asked in a television interview to explain the looting: "Because we hate 'em. Everybody hates them" (Frontline, April 27, 1993).

This racial animosity springs from a number of sources within the context of both African American and Asian American's efforts to construct their racial identities. Like all other members of United States society, each of these groups perceives the other through the lens of cultural mythologies. Asians see African Americans as criminals, as welfare cheats, as threats to their economic and physical well-being. African Americans see Asians as 'perpetual foreigners', as unsavory business people. These tendencies are exacerbated by ongoing media coverage which highlights the tensions rather than efforts at reconciliation. This was the case in the context of the 1990 boycott of Korean businesses in Flatbush, N.Y., where the media coverage was deemed to be "inflammatory and polarizing" and "overly simplistic and in some cases blatantly racist" (U.S. Commission on Civil Rights, 1992: 37). And it was also the case in the treatment of the Los Angeles riots in 1992, in which the mainstream media contributed to the animosity by

> . . . spotlighting tensions between African Americans and Koreans above all efforts to work together . . . (and) by exploiting racist stereotypes of Koreans as unfathomable aliens, this time wielding guns on rooftops and allegedly firing wildly into crowds (Kim, 1993: 221).

Such divisive mechanisms encourage blacks and Asians to adopt an oppositional stance relative to one another – they are portrayed as inevitable enemies rather than allies. And unfortunately, they all too often accept these externalized interpretations of their relationship, as evidenced by their long term animosities. Jo (1992) catalogs the charges and countercharges levied by blacks and Asians as each group attempts to construct itself as the wronged party. Korean shopowners in particular are said to be rude, exploitive, unwilling to hire blacks, and unfairly advantaged by government programs. African Americans are said to be unwilling to co-operate or understand, unreliable workers, criminal threats and unfairly advantaged by government programs (see also McClain & Stewart, 1995).

Black hostility toward Asian shop-owners has apparently inspired anti-Asian violence, as a means of recouping their lost prominence in the community. A seven block black neighborhood in Washington D.C. has seen nine fire-bombings of Korean businesses since 1984. In Los Angeles, more than 20 Korean shop-owners have been killed in black communities since 1990. During the Los Angeles riots, 300 Chinese businesses were looted and burned; in all, 40% of the businesses lost were Asian owned (Cho, 1993). Moreover, anti-Asian violence and harassment are endemic in these communities (U.S. Commission on Civil Rights, 1992).

In large measure, black perpetrators of anti-Asian hate crime are reacting to the particular way in which Asian Americans construct their racial identity in the context of a white supremacist culture. Asians' assigned roles as 'middle-men' is taken as a sign of their 'preferred status' in United States society. They are held accountable – and found guilty – as illegitimate interlopers, who entered 'the game' very late, yet nonetheless managed to vault over the heads of native-born African Americans. The 'appropriate' place of Asian immigrants – according to native-born African Americans – is at the end of the queue. In contrast, they are perceived to have relatively little problem in assimilating into the mainstream relative to the historical experiences of blacks. In spite of the fact that they are racially and ethnically 'not white', Asians are *seen to be* readily accepted by the white community. A community worker interviewed by Jo (1992: 405) declaimed that "Koreans think that they are white!"

The dual image of Asians as middle-men and as the model minority further divides them from African Americans. Freer (1994: 191) attributes the hostility of blacks toward Asians to the perception that Asian Americans collectively attain the American dream at the expense of black success. Asians' efforts to assimilate and advance are taken as arrogance that must be met with hostility, even violence. Perhaps even more than the dominant white culture, African American inner city poor are threatened by the apparent success of Asian Americans. In addition to the economic competition represented by the latter is the competition for place and status relative to the white power structure. Asian Americans are exploitable as a wedge against native born blacks who are told to look to Asians as an indication that the American dream is open to all, *if only you are willing to work for it.* Korean shopowners, Chinese entrepreneurs, and Japanese executives are held up as models of the potential for assimilation and advancement – with the implied message that African Americans just don't work hard enough. To the extent that Asian Americans are seen to enact this role, they are met with the hostility of blacks who do not fare well in the implied comparison. Consequently, Asians become caught between the racism of whites and the racialized resentment of blacks:

> The model minority myth plays a key role in establishing a racial hierarchy which denies the oppression of Asian Americans while simultaneously legitimizing the oppression of other racial minorities and poor whites (Chang, 1995: 329).

Violence is a readily available outlet for this misdirected hostility. It is a means of empowerment for blacks who see themselves losing ground relative to newly arrived immigrants – losing ground, that is, in economic, political and cultural terms. In light of the impoverishment and structural unemployment of so many black youths, in particular, alternative mechanisms for empowerment are rare. Consequently, violence directed towards Asian Americans is a simultaneous effort to reclaim some of this loss in status, while seeking to remind Asians of their appropriate place in the racialized pecking order.

JEWS AND AFRICAN AMERICANS

Joshua Price poignantly expresses one source of inter-group hostility among oppressed groups when he states that "Although Jewish, I understand myself as an almost-insider to Anglo culture in the United States" (cited in Lugones & Price, 1995: 113). He is not alone in seeing himself and American Jews as 'almost insiders', quite like Asians. And this provokes anger toward what many perceive as a 'privileged minority'. Price further admits to complicity in white dominance, in exchange for inclusion in their ranks. He too is asked – and often agrees – "to maintain solidarity and loyalty – often in order to break, exclude, violate, exploit and deny those people who are outside the inner circle" (Lugones & Price, 1995: 113). Much like African Americans throughout their history in the United States, Jews experience a 'double consciousness' described by Du Bois as the simultaneous construction of identity within the context of both dominant and subordinate cultures. Understandably, this causes insecurity and discomfort all around. Cornel West (1994: 111) also speaks to this uneasy positioning of Jews relative to subordinate blacks and dominant whites when he observes that

> The images of black activists yelling "Where's Hitler when we need him?" and "Heil Hitler," juxtaposed with those of David Duke celebrating Hitler's birthday seem to feed a single line of intolerance burning on both ends of the American candle, that threatens to consume us all.

More so even than Asians, American Jews occupy a paradoxical position in the racialized hierarchy of power and place. And, more so than Asians, they are seen as co-conspirators in the plot to maintain the subordination of blacks. Consequently, they are held accountable by black Americans to a construction of racial identity that is seen as oppositional and threatening. To an alarming extent in some quarters, Jews are held to be complicit in the formation and maintenance of a racialized hierarchy which subordinates blacks. Anti-Semitic violence then becomes a mechanism for transcending racial domination, and an important resource for constructing relational identities.

This uneasy relationship between blacks and Jews has a lengthy history, punctuated by periods of conciliation. In many ways, black anti-Semitism has paralleled that of the white Christian majority. Nineteenth century religious teachings portrayed Jews as protagonists, as in black catechisms, for example:

> Q. Who killed Jesus?
> A. The wicked Jews.
> Q. The wicked Jews grew angry with our Savior, and what did they do to Him?
> A. They crucified Him.

Similar sentiments are expressed in many black spirituals of the time, which make such claims as "Virgin Mary had one son, The cruel Jews had him hung" (Dinnerstein, 1994: 198). As often as not, prejudices grounded in religious teachings were accompanied by secular stereotypes which further vilified the 'wicked Jews' as greedy, insatiable and conniving in their quest for wealth. A turn of the century article in *Colored America* insisted that Jews were

> . . . parasitical and predatory rather than conservatory and constructive in tendencies – preying upon and devouring the substance of others (cited in Dinnerstein, 1994: 199).

Little has changed in this century. Black Americans still share with white Americans the perception of Jews as Christ killers, predators and greedy financiers:

> This is part of the way racism works – it is easier to scapegoat Jews . . . than to target larger structures of white supremacy. . . . It is a distortion of reality to act as though any form of

> black anti-Semitism, however virulent, exists in isolation from the anti-Semitism that is learned whenever anyone absorbs without question the values of mainstream white culture (hooks, 1995: 210).

Yet the relationship between blacks and Jews in the United States retains its own specificity in light of the relative economic and political positions of the two groups. On the one hand, both are 'not white' and therefore outsiders. But Jews are both 'not white' and 'white', or at least 'almost white'. By virtue of the latter, they are also insiders, sharing the white skin privilege – something African Americans can never accomplish.

In this vein, West (1993) identifies the predominant cultural and structural dynamics which condition relations between blacks and Jews, and which set the stage for interethnic violence. First, he contends, black anti-Semitism reflects black anti-whitism. Jewish complicity in the politics and economics of racism in the United States is seen to reinforce the subordination of the nation's black community. There is a sense among some blacks that Jews could only become 'white' in America because of the existence of blacks. Thus, Jewish-black relationships reproduce broader white-black relationships of power which assume a paternalistic, often exploitive face, in that they

> ... have almost always been as philanthropist to recipient, shopkeeper to customer, landlord to tenant, employer to employee, teacher to student, welfare worker to client, and so forth (Dinnerstein, 1994: 224).

The perceived role of Jews in the continued oppression of black America is taken as a sign of betrayal between 'natural allies'. West (1994) holds that blacks have heightened expectations of Jews in light of the similarity of their historical experiences of oppression, and earlier coalitions around civil rights. Blacks and Jews alike have experienced (at different times and places) enslavement, ghettoization, subjugation, diaspora and violence. In light of these similarities, then, black Americans are embittered by the perception that their allies have become their enemies. As evidence of this betrayal, they point to Jewish resistance to affirmative action and state social security provisions, for example. In a dialogic paper, Pogrebin & Hutchinson (1994: 219) declare:

> Now we often march in opposite directions or face each other across an abyss. Now our two communities clash regularly over issues of power, priorities, competitive oppression and conflicting self-interest.

This sense of betrayal is enhanced by the corresponding fear that Jews, like Asians, like Hispanics, like all other subordinate groups, have vaulted past African Americans in economic and political strength. The result is a case of "underdog resentment and envy directed at another underdog who has 'made it' in American society" (West, 1994: 151). Jewish appeals to the aforementioned similarity of oppression, then, ring hollow in the ears of African Americans who have been surpassed in power, left behind by Jewish success. Moreover, whatever measure of success achieved by Jews is seen as a further evidence of Jews as co-conspirators in white racism. It is success won, not through effort and initiative, but by white nepotism.

While the social and economic malaise of black Americans provides the background for black anti-Semitism, the Nation of Islam in recent years has provided its public forum. The rhetoric of Islam – as preached by such leaders as Farrakhan and Muhammad, for example – provides a menu of ideologies which co-construct blacks as the chosen people and Jews as the worst of all 'white Devils'. Just as white supremacist groups condition and encourage hate motivated sentiment and activity, so too does the anti-Semitism of Farrakhan and Muhammad. Jews are to be blamed for alcohol and drug abuse in black communities, for negative stereotypes of blacks in entertainment media, for black poverty. *The Secret Relationship Between Blacks and Jews* provides much of the fodder for this virulence, holding Jews accountable for slavery and black genocide in general. Muhammad's and Farrakhan's frequent references to 'Jew York City', 'Jewnited Nations' and 'Columbia Jewniversity' highlight the perception of the breadth of Jewish control and manipulation. In a recent interview, Farrakhan exploits the image of Jews as usurious leeches:

> In the '20s, '30s and '40s, up into the '50s, the Jews were the primary merchants in the black community. Wherever we were, there they were. What was their role? We bought food from them; we bought clothing from them; we bought furniture from them; we rented from them . . . Sucking the lifeblood from our own community (Farrakhan, 1996: 53).

In the context of the structural and cultural relationship between blacks and Jews, interethnic violence persists as a resource for the establishment and re-establishment of racial and ethnic identity. This is especially important for poor African American youth who lack access to alternative means by which to compete with Jews. Resentment toward Jewish progress is manifest in bias-motivated harassment and victimization which enhances that status of the perpetrator(s), while seeking to disempower the victim(s). It attests to the group alliance of the perpetrators, and especially to their 'insider' status. Conversely, it reconstructs the Jewish victims as perpetual 'outsiders'. This is a curious – but important – inversion of the groups' relative positionality in the broader culture, where blacks are always the outsiders, while Jews are seen to travel the boundaries between insider and outsider at will. This is at the heart of what Steele (1994: 180) refers to as the 'unseen problem' between blacks and Jews: the "presumption by the larger society that we make up a brotherhood of outsiders . . . and we fight against each other to prove it wrong, to show that we have no such brotherhood."

These dynamics were readily apparent in the violence that followed the death of Gavin Cato in Crown Heights, N.Y. The long simmering antagonism toward the neighborhood's Hasidic Jews boiled over into a week of violence, which included the stabbing death of Yankel Rosenbaum. Here, too, the black community espied evidence of preferential treatment and white racism, as when Hasidim were accompanied by police escorts, or allowed to organize street patrols which harassed black community members. In response, spurred by Cato's death, blacks sought to preserve both the geographical and racialized boundaries between themselves and the 'Other' by violence, harassment and vandalism accompanied by exclusionary messages that included "Heil Hitler" and "Get the Jews Out."

For the black community of Crown Heights, the Hasidim were "out of place" both in geographical and political terms. They had betrayed their 'allies' by siding with the white majority. Consequently, the killing of Cato lit a fuse that had long laid in wait. His death provided the context and the motive for black demonstrations of racial solidarity, even across ethnic groups (e.g. Haitians, Jamaicans).

To humiliate, devalue and victimize the identifiable 'white Jewish devil' is to simultaneously proclaim the positive collective identity of the African American, very much in contrast to the negative identity generally assigned blacks in this culture. It is, moreover, a means of distancing oneself from the Other. Anti-Semitic violence reinforces the differences rather than the similarities between these two differently oppressed groups. It reasserts the particular and unique suffering of blacks in the United States. There is a certain resentment toward the Jewish assumption of share oppression. The language and activities of anti-Semitism make that resentment clear (Lester, 1994: 172).

GAY MEN AND COMMUNITIES OF COLOR

Intercultural violence is not necessarily restricted to conflict between ethnic and racial groups. Anti-gay violence cross-cuts ethnicity. Violence perpetrated against gays by men of color crosses the axes of race, gender and sexuality. In other words, regardless of race or ethnicity, masculinity, in particular, assumes heterosexuality. For example, Hooks (1990, 1992, 1994, 1995) returns again and again to the tendency of black males to demand compulsory heterosexuality. In this sense, gay men, especially black and Chicano gay men, are further marginalized within their racial communities, since they are simultaneously 'race traitors' and 'gender traitors'. Hooks (1992: 113) cites the lamentation of a black gay writer, who claims that "nobody wants to know my name, or hear my voice." Gay men of color are 'outsiders' on both the axes of gender and racial identity.

The complexity of attempts to simultaneously negotiate ethnic and sexual identities remains dramatically underdeveloped both theoretically and empirically. What follows, then, is a tentative exploration of how men of color may or may not use 'gay-bashing' as a resource in their own performance of a racialized gender identity. Again, as with interracial violence, it is imperative to consider anti-gay violence by people of color in the context of broader patterns of patriarchal white supremacy. It is these interlocking structures of inequality that condition and delimit the resources available to men of color as they seek to establish themselves in United States society. Yet the impact of these broader patterns on each cultural group retains a specificity, depending on historical and cultural features of each. For far too many gay men and women, being black or Jewish or Latino among like others offers no refuge when one is gay. Where they might expect solidarity on the basis of their ethnic identity, they may instead suffer stigmatization, persecution and violence on the basis on their performance of their gender and sexual identity.

Many traditional Native American cultures would have no call to denigrate homosexuality in their midst. More so than most groups in the United States, Native Americans often hold to a flexible, fluid conceptualization of sexuality. Behavior is more likely to be evaluated according to the appropriateness of the context, rather than the behavior itself. Thus, there is no rigid prescription against homosexuality which might otherwise result in its violent suppression. On the contrary, many Native traditions refer to the 'Two Spirited' as one who is valued because of his/her inherent combination of both the male and female spirits. This dualism allows the Two Spirited to see and perceive the world from a much clearer, much more complete perspective. Rather than defining sexuality in binary terms then, many Native cultures see one's sexual identity along a mutable continuum:

> If one takes the line of male/female, gay/straight, and bends it into a circle, there are an infinite number of points. Just so, there are theoretically an infinite number of points of gender and sexual identity for an individual that can shift and differ over time and location (Tafoya, 1997: 8).

Some Asian cultures share elements of this tolerant outlook on sexual diversity. Historically, same sex relationships have permeated the upper echelons of Japanese society, including the wealthy urban classes, Buddhist clergy and the military. In fact, *nanshoku* ('male colors') or *shudo* ('way of companions') was so intimately connected to the warrior society that it was often refered to as the 'past-time' of the Samurai – a manly bunch if ever there was one! Clearly, then, masculinity did not require unfailing performances of heterosexuality.

However, while this tradition was readily accepted for centuries, it seems to have become latent since the turn of the century, a phenomenon Miller (1995) attributes to the Westernization of Japan – a process which included a transition in sexual morality.

Nonetheless, there remains a significant distinction between Asian and Western reactions to homosexuality. Miller (1995) and Greene (1997) both assert that Asian resistance is grounded, not in homophobia, nor in heterosexism, but in pressure to marry. Asian American men, in particular, are held accountable to family rather than gender expectations. Homosexuality is thus a punishable threat to the family line and name. Where a male renounces this obligation and traditional obedience to the family, efforts to regulate his behavior may include violence, even death. Hence, this violence constitutes an attempt on the part of male family members (not judgmental strangers) to reassert their dominance in the familial relationship; at the same time, it is a penalty intended to ensure compliance with the interests of the family. Consequently, anti-gay violence in this context is not an attempt by strangers to assert their aggressive masculinity. Rather it is a weapon for ensuring family honor, and thereby ensuring cultural continuity. This may be especially important for Asians living in the United States, where the Western culture generally poses a threat to the continuation of the traditional Asian family line and values. Anti-gay violence, then, enforces conformity to cultural norms of both sexuality and family.

Family also provides a context for anti-gay violence among Latinos, yet in a different – and generally more intense – manner (Greene, 1997). In traditional Latino cultures, masculinity is rigidly enacted through the patriarchal family, wherein masculinity is associated with roles of provider and protector. Homosexual men, on the other hand, are regarded as effete, and both incapable and unwilling to assume these roles. Consequently, they are labeled as traitors to the family as well as to the culture itself. Moraga (1996: 299) argues that the rigidity of Latino conceptions of masculinity is stronger than virtually any other culture in the United States:

> Because they (gay Latino men) are deemed inferior for not fulfilling the traditional role of men, they are more marginalized from mainstream heterosexual society than other gay men and are especially vulnerable to male violence.

As a result, gay or effeminate Latino men are subject to persistent violence as punishment for their betrayal. They have transgressed an inviolate boundary, thereby threatening the solidarity and dominance of Latino men as a group.

However, this virulent homophobia is not without its contradictions, especially among Chicano men. Drawing on anthropological evidence, Almaguer (1995) asserts that the Chicano understanding of homosexuality revolves around sexual acts rather than sexual preferences per se. Specifically, a distinction is drawn between *activo* and *pasivo*, with stigmatization and ridicule reserved for the *pasivo*. The latter is deemed to be enacting a passive, subservient, feminine identity, very much out of line with the favored *activo*, who is by definition active, aggressive and very 'male'. The *activo* may in fact gain status through his dominance of the weaker recipient who is judged to be a feminized man: biologically male, but not really a man (Almaguer, 1995).

The *pasivo*, or more vulgarly termed *joto* or *puto*, does not meet the standards established for the construction of the aggressive, heterosexual dominant male. On the contrary, he is accused of having betrayed the Chicano's prescribed gender and sexual performance. As a consequence of this, he is constantly vulnerable to the violence of other Chicanos, who are simultaneously affirming their solidarity for all to see. It is left to the perpetrators of anti-gay violence to reassert masculine privilege and heterosexuality by the negation of the threat:

> The openly effeminate Chicano gay man's rejection of heterosexuality is typically seen as a fundamental betrayal of Chicano patriarchal cultural norms. He is viewed as having turned his back on the male role that privileges Chicano men . . . Those who reject these male prerogatives are viewed as non-men (Almaguer, 1995: 425).

The dynamics of Latino intolerance for homosexual men take on a special significance in the United States. Here, traditional resources for enacting masculinity are limited by structures of inequality – racism and classism – which inhibit the Latino male's ability to express his manhood through the familial roles of provider and protector. Elevated rates of unemployment, underemployment and impoverishment have meant that many of Mexican, Puerto Rican or Cuban heritage, for example, continue to find that their ability to support a family is dramatically undermined (Feagin & Feagin, 1996). This predisposes them to judge harshly those who 'choose' not to struggle beside them to preserve the family, and concomitantly, the culture. Moreover, the curtailed economic success leaves aggressive heterosexuality as a paramount residual means through which to construct their manliness. As an activity, anti-gay violence is tailor-made for this construction of masculinity, since it allows the visible demonstration of the most salient features of manliness: aggression, domination and heterosexuality.

If this holds true for Latino males, it is perhaps doubly true for African American men whose capacity to enact traditional patriarchal masculinity has been even more thoroughly circumscribed. Greene (1997) postulates that homophobia among African Americans generally may be more pronounced since it is multiply determined by sexism, Christian religiosity (especially Southern Baptist), and external and internalized racism. By virtue of their long experiences within a white Christian society, African Americans have also internalized the norms and values associated with white patriarchal notions of masculinity; yet by virtue of their class and race subordination, poor black male youths, in particular, do not have access to the resources by which they might 'appropriately' enact the hegemonic form of masculinity. Consequently, African American males have adopted alternative versions of what it is to be 'a man'. This perspective is aptly summed up in Majors & Billson's (1992: 1) analysis of the 'cool pose' as a resource for constructing an aggressive masculinity in the face of ongoing racism:

> African American men have defined manhood in terms familiar to white men: breadwinner, provider, procreator, protector. Unlike white men, however, blacks have not had consistent access to the same means to fulfill their dreams of masculinity and success.

Under advanced capitalism, the ability of poor black males to perform as breadwinners or providers is limited by wage and employment structures which marginalize them. They must, then, search elsewhere for the resources through which they can assert their manliness. Disempowered politically, economically and socially, young black males must express their capacity, their power as males' through alternate means. Having been denied the typical avenues by which to establish masculinity, they must nonetheless constantly prove to themselves and to others that they are men. Hooks (1992), Oliver (1988) and Majors & Billson (1992) all concur that for many, this is achieved through the performance of 'compulsive masculinity' wherein

> . . . typical masculine values become a rigid prescription for toughness, sexual promiscuity, manipulation, thrill-seeking and a willingness to use violence to resolve interpersonal conflict (Majors & Billson, 1992: 34).

Especially for marginalized black men on the street, violence becomes an important resource for "masculine expression and validation." Consequently, an aggressive stance relative to other marginalized groups – including gay men – is a "means by which to gain recognition and to reward one's masculinity, a means with which to solve the gender problem of accountability" (Messerschmidt, 1993: 111).

Moreover, a central facet of this form of identity construction is adherence to a similarly 'compulsive' and compulsory heterosexuality. As Messerschmidt (1993: 115) asserts, for white and black men alike, "heterosexual intercourse is the hallmark of one's identity as a man and inadequate performance severely threatens one's masculinity." We might contrast this to masculinity enacted through 'patriarchal status' or a generalized assertion of power on the basis of maleness. The alternative 'phallocentric masculinity' implies that "what a male does with his penis becomes a greater and certainly more accessible way to assert masculine status" (Hooks, 1992: 94). For many young men, black masculinity demands the performance and policing of a narrowly defined version of manliness which requires aggressive heterosexuality. This form of masculinity might find expression in sexual promiscuity, sexual assaults, domestic violence, and, most notably for present purposes, anti-gay violence. The latter, in this context, becomes a means of performing accountable gender identity.

Violence against gay black men is a corollary of the cool pose. As such, it is "about how black males have created a tool for hammering masculinity out of the bronze of their daily lives" (Majors & Billson, 1992: 1). It is a means of asserting the perpetrator's identity for all to see and evaluate – he is a manly man, virile, strong, heterosexual and in control of at least these aspects of his life. If he is to be held accountable to his gender identity – given the limits imposed by his racial identity – at least he can make the claim to have acted in accordance with the prerequisites of aggressive heterosexuality by simultaneously engaging in violence himself, and punishing the 'unmanly' other. I am not implying here that all straight black men beat gay men of color. Rather, what I am suggesting is that this is but one alternative resource for achieving gender accountability when other 'legitimate' means such as employment, professional status, or political office holding are closed off. It is a situational option, dependent upon ease of access to other equally illegitimate means such as drug sales, or domestic violence. Anti-gay violence is most likely to emerge in a group setting where homosocial bonding paradoxically requires reaction against homosexual threats.

Conversely, violence against gay black males also shapes the victim's identity. James Baldwin once claimed that being attacked by white people – on account of his refusal to be a 'good Negro' – only made him angry, whereas being attacked by black people – on account of his refusal to be a 'real man' – made him want to cry. Such victimization is a rejection, an ostracism, a penalty. This is echoed by West's (1994: 129) observation that gay black men

> ... reject the major stylistic option of black machismo identity, yet are marginalized in white America and penalized in black America for doing so. In their efforts to be themselves, they are not really 'black men', not machismo-identified.

The youth on the street who punishes the gay black male is holding his victim accountable to a particular racialized gender performance. He is seeking to police the boundaries of sexuality that are defined for him in the dominant black institutions. He is reading the script written by the A. M. E. Church, which has publicly denounced homosexuality, or the Howard University newspaper which referred to black gay males as 'freaks' engaged in 'depravity'. Gay black men are seen to have rejected these censures, and are therefore vulnerable to violent attempts to make them repent and realign themselves with their brothers. Failing this, anti-gay violence alternatively reminds gay black males that they are traitors to their race and their gender, and therefore reviled.

CONCLUSION: BUILDING BRIDGES

These preliminary remarks were intended to draw attention to the reality of intercultural violence among and between oppressed groups. I have suggested that such intercultural violence may be seen as efforts to police boundaries between groups, thereby enhancing the solidarity and privilege of the perpetrator's reference community, and simultaneously minimizing the power of the victim group. Deprived of the unobstructed right to enact a forceful identity within view of the majority culture, members of subordinate groups may turn to hate crime as an alternative resource for constructing their collective identities. Disempowered economically, socially and politically, minority groups struggle between themselves for ascendancy, often in violent and repressive ways.

The irony of this particular manifestation of identity construction is that the perpetrator and the victim have often experienced a similarity (but not sameness) of oppression. In other words, blacks, Jews, Asians, homosexuals and others not explicitly noted herein have all suffered various degrees of discrimination and victimization. Yet rather than acknowledging this and forming coalitions, they have often resorted to conflict among themselves. Freer (1994: 174) provides an apt summary:

> Competition for scarce resources is akin to metaphorical competition over a single slice of an entire pie. Too often the fight between two individuals, or in this case groups, over that single slice detracts from any efforts to enlarge the pie itself, or to question the distribution scheme in the first place.

Contending minority groups have so internalized the dominant aspects of white masculine supremacy that this is the only lens through which they can view one another. In such a context, it becomes difficult to imagine how the 'pie' might otherwise be shaped and shared.

Consequently, the interethnic alliances necessary to minimize interethnic violence rest on practices that empower all minority groups in such a way as to lift all together, rather than some at the expense of others. In other words, such strategies must be "transformative rather than simply effective in reducing tensions or addressing particular problems" (Okazawa-Rey & Wong, 1996: 35). Energies must be devoted to the identification and acknowledgement of what these communities share rather than what divides them. All too often this will mean recognizing the common economic, political, and social oppression have historically threatened minority groups in the United States. For example, while the Los Angeles riots of 1992 revealed the multiple fractures among and between minority groups, it was also a pivotal point for the mobilization of inter-ethnic coalitions. Unlike any prior urban 'race riot', the events leading up to and following the Los Angeles made visible the antipathy of white citizenry and public officials toward all minority groups. Consequently, they renewed and gave birth to an array of – albeit tenuous – coalitions, such as the Multicultural Collaborative. It is of course unfortunate that it took such a tragic series of events to inspire this co-operation; nonetheless, these organizations suggest models for future collaboration that crosses racial, ethnic, gender and class lines. If a city so fraught with conflict and violence can engender inter-group collaboration, then coalitions elsewhere must not be inconceivable. Moreover, coalition building can go far in establishing a dynamics of identity construction that is less competitive, less of a zero-sum game.

Successful transformation also requires that scholars and activists consistently look "beyond black and white" to recognize the layers of inter-racial and inter-ethnic relationships that characterize the contemporary United States. Changing demographics and shifting power bases demand that we widen our understanding of difference in this country. Both inter- and intra-group cleavages attest to the complexity of ethnic relations as we enter the twenty-first century. Consequently, "a serious effort must be made to incorporate the histories and contemporary experiences of people of color between the two poles of Black and white on the racial spectrum" (Cho, 1993: 207). Moreover, as argued herein, this incorporation must move beyond even the inclusion of diverse racial identities to account for the experiences of those whose identities are simultaneously grounded in gender and class categories, since 'doing difference' subsumes multiple identities.

ACKNOWLEDGMENTS

The author wishes to thank Alex Alvarez for his thoughtful feedback and advice, and an anonymous reviewer for *Sociology of Crime, Law and Deviance*.

REFERENCES

Almaguer, T. (1995). Chicano Men: A Cartography of Homosexual Identity and Behavior. In: M. Kimmel & M. Messner (Eds), *Men's Lives* (pp. 418–431). Boston, MA: Allyn and Bacon.

Chang, R. (1995). Toward and Asian American Legal Scholarship: Critical Race Theory, Post-Structuralism and Narrative Space. In: R. Delgado (Ed.), *Critical Race Theory* (pp. 322–336). Philadelphia PA: Temple University Press.

Cho, S. (1993). Korean Americans vs. African Americans: Conflict and Construction. In: R. Gooding-Williams (Ed.), *Reading Rodney King, Reading Urban Uprisings* (pp. 196–211). New York NY: Routledge.

Chun, Je. (1996). Color of Racism. *WE Magazine* (online), www.asiandir.com/we.march96/color.html.

Connell, R. (1987). *Gender and Power.* Stanford, CA: Stanford University Press.

Dinnerstein, L. (1994). *Anti-Semitism in America.* New York, NY: Oxford University Press.

Farrakhan, L. (1996). Black Muslims are Telling the Truth About Jews. In: P. Winters (Ed.), *Race Relations: Opposing Viewpoints* (pp. 50–54). San Diego, CA: Greenhaven Press.

Feagin, J., & Feagin, C. B. (1996). *Racial and Ethnic Relations.* Upper Saddle River, NJ: Prentice-Hall.

Fine, M. (1997). Witnessing Whiteness. In: M. Fine, L. Weis, L. Powell, & L. Mun Wong (Eds), *Off-White: Readings on Race, Power and Society* (pp. 57–65). New York, NY: Routledge.

Freer, R. (1994). Black-Korean Conflict. In: M. Baldassare (Ed.), *The Los Angeles Riots* (pp. 175–203). Boulder, CO: Westview.

Frontline (1993). *Is LA Burning?*

Greene, B. (1997). Ethnic Minority Lesbians and Gay Men: Mental Health and Treatment Issues. In: B. Greene (Ed.), *Ethnic and Cultural Diversity Among Lesbians and Gay Men* (pp. 216–239). Thousand Oaks, CA: Sage.

Herdt, G. (1997). *Same Sex, Different Cultures.* Boulder, CO: Westview.

Holmes, S. (1994). Survey Finds Minorities Resent One Another Almost As Much as They Do Whites. *New York Times*, B8.

Hooks, B. (1995). *Killing Rage.* New York, NY: Henry Holt and Co.

Hooks, B. (1994). *Outlaw Culture.* New York, NY: Routledge.

Hooks, B. (1992). *Black Looks.* Boston MA: South End Press.

Hooks, B. (1990). *Yearning.* Boston, MA: South End Press.

Ikemoto, L. (1995). Traces of the Master Narrative in the Story of African American/Korean American Conflict: How We Constructed Los Angeles. In: R. Delgado (Ed.), *Critical Race Theory* (pp. 305–315). Philadelphia, PA: Temple University Press.

Jo, M. H. (1992). Korean Merchants in the Black Community: Prejudice Among the Victims of Prejudice. *Ethnic and Racial Studies, 15*(3), 395–411.

Kim, E. (1993). Home is Where the *Han* Is: A Korean American Perspective on the Los Angeles Upheavals. In: R. Gooding-Williams (Ed.), *Reading Rodney King, Reading Urban Uprisings* (pp. 215–235). New York, NY: Routledge.

Lester, J. (1994). The Lives People Live. In: P. Berman (Ed.), *Blacks and Jews: Alliances and Arguments* (pp. 164–177). New York, NY: Delta.

Lugones, M., & Price, J. (1995). Dominant Culture: El Deseo por un Alma Pobre. In: D. Harris (Ed.), *Multiculturalism From the Margins* (pp. 103–128). Westport, CT: Bergin and Garvey.

Majors, R., & Billson, J. M. (1992). *Cool Pose.* New York, NY: Touchstone Books.

McClain, P., & Stuart, J. (1995). *Can We All Get Along? Racial and Ethnic Minorities in American Politics.* Boulder, CO: Westview.

Messerschmidt, J. (1997). *Crime As Structured Action.* Thousand Oaks, CA: Sage.

Messerschmidt, J. (1993). *Masculinities and Crime.* Lanham, MD: Rowman & Littlefield.

Miller, N. (1995). *Out of the Past: Gay and Lesbian History from 1969 to the Present*. New York, NY: Vintage Press.

Moraga, C. (1996). Queer Aztlán: The Reformation of Chicano Tribe. In: D. Morton (Ed.), *The Material Queer* (pp. 297–304). Boulder, CO: Westview.

Oliver, M., & Johnson, J. (1984). Interethnic Conflict in an Urban Ghetto: The Case of Blacks and Latinos in Los Angeles. *Research in Social Movements, Conflict and Change*, *6*, 57–94.

Oliver, W. (1988). The Symbolic Display fo Compulsive Masculinity in the Lower-class Black Bar. Unpublished manuscript, University of Delaware, Newark, DE.

Pogrebin, L. C., & Hutchinson, E. O. (1994). A Dialogue on Black-Jewish Relations. In: F. Pincus & H. Ehrlich (Eds), *Race and Ethnic Conflict* (pp. 219–226). Boulder, CO: Westview.

Rothenberg, P. (1992). The Construction, Deconstruction and Reconstruction of Difference. In: R. Baird & S. Rosenbaum (Eds), *Bigotry, Prejudice and Hatred* (pp. 47–64). Buffalo, NY: Prometheus Books.

Sheffield, C. (1995). Hate Violence. In: P. Rothenberg (Ed.), *Race, Class and Gender in the United States* (pp. 432–441). New York, NY: St. Martin's Press.

Steele, S. (1994). Breaking Our Bond of Shame. In: P. Berman (Ed.), *Blacks and Jews: Alliances and Arguments* (pp. 178–180). New York, NY: Delta.

Tafoya, T. (1997). Native Gay and Lesbian Issues: Two Spirited. In: B. Greene (Ed.), *Ethnic and Cultural Diversity Among Lesbians and Gay Men* (pp. 1–10). Thousand Oaks, CA: Sage.

United States Commission on Civil Rights (1992). *Civil Rights Issues Facing Asian Americans in the 1990s*.

West, C., & Fenstermaker, S. (1995). Doing Difference. *Gender and Society*, *9*(1), 8–37.

West, C., & Fenstermaker, S. (1993). Power, Inequality and the Accomplishment of Gender: An Ethnomethodological View. In: Paula England (Ed.), *Theory on Gender/Feminism on Theory* (pp. 151–174). New York, NY: Aldine de Gruyter.

West, C., & Zimmerman, D. (1987). Doing Gender. *Gender and Society*, *1*(2), 125–151.

West, C. (1994). *Race Matters*. New York, NY: Vintage Books.

West, C. (1993). Learning to Talk Race. In: R. Gooding-Williams (Ed.), *Reading Rodney King, Reading Urban Uprisings* (pp. 255–260). New York, NY: Routledge.

Winant, H. (1998). Racial Dualism at Century's End. In: W. Lubiano (Ed.), *The House that Race Built* (pp. 87–115). New York NY: Vintage.

Winant, H. (1997). Where Culture Meets Structure: Race in the 1990s. In: D. Kendall (Ed.), *Race, Class and Gender in a Diverse Society* (pp. 26–38). Boston, MA: Allyn and Bacon.

Young, I. M. (1990). *Justice and the Politics of Difference*. Princeton, NJ: Princeton University Press.

EXPLAINING VIOLENCE, SUBSTANCE ABUSE, AND PERSISTENT VIOLENCE AMONG MEN: ELABORATING A SIDE-BY-SIDE INTEGRATIVE MODEL OF FOUR THEORETICAL PERSPECTIVES

Rebecca S. Katz

ABSTRACT

This chapter elaborates on the continuing proposal for the theoretical integration of a number of perspectives traditionally not associated with one another (see Katz, 1999). Separately, each perspective intuitively maintains a great deal of interest although only one has been consistently tested by criminologists. This integrative approach links James Messerschmidt's structured action theory, the age-graded life course theory (Sampson & Laub, 1993), Gottfredson & Hirschi's general theory of crime (1990) and John Bowlby's attachment theory (Bowlby, 1949). Such an integrative approach may explain Sampson & Laub's qualitative findings that marital attachments failed to lead substance-abusing men towards desistance but rather led them toward continued violent offending against their wives. This proposed side-by-side integration posits that substance abusing men will remain unaffected by marital attachments as the result of early gendered socialization and attachment processes which promote

Sociology of Crime, Law and Deviance, Volume 2, pages 325–342.

ISBN: 0-7623-0680-7

doing hegemonic masculinity (Bird, 1996; Connell, 1995; Messerschmidt, 1993, 1997; Stark & Flitcraft, 1996).

INTRODUCTION

R. W. Connell's theoretical work explains that gender is not fixed but "constructed in interaction." Connell's model of masculinity is a developmental approach to the understanding of gendered behavior as it occurs within specific contexts (Connell, 1995, p. 35). Such a gendered approach to explaining crime is necessary since men and boys commit most crime (Messerschmidt, 1993, 1997). Connell's model may be integrated in a side-by-side fashion to a number of other theoretical perspectives already integrated in such a fashion (see Katz, 1999). These other theories include Gottfredson & Hirschi's theory of self-control, Bowlby's attachment theory, and Sampson & Laub's life-course perspective. This integrated model may lead criminologists toward a clearer understanding of the relationship between the gendered nature of early and later attachment relationships and the precise nature of the nexus between violence and substance abuse (Chodorow, 1978). Finally, such an approach may bridge the gap between work examining violence committed by men against women and violence committed by men against other men (Russell, 1998; Schwartz & DeKeseredy, 1998; Websdale, 1998; Holtzworth-Munroe, Stuart & Hutchinson, 1997)

SIDE-BY-SIDE INTEGRATION OF ATTACHMENT THEORY, GOTTFREDSON AND HIRSCHI'S GENERAL THEORY OF CRIME, AND STRUCTURED ACTION THEORY

John Bowlby's attachment theory assumes that the development of an attachment between the primary caretaker and the child is the basic foundation of all future development (Bowlby, 1969, 1988; Horner, 1991; Katz, 1999). The primary caretaker facilitates this attachment process through the provision of safety and security to the infant through meeting the child's needs (Bowlby, 1988: 27). A number of studies substantiate this process beginning at birth which results in a secure attachment style by the end of the second year of life (Bowlby, 1988: 11; Isabella, Belsky & von Eye, 1989; Ainsworth, Walters & Wall, 1978; Greenberg, Cicchetti & Cummings, 1992, Howing, Wodarski, Kurtz & Gaudin, 1993; Jacobson, Huss & Fendrich, 1997; Moffitt, 1997). Unfortunately, most sociological research ignores early infant attachment in attempting to explain deviant behavior. However, Gottfredson & Hirschi's general theory of crime contains a similar construct, the emotional investment of the parent in the child (see Katz, 1999). This emotional investment sets into motion other parenting behaviors such as, monitoring the child's behavior, recognizing deviant behavior when it occurs and then punishing the child (Gottfredson & Hirschi, 1990: 97). Although Gottfredson & Hirschi do not explain this emotional investment process, Bowlby's attachment theory clearly does.

Extant research supports this connection between the above two perspectives. Specifically, recent longitudinal research in Denmark substantiates that early secure attachment leads to the ability to delay gratification, a key component of self-control (Jacobson, et al., 1997). Furthermore, an insecure attachment style has been clearly linked to later involvement in violent behavior (Raine, Brennan & Mednick, 1997). Thus an insecure attachment fails to engage the other parental management behaviors discussed in Gottfredson and Hirschi's theory and results in low self-control. First linking Bowlby's attachment theory to Gottfredson and Hirschi's perspective by using a side-by-side model may provide a more succinct explanation for the development of self-control. Such a goal in integration allows partial theories to be placed side-by-side and segregates the cases to which they are considered applicable (Elliot, Ageton & Canter, 1979; Hirschi, 1979). It is not considered comprehensive integration since the compatibility of theoretical assumptions is not addressed (Liska, Krohn & Messner, 1989). This side-by-side approach links theories through their common theoretical constructs, a secure infant-parent attachment, and the emotional investment of the parent in the child. Secondly, such integration may eliminate some noted problems with Gottfredson and Hirschi's theory. One of problems with their perspective, illustrated by Longshore and colleagues recent criticism, is that although a general theory of crime predicts crimes of force and fraud, especially among men 'self-control seems not to improve predictive power or conceptual clarity regarding the etiology of crime' (Longshore et al., 1998; Grasmick, Tittle, Bursik & Arneklev, 1993; Nagin & Paternoster, 1993; Gibbs, Giever, Giever & Martin,1998; Keane, Maxim & Teevan, 1993; Evans, Cullen & Burton, 1997; Lynam, Moffitt & Stouthamer-Loeber, 1993; Krueger Avashalom, Caspi, Moffitt, White & Stouthamer-Loeber, 1996; Longshore, Turner & Stein, 1996; Marenin & Reisig, 1995; Piquero & Rossay, 1998). This seems to be true even when self-control is measured as it interacts with the opportunity to commit crime (Longshore, 1998). Similarly, LaGrange & Silverman's recent analysis reveals that while some dimensions of self-control have variant effects on males and females, other dimensions have invariant effects (LaGrange & Silverman, 1999). But even LaGrange & Silverman's work, which discussed the need to take into account differential gender socialization practices, did not do so. Further they failed to account for the emotional investment of the parent in the child and for the personality characteristic of empathy. However, this is not necessarily any fault of their own because the second problem in Gottfredson & Hirschi's theory is the failure to account for the gendered nature of the development of self-control. Furthermore Gottfredson & Hirschi also fail to take into account the gendered nature of the emotional investment of the parent in the child (see LaGrange & Silverman, 1999). In our society, mothers continue to be the primary caretakers of children, and this emotional investment process or the early infant-parent attachment process clearly differs in accordance with the gender of the child. These gendered early infant-parent attachment processes differentially effect men's and women's later attachments, empathy and criminal behavior (Connell, 1995; Chodorow, 1978; Messerschmidt, 1993, 1997; Holtzworth-Munroe, et al., 1997). Linking the gendered attachment process to the ideology of masculinity, may provide fertile ground for the development of an improved theory of the gendered nature of all criminal behavior.

GENDERED ATTACHMENT PROCESSES AND IDENTITY DEVELOPMENT

Nancy Chodorow theorized that the attachment to a primary parental figure in infancy was a distinctively unique process for males. She hypothesized that males were pushed to separate early from the mother in order to development what is defined as an appropriate masculine identity while females separated and individuated from mothers much later (Chodorow, 1978; Erickson, 1950). Compelling evidence reveals some truth to these claims. Specifically, research shows that stages five and six of Eric Erickson's model of psychosocial development, are different for each gender. Males form identity first (as in Erikson's original model) and then move on to experience intimacy. Whereas females experience the intimacy stage first and then maneuver through identity development (Cosse, 1992). Cosse reviews the work of Hodgson & Fischer who studied identity development in one hundred males and females and found three distinctive pathways of development. The first was occupied only by females and was referred to as the interpersonal path. This pathway was characterized by concerns with roles and values. The second pathway was followed predominantly by boys, this was called the intrapersonal path and was characterized by concerns about achievement, religious, and political ideology. The third pathway was described as androgynous which was a combination of the characteristics of the first two (Cosse, 1992). Similarly other studies report that female development is characterized by high levels of empathy and an ethic of care; while male development is characterized by concerns about individual competency, mastery of the environment and an ethic of justice (Cosse, 1992; Olweus & Endresen, 1998). It seems then that only among girls do early attachment processes facilitate the development of empathy. Evidence of this is illustrated in Roberts & Strayer's empirical analysis revealing that girls demonstrate more emotional expressiveness, insightfulness, and empathy than boys. However, when boys exhibited more emotional expressiveness, insight, and role taking ability; they were more likely to exhibit empathy. Thus role-taking ability among boys develops empathy which then predicts pro-social behavior. Therefore, boys must develop role-taking abilities to become empathetic in order to avoid deviant behavior, while empathy develops early on for most girls as the result of the secure attachment to their mothers.

A variety of additional research supports the relationship between impairment in the ability to demonstrate empathy and the capacity for violent behavior (Richardson, Hammock, Smith & Gardner, 1994; Ken-Ichi & Mukai, 1993; Born, Chevalier & Humblet, 1997). Thus an absence of the gendered infant-parent attachment may impair a male's ability to later take on the role of others, resulting in low levels of empathy and an increased propensity for crime and violence. Conversely, new evidence also reflects that an exaggerated attachment to the opposite sexed parent or adult and the lack of attachment to the same sexed parent or adult significantly increases serious delinquency involvement (Born, Chevalier & Humblet, 1997). Thus an exaggerated mother-son secure infant-parent attachment resulting from the failure of the infant to separate or individuate from the mother may also lead to the development of low levels of empathy. Moreover, the lack of development a secure father-son attachment may interact with the failure to separate from the mother to increase the likelihood of the development of low levels of empathy. Either of these types of attachment may then interact with the ideological or structural underpinnings of hegemonic masculinity, increasing the strength of the relationship between low levels of empathy and violence or substance abuse. Such a link may provide the explanation for Sampson & Laub's qualitative findings illustrating that substance abusing men failed to desist from criminal behavior as the result of marital attachments and beat their wives (Sampson & Laub, 1993; Laub, Nagin & Sampson, 1998). Even more recent work by Laub & Sampson illustrates that 'adolescence competence', a construct which resembles Gottfredson and Hirschi's construct of high self-control, fails to lead to desistance among substance abusing men (Laub & Sampson, 1998).

SUBSTANCE ABUSE AND VIOLENT CRIME

Although both the theoretical and empirical links between alcohol and violent behavior have been controversial issues in criminology, evidence supports the existence of a significant relationship between the two (Buss & Abdu, 1995; Nash-Parker, 1995; Browne, 1993; Collins, 1988; Walters, 1994; Huizinga, Loeber & Thornberry, 1993; Horney et al., 1995). Macro level research reflects that the increased consumption of alcohol predicts family homicides as well as primary non-intimate homicides even when controlling for poverty, the social control of alcohol, and capital punishment (Nash-Parker, 1995). Even experimental studies in the laboratory substantiate that alcohol consumption increases the intensity of shocks willingly administered by male subjects to both males and females (Giancola & Zeichner, 1995). Other evidence illustrates that husbands high on drugs were more likely to commit minor assaults on wives although wives' own drunkenness also predicted these assaults. However, wives' drunkenness did not predict severe violence perpetrated by the husband while the males' drunkenness did (Kaufman, Kantor & Straus, 1989). The longitudinal three cite study of delinquency, currently on-going in Rochester, Denver & Pittsburgh, reveals that substance abuse increases serious delinquency more than involvement in serious delinquency increases the likelihood of substance abuse (Huizinga, Loeber & Thornberry, 1993). However, other work by Horney and colleagues found that an underlying propensity (like low self-control) did not explain the relationship between drug and alcohol use and crime (Horney, et al., 1995; Gottfredson & Hirschi, 1990). Perhaps other individual level personality traits or states, or macro level constraints may explain the nature of this relationship. Exploring the link between heavy drinking and violence and perhaps their common causal predictors may lead us to a clearer explanation of the violence in our culture. Moreover, such a model may also explain the missing link in patterns of desistance already noted by other developmental theorists.

WHAT IS THE MISSING LINK IN PATTERNS OF PERSISTENCE OR LACK OF DESISTANCE AMONG VIOLENT AND SUBSTANCE ABUSING MEN?

Laub & Sampson's quantitative work and other subsequent research reveals that men's attachment to a spouse, work, or other family members, leads toward desistance from most types of crimes (Sampson & Laub, 1993; Laub et al., 1998; Horney et al., 1995; Farrall & Bowling, 1999). However, Sampson & Laub's qualitative work revealed that substance abusing men did not desist as the result of marital attachments, but rather committed violence against their wives (Sampson & Laub, 1993; Laub et al., 1998; Horney et al., 1995). On the surface this may be explained by an insecure attachment developed in early infancy, leading to the inability to behave empathetically. As mentioned previously, a variety of psychological research exists which substantiates the link between an insecure attachment style formed in early childhood and an increased likelihood of committing intimate violence in adulthood (Kesner, Julian & McKenry, 1997; Raine et al., 1997; Moffitt, 1997; Soko-Katz, Dunham & Zimmerman, 1997; Roberts & Noller, 1998; Holtzworth-Munroe et al., 1997). Even more recent tests of Sampson and Laub's theory among married African American and Hispanic men discovered a similar pattern. That is frequent drunkenness interfered with desistance processes (Nielsen, 1999). Unfortunately, Nielsen did not take violent behavior into account in re-examining Sampson & Laub's earlier findings. Thus these results offer little in the way of explaining the exact nature of the desistance process from violence or substance abuse. However, accounting for the early gendered nature of the parent-infant attachment relationship and the gendered nature of the development of empathy may explicate the exact nature of the desistance and persistence process among males. The key to these relationships lies in how early infant-parent attachments are replicated in the later lives of adult men. Research exists substantiating that secure attachment patterns from infancy are related to later adult attachments. For example, Holtzworth and colleagues review of the extant literature clearly illustrates that early parent-infant attachment patterns predict later marital satisfaction, better marital adjustment, improved awareness of and kindness toward others, and non-violent behavior (Holtzworth et al., 1997). However, in light of the previous discussion, even males with secure attachments will demonstrate less empathy than females with secure attachments, and thus will be more likely to behave violently. A consideration of Terrie Moffitt's work may provide enlightening at this juncture.

Moffitt hypothesizes that men who desist are simply adolescent-limited offenders. This is congruent with the above literature regarding males entering the intimacy stage later than girls' and thus desisting in late adolescence. For some males, that means marriage is occurring simultaneously simply as a part of the normal course of development. Thus males marry after they develop role taking abilities and thus empathy. Similarly, their premarital and marital attachments to women lead to their desistance from adolescent limited offending. However, Moffitt's description of the second group of males, the life-course persistent males, reflects that these males are often involved in substance abuse as well as violent behavior and are least likely to desist. Thus it could be stated that these men most closely resemble Laub and Sampson's substance abusing males who beat their wives. Yet Moffitt's work clearly identifies the persistent male offenders as suffering from a variety of interpersonal and intellectual deficits, including neurological problems. But this theoretical work proposes that Moffitt's alleged life-course-persistent males really consist of several groups of male offenders. These are defined below:

(1) Men who are insecurely attached and will never develop role taking abilities or empathy. These are the truly life-course persistent males. They also abuse chemicals. For these men low self-control will be the best predictor of substance abuse and violence.
(2) Men who are securely attached who will develop empathy later in life after developing the ability to take the role of others. These men will be most likely to desist from crime and deviance.
(3) Men who are securely attached but in early childhood develop a hegemonic masculine identity from a significant adult male or from other males in their social world (media influences, neighborhood, school, etc.) which will prevent them from developing role taking abilities and thus empathy. For these men the best predictor of violence and substance abuse will be hegemonic masculinity.
(4) Men who are insecurely attached and will not develop role taking ability nor the capacity for empathy. These men may also develop a hegemonic masculinity identity, which will act to further increase their propensity for involvement in violence and substance abuse. These should be the most serious violent offenders, and the group least likely to desist.

Thus the most persistent offending will come from group four, men who are insecurely attached and who practice hegemonic masculinity (Messerschmidt, 1993; 1997; Schwartz & DeKeseredy, 1997).

HEGEMONIC MASCULINITY

Messerschmidt theorizes that the patriarchal nature of our social structure facilitates a common ideological definition of masculinity called hegemonic masculinity. Messerschmidt explains that the greater likelihood of involvement in a variety of deviant behaviors among men, including violent behavior in general, violence against women and substance abuse results from this ideology of masculinity which is built into our social structures, institutions, and everyday social practice (Messerschmidt, 1993, 1997). Schwartz & DeKeseredy's peer model of sexual assault posits that patriarchy develops as the result of homosocial male groups, rather than resulting from a macro level imperative or existing within the institutions of our social structure. In other words, Schwartz & DeKeseredey's model unlike Messerschmidt's, does not imply that all men adhere to patriarchal beliefs. Rather, only those men who participate in social relationships with other males where patriarchy is openly discussed, peer supported, verbally reinforced, and practiced will adhere to hegemonic masculine ideals. Moreover, Schwartz & DeKeseredy claim that it is only among those men, that crime will be committed, particularly rape (Schwartz & DeKeseredy, 1997). This paper posits that Schwartz & DeKeseredey's model is clearly subsumed within Messerschmidt's conceptualization of structured action theory. Thus empirically testing one perspective is clearly also testing the other.

Messerschmidt discusses the importance of male dominance bonding and the necessity of appearing heterosexual as well as performing heterosexually to demonstrate masculinity to other men. Messerschmidt cites R. W. Connell's 1987 definition of hegemonic masculinity as "the ascendancy of a certain form of masculinity that is embedded in religious doctrine and practice, mass media content, wage structures, the design of housing, and welfare/taxation policies. This is the dominant form of masculinity to which other types of masculinity are subordinated not eliminated, and it provides the primary basis for relationships among men. This ideology emphasizes practices toward authority, control competitive individualism, independence, aggressiveness and the capacity for violence". Finally, he states, "force and threat of force may be used to help maintain hegemonic masculinity" (Messerschmidt, 1993, p. 81–82). He explains that the key to understanding hegemonic masculinity is the gender division of labor, which organizes relations of power, authority and control by men and the subordination of women. This ideological hegemony pushes men to feel obliged to have influence or control over other men and women. Thus men do gender by doing hegemonic masculinity and those who cannot do hegemony in traditionally defined ways will do hegemonic masculinity by doing crime.

Traditional methods of achieving hegemony (accommodating masculinity), according to Messerschmidt's perspective include, work in the paid labor market, the subordination of women in intimate relationships, heterosexism, a driven and uncontrollable sexuality, practicing authority, control, drinking alcohol, competitiveness, individualism, independence and aggressiveness. Thus men do gender in order to ascend a hierarchy of power among men as organized across race and class, by subordinating and controlling other men and women in legitimate or illegitimate ways. Oppositional hegemonic masculinity is accomplished by doing violence, rape, threatening to do violence and rape, and the public and private denigration of other men and women (Messerschmidt, 1993; Messerschmidt, 1997; Bird, 1996; Connell, 1995). Clearly, Schwartz & DeKeseredey's peer support model, which explains college male sexual assault of women, is subsumed within Messerschmidt's structured action theory.

A variety of existing research, although all of it not directly testing structured action theory, is supportive of these major tenets. For example, Yllo & Straus demonstrated that across thirty states, economic gender inequality and social structural gender inequality increased the rate of severe violence perpetrated against women by their husbands (backlash effects were also present) (Yllo & Straus, 1990). And although their work found no relationship between beliefs in patriarchal norms and gender inequality, belief in patriarchal norms increased violence against wives. Furthermore, where women were achieving greater economic and structural equality, belief in patriarchy increased and wife beating was at it's highest. This too is supportive of Messerschmidt's perspective, which suggests that the social structure and the ideology of masculinity may both operate separately as well as interact to increase male violence. Moreover, theoretical work by Susan Faludi succinctly explicated the relationship between women's economic gains and the ideological and sometimes institutional backlash against those gains. Similarly, backlash effects have also been illustrated in Peterson & Bailey research on the effects of women's economic gains on rape rates (Peterson & Bailey, 1992). Unfortunately, Yllo & Straus's indicator of patriarchal beliefs was not derived from the men who were beating their wives. If we are to accurately test the influence of hegemonic masculinity or patriarchy both on inequality and violence, we must measure the belief in patriarchy or the practice of patriarchy (or hegemonic masculinity) from the same men who are beating their wives or partners. Nonetheless, Yllo & Straus's research clearly supports the tenets of Messerschmidt's perspective in that both the social structure and the ideology of masculinity (hegemonic masculinity or patriarchy) predicted the violence that men committed against their women partners. Other research also supports the influence of patriarchy on male violence against women partners. Crossman et al., illustrate that as males weld more power in marital relationships or as relationships become less egalitarian, male violence against their wives increases (Crossman, Stith & Bender, 1990). Thus as hegemonic masculinity becomes less valued, subordinating women and drinking heavily or using substances heavily becomes less valued as well. Although limited, other research supports the existence of hegemonic masculine ideology, especially among male street offenders (Collison, 1996). Further, although Swartz & DeKeseredy mention Lee Bowker's early work on male violence and adherence to a patriarchal ideal; Bowker's more recent work, 'The Violences of Men', also supports Messerschmidt's conceptualization of hegemonic masculinity. That is Bowker finds that men who physically assault their wives believe in the importance of male privilege and male dominance and abuse their wives when that authority and dominance become threatened (Bowker, 1997). Both Sharon Bird's research and Swartz & DeKeseredy's review of the literature through analyses of male homosocial conversations illustrates support for the verbal and social maintenance of hegemonic masculine ideals and behavior. Bird's findings illustrated that men discussed topics including the subordination of other men and the objectification and denigration of women (Bird, 1996). Swartz & DeKeseredy also discuss their work on the Canadian National Survey of colleges and other research revealing that male dominance over women includes the sexual objectification of women (Crowell & Burgress, 1996; Itzin, 1992). This objectification of women is illustrated not only among males in homosocial interactions but has also been found in both television and films, as well as normative publications outside the realm of what could be termed soft and hard core pornography (Linz & Donnerstein, 1994; Molitor & Sapolsky, 1994). These images are linked to support of rape myths and are used by convicted rapists to enhance their deviant fantasies (Swartz & DeKeseredy, 1998; Linz, Wilson & Donnerstein, 1992; Donnerstein & Linz, 1994; Russell, 1998). Messner's work also reveals that a part of the development of a masculine identity includes male ambivalence toward intimacy, an affinity to rule-bound, competitive, and hierarchical sport, and instrumental relationships with others (Messner, 1990). Simpson & Ellis also examined hegemonic masculinity's effect on crime through an examination of the National Youth Survey. Their work revealed that the factors explaining violent crime among males and females vary (Simpson & Ellis, 1995). Unfortunately their only indicator of hegemonic masculinity was one question which asked males about their desire for marriage and family in the future. This indicator negatively affected property offending but had no effect on violent offending. Their findings also reflected that among white males both peer influences and school influences were more powerful predictors of violence. Although they didn't identify them as such, these may have been better indicators of hegemonic masculinity. Thus measuring adherence to a hegemonic masculine ideology and peer influences among males, may provide a clearer theoretical foundation from which to explain the relationship between violent behavior in general, violence against women and its relationship to the consumption of alcohol and other drugs. Also this theory may facilitate a clearer understanding of Sampson & Laub's qualitative findings that substance abusing men were less likely to desist as the result of marital attachments and were more likely to beat their wives (Sampson & Laub, 1993).

THE INTEGRATED THEORETICAL MODEL

This model is a side-by-side integration of Bowlby's attachment theory, Gottfredson & Hirschi's theory of self-control, and Sampson & Laub's developmental theory (see Katz, 1999). Further these perspectives are also linked to structured action theory which subsumes the male-peer support model as well as the construct of the gendered nature of the attachment process. Bowlby's theory is linked side-by-side to Gottfredson & Hirschi's theoretical perspective through the construct of the secure infant attachment relationship or what Gottfredson & Hirschi referred to as the emotional investment of the parent in the child. This attachment, whether secure or insecure, is understood as a clearly gendered process. During this attachment process the primary caretaker or the mother, separates from the male child early in his development. Among some males this may begin the process of his inculcation into a hegemonic masculine identity and a patriarchal society which is verbally supported and reinforced through homosocial male groups, or initially by his father or brother(s). Thus initial support for this ideology may occur through primary male relationships in the family. This in turn may facilitate the gendered nature of delinquency and adult crime. However, those males who have established a secure attachment and are not exposed to hegemonic masculinity, may be more likely to age-out of crime much earlier, as they develop the ability to take on the role of the other and thus empathy. Moreover these males may more easily become attached to a female and desist from crime. Insecurely attached men will develop low self-control and some of these males may be constrained to perform hegemonic masculinity. These men will be least likely to desist and most likely to be abusive to their wives. However, men who are securely attached and who abuse alcohol or drugs will also be unable to reach the intimacy stage of development and may well become embedded in practicing hegemonic masculinity. These men will be unable to desist and will be more likely to abuse their wives or partners. Finally, men who are securely attached but who become embedded in hegemonic masculinity whether through family or peers will also be more likely than other securely attached males to be involved in violence and substance abuse and will also be less likely to desist. This proposed theoretical integration is critically important, particularly with regard to ending America's reign as one of the most violent industrialized societies in the world. Moreover, recognizing the gendered nature of most violence may lead to changes in the way American mothers and fathers parent boys and girls and the recognition by policy makers that social institutions which foster hegemonic masculinity are partly responsible for male involvement in violent crime and alcohol abuse.

REFERENCES

Agnew, R. (1992). Foundation For a General Strain Theory of Crime and Delinquency. *Criminology, 30*, 47–87.

Ainsworth, M. D., Blehar, M., Waters, E., & Wall, S. (1978). *Patterns of Attachment: A Psychological Study of the Strange Situation*. Hillsdale, NJ: Erlbaum Press.

Arneklev, B. J., Grasmick, H. G., Tittle, C. R., & Bursik, R. J. Jr (1993). Low Self-Control and Imprudent Behavior. *Journal of Quantitative Criminology, 9*, 225–247.

Bahr, S. J., Hawks, R. D., & Wang, G. (1993). Family and Religious Influences on Adolescent Substance Abuse. *Youth and Society, 24*, 443–465.

Brannigan, A. (1997). Self Control, Social Control and Evolutionary Psychology: Towards an Integrated Perspective on Crime. *Canadian Journal of Criminology, 39*, 403–431.

Bartusch J. D. R., Lynam, D. R., Moffitt, T., Silva, P. A. (1997). Is Age Important? Testing a General Versus a Developmental Theory of Antisocial Behavior. *Criminology, 35*, 13–48.

Benson, M. L., & Moore, E. (1992). Are White Collar and Common Offenders the Same? An Empirical And Theoretical Critique of a Recently Proposed General Theory of Crime. *Journal of Research in Crime and Delinquency, 29*, 251–272.

Bird, S. (1996). Welcome to the Men's Club Homosociality and the Maintenance of Hegemonic Masculinity. *Gender and Society, 10*, 120–132.

Born, M., Chevalier, V., & Humblet, I. (1997). Resilience, desistance and delinquent career of adolescent offenders. *Journal of Adolescence, 20*, 679–694.

Bowlby, J. (1947). *Forty-four Juvenile Thieves: Their Characters and Home Life*. First Published December 1946. Bailliere, Tindall and Cox, London.

Bowlby, J. (1988). *A Secure Base. Parent-Child Attachment and Healthy Human Development*. Basic Books Inc.

Bowlby, J. (1969). *Attachment and Loss, Volume I*. Basic Books.

Brown, A., & Williams, K. R. (1993). Gender, Intimacy, and Lethal Violence: Trends from 1976 through (1987). *Gender and Society, 7*, 78–98.

Buss, T. F., & Abudu, R. (1995). Repeat Victims of Violence in An Urban Trauma Center. *Violence and Victims, 10*, 183–194.

Caspi, A., Moffitt, T., E., Silva, P., A., Stouthamer-Loeber, M., Krueger, R. F., & Schumtte, P. S. (1994). Are Some People Crime-Prone? Replications of the Personality-Crime Relationship Across Countries, Genders, Races and Methods. *Criminology, 32*, 163–195.

Cernkovich, S. A., & Giordano, P. C. (1987). Family Relationships and Delinquency. *Criminology, 25*, 295–313.

Chibucos, T. R. & Kail, P. R. (1981). Longitudinal Examination of Father-Infant Interaction and Infant Father Attachment. *Merrill-Palmer Quarterly, 27*, 81–96.

Cohen, T. F. (1987). Remaking Men: Men's Experiences Becoming and Being Husbands and Fathers and Their Implications for Reconceptualizing Men's Lives. *Journal of Family Issues, 8*, 57–77.

Collins, J. J. (1988). Suggested Explanatory Frameworks to Clarify the Alcohol Use/Violence Relationship. *Contemporary Drug Problems, 15*, 107–121.

Connell, R. W. (1995). *Masculinities*. University of California Press. Berkeley Los Angeles.

Crossman, R. K., Sith, S. M., & Bender, M. M. (1990). Sex Role Egalitarianism and Martial Violence. *Sex Roles, 22*, 293–304.

Dutton, D. G., Saunders, K. S., & Bartholomew (1994). Intimacy-Anger and Insecure Attachment as Precursors of Abuse in Intimate Relationships. Journal of Applied Social Psychology, *24*, 1367–1386.

Erikson, E. H. (1964). Childhood & Society. New York, W. W. Norton and Company Inc.

Evans, T. D., Cullen, F. T., & Burton, V. S. (1997). The Social Consequences of Self-Control: Testing the General Theory of Crime. *Criminology, 35*, 475–504.

Farrall, S., & Bowling, B. (1999). Structuration, Human Development and Desistance from Crime. *British Journal of Criminology, 39*(2), 253–268.

Farrington, D. P., & West, D. J. (1995). Effects of Marriage, Separation, and Children on Offending by Adult Males. *Current Perspectives on Aging and the Life Cycle, 4*, 249–281.

Giancola, P. R., & Zeichner, A. (1995). An Investigation of Gender Difference in Alcohol-Related Aggression. *Journal of Studies on Alcohol, 56*, 573–579.

Gibbs, J. J., Giever, D. G., & Martin, J. S. (1998). Parental Management and Self-Control: An Empirical Test of Gottfredson and Hirschi's General Theory. *Journal of Research in Crime and Delinquency, 25*, 40–70.

Gottfredson, M., & Hirschi, T. (1990). *A General Theory of Crime*. Stanford, California Stanford University Press.

Grasmick, H. G., Tittle, C. R., Bursik, R. J. Jr., & Arneklev, B. J. (1993). Testing the Core Empirical Implications of Gottfredson and Hirschi's General Theory of Crime. *Journal of Research in Crime and Delinquency, 30*, 5–29.

Greenberg, M. T., Cicchetti, D., & Cummings, M. E. (1992). *Attachment in the Preschool Years*. The University of Chicago Press, Chicago and London.

Hawks, R. D., Bahr, S. J., & Wang, G. (1994). Adolescent Substance Use and Codependence. *Journal of Studies on Alcohol, 55*, 261–268.

Hirschi, T. (1979). Separate and Unequal is Better. *Journal of Research in Crime and Delinquency 16*(5), 35–38.

Hirschi, T., & Gottfredson, M. (1993). Commentary: Testing The General Theory of Crime. *Journal of Research in Crime and Delinquency, 30*, 47–54.

Hochschild, A., & Machung, A. (1989). *The Second Shift*. New York, Viking Penguin Press.

Holtzworth-Munroe, S., Gregory, L., & Hutchinson, G. (1997). Violent Versus Nonviolent Husbands: differences in Attachment Patterns, Dependency, and Jealousy. *Journal of Family Psychology, 11*, 314–331.

Horney, J., Osgood, W. D., & Marshall, H. I. (1995). Criminal Careers in the Short-Term: Intra-Individual Variability in Crime and Its Relation to Local Life Circumstances. *American Sociological Review, 60*, 655–673.

Horner, A. J. Ph.D. (1991). *Psychoanalytic Object Relations Therapy*. Janson Aronson Inc. Northvale, New Jersey, London.

Howing, P. T. MSW, Wodarski, J. S. Ph.D., Kurtz, D. P. Ph.D., & Gaudin, M. Jr. Ph.D. (1993). *Maltreatment and the School-Age Child., The Haworth Press Inc.*

Huizinga, D., Loeber, R., Thornberry, T. (1993). Urban Delinquency and Substance Abuse. Office of Juvenile Justice and Delinquency Prevention. Office of Justice Programs. U.S. Department of Justice.

Isabella, R. A., Belsky, J., & Von Eye, A. (1989). Origins of Infant-Mother Attachment: An Examination of Interactional Synchrony During the Infant's First Year. *Developmental Psychology, 25*, 12–21.

Jacobson, T., Huss, M., & Fendrich, M. (1997). Children's Ability to Delay Gratification: Longitudinal Relations to Mother-Child Attachment. *The Journal of Genetic Psychology, 158*, 411–426.

Johnson, V., & Pandina, R. J. (1993). Affectivity, family drinking history, and the Development of problem drinking: A Longitudinal Analysis. *Journal of Applied Social Psychology, 23*, 2055–2073.

Katz, R. (1999). Building the Foundation for a Side-by-Side Explanatory Model: A General Theory of Crime, the Age-Graded Life Course Theory and Attachment Theory. *Western Criminology Review, 1:2*. http://wcr.sonoma.edu/v1n2/katz.html

Kaufman Kantor, G., & Straus, M. A. (1989). Substance Abuse as Precipitant of Wife Abuse Victimization's. *American Journal of Drug and Alcohol Abuse, 15*, 173–189.

Keane, C., Maxim, P. S., & Teevan, J. J. (1993). Drinking and Driving, Self-Control, and Gender: Testing a General Theory of Crime. *Journal of Research in Crime and Delinquency, 30*, 30–46.

Kesner, J. E., Julian, T., & McKenry, P. C. (1997). Application of Attachment Theory to Male Violence Toward Female Intimates. *Journal of FamilyViolence, 12*, 211–228.

Krueger, R. F., Avashalom, C., Moffitt, T. E., White, J., & Stouthamer-Loeber, M. (1996). Delay of Gratification, Psychopathology, and Personality: Is Low Self-Control Specific to Externalizing Problems? *Journal of Personality, 64*, 107–129.

LaGrange, T. C., & Silverman, R. A. (1999). Low Self-Control and Opportunity: Testing the General Theory of Crime As An Explanation for Gender Differences in Delinquency. *Criminology, 37*, 41–72.

Larzelere, R. E., & Patterson, G. (1990). Parental Management: Mediator of the Effect of Socioeconomic Status on Early Delinquency. *Criminology, 28*, 304–324.

Laub, J. H, Nagin, D. S., & Sampson, R. J. (1998). Trajectories of Change in Criminal Offending: Good Marriages and the Desistance Process. *American Sociological Review, 68*, 225–238.

Laub, J. H., & Sampson, R. J. (1993). Turning Points in the Life Course: Why Change Matters to the Study of Crime. *Criminology, 31*, 301–325.

Laub, J. H., & Sampson, R. J. (1998). The Long-Term Reach of Adolescent Competence: Socioeconomic Achievement in the Lives of Disadvantaged Men. In: A. Colby, J. James, & D. Hart (Eds), *Competence and Character Through Life*. The University of Chicago Press, Chapter Four.

Loeber, R. (1990). Development and Risk Factors of Juvenile Antisocial Behavior and Delinquency. *Clinical Psychology Review, 10*, 1–41.

Loeber, R., Stouthammer-Loeber, M., Van Kammen, W., & Farrington, D. P. (1991). Initiation, Escalation and Desistance in Juvenile Offending and their Correlates. *The Journal of Criminal Law and Criminology, 20*, 38–82.

Loeber, R., Green, S. M., Lahey, B. B., Christ, M. A. G., & Frick, P. J. (1992). Developmental Sequences in the Age of Onset of Disruptive Child Behaviors. *Journal of Child and Family Studies, 1*, 21–41.

Loeber, R., Wung, P., Keenan, K., Girous, B., Stouthammer-Loeber, M., Van Kammen, W. B., & Maughan, B. (1993). Developmental Pathways in Disruptive Child Behavior. *Developmental and Psychopathology, 5*, 103–133.

Longshore, D., Turner, S., & Stein, J. A. (1996). Self-Control in A Criminal Sample. An Examination of Construct Validity. *Criminology, 34*, 209–228.

Longshore, D., Stein, J., & Turner, S. (1998). Reliability and Validity of Self-Control Measure: Rejoinder. *Criminology, 34*, 175–182.

Longshore, D. (1998). Self Control and Criminal Opportunity: A Prospective Test of the General Theory of Crime. *Social Problems, 45*(1), 102–113.

Lynam, D., Moffitt, T., & Stouthamer-Loeber, M. (1993). Explaining the Relation Between IQ and Delinquency: Class, Race, Motivation, School Failure, or Self-Control? *Journal of Abnormal Psychology, 102*, 187–196.

Marenin, O., & Reisig, M. D. (1995). A General Theory of Crime and Patterns of Crime in Nigeria: An Exploration of Methodological Assumptions. *Journal of Criminal Justice, 23*, 501–518.

McCord, J. (1991). Family Relationships, Juvenile Delinquency, and Adult Criminality. *Criminology, 29*, 399–417.

McCord, J. (1979). Some Child Rearing Antecedents of Criminal Behavior in Adult Men. *Journal of Personality and Social Psychology, 37*, 1471–1486.

McDevitt, T. M., Lennon, R., & Kopriva, R. J. (1991). Adolescents' Perceptions of Mothers' and Fathers' Prosocial Actions and Empathic Responses. *Youth and Society, 22*, 387–409.

McGee, L., & Newcomb, M. D. (1992). General Deviance Syndrome: Expanded Hierarchical Evaluations at four Ages from Early Adolescence to Adulthood. *Journal of Consulting and Clinical Psychology, 60*, 766–776.

Messerschmidt, J. W. (1993). Masculinities and Crime Critique and Reconceputalization of Theory.Rowman and Littlefield Publishers, Inc.

Messerschmidt, J. W. (1997). *Crime as Structured Action Gender, Race, Class, and Crime in the Making*. Sage Publications.

Minty, B. (1988). Public Care or Distorted Family Relationships: The Antecedents of Violent Crime. *The Harvard Journal, 27*, 172–187.

Nagin, D. S., & Farrington, D. P. (1992). The Onset and Persistence of Offending. *Criminology, 30*, 501–528.

Nagin, D. S., & Paternoster, R. (1991). On the Relationship of Past to Future Participation in Delinquency. *Criminology, 29*, 163–189.

Nagin, D. S., & Paternoster, R. (1993). Enduring Individual Differences and Rational Choice Theories of Crime. *Law and Society Review, 27*, 467–496.

Nagin, D. S., & Paternoster, R. (1994). Personal Capital and Social Control: The Deterrence Implications of a Theory of Individual Differences in Criminal Offending. *Criminology 32*, 581–606.

Nagin, D. S., Farrington, D. P., & Moffitt, T. E. (1995). Life-Course Trajectories of Different Types of Offenders. *Criminology, 33*, 111–139.

Nash-Parker, R. (1995). *Bring Booze Back*. In: The Relationship Between Alcohol and Homicide. *Journal of Research in Crime and Delinquency, 32*, 3–38.

Nielson, A. L. (1999). Testing Sampson and Laub's Life Course Theory: Age, Race/Ethnicity, and Drunkeness. *Deviant Behavior, 20*, 129–151.

Ohbuchi, K.-I., Ohno, I., & Mukai, H. (1993). Empathy and Aggression: Effects of Self-Disclosure and Fearful Appeal. *The Journal of Social Psychology, 133*, 243–353.

Paternoster, R., & Brame, R. (1997). Multiple Routes to Delinquency? A Test of Developmental and General Theories of Crime. *Criminology, 35*, 49–84.

Patterson, G. R., & Yoerger, K. (1993). Developmental Models for Delinquent Behavior. In: S. Hodgins (Ed.), *Mental Disorder and Crime*. Sage Publications, Chapter 8.

Piquero, A. R., & Rosay, A. B. (1998). The Reliability and Validity of Grasmick et al.'s Self-Control Scale: A Comment on Longshore et al.* *Criminology, 36*, 157–173.

Piquero, A., & Tibbets, S. (1996). Specifying the direct and indirect effects of low self-control and situational factors in offenders' decision making: Toward a more complete model of rational offending. *Justice Quarterly, 13*, 481–510.

Pulkkinen, L., & Pitkanen, T. (1993). Continuities in Aggressive Behavior From Childhood to Adulthood. *Aggressive Behavior, 19*, 249–263.

Raine, A., Brennan, P., & Mednick, S. A. (1997). Interaction Between Birth

Complications and Early Maternal Rejection in Predisposing Individuals to Adult Violence: Specificity to Serious, Early-Onset Violence. *American Journal of Psychiatry, 154*, 1265–1271.

Rankin, J. H., & Wells, E. L. (1990). The Effect of Parental Attachments and Direct Control on Delinquency. *Journal of Research in Crime and Delinquency, 27*, 140–165.

Rankin, J. H., & Kern, R. (1994). Parental Attachments and Delinquency. *Criminology, 32*, 495–515.

Reed, G. E., & Yeager C. P. (1996). Organizational Offending and Neoclassical Criminology: Challenging the Reach of a General Theory of Crime. *Criminology, 34*, 357–382.

Richardson, D. R., Hammock, G. S., Smith, S. M., Gardner, M. (1994). Empathy as a Cognitive Inhibitor of Interpersonal Aggression. *Aggressive Behavior, 20*, 275–290.

Rowe, D. C., Rogers, J. L., & Meseck-Bushey, S. (1992). Sibling Delinquency and the Family Environment: Shared and Unshared Influences. *Child Development, 63*, 59–67.

Rowe, D. C., & Farrington, D. P. (1997). The Familial Transmission of Criminal Convictions. *Criminology, 35*, 177–201.

Rosen, L. (1985). Family and Delinquency: Structure or Function? *Criminology, 23*, 553–573.

Rubin, L. (1983) Intimate Strangers. New York, HarperCollins Publishers.

Russell, G. (1982). Highly Participant Australian Fathers: Some Preliminary Findings. *Merrill–Palmer Quarterly, 28*, 127–156.

Sampson, R. J., & Laub, J. H. (1990). Crime and Deviance Over the Life Course: The Salience of Adult Social bonds. *American Sociological Review, 55*, 609–627.

Sampson, R. J., & Laub, J. H. (1993). *Crime in the Making Pathways and Turning Points, Through Life.* Harvard University Press, Cambridge, Massachusetts.

Sampson, R. J., & Laub, J, H. (1996). Socioeconomic Achievement in the Life Course of Disadvantaged Men: Military Service as a Turning Point, Circa 1940–1965. *American Sociological Review, 61*, 347–367.

Schwartz, M. D., & DeKesredy, W. (1997)., *Sexual Assault on the College Campus The Role of Male Peer Support.* Thousand Oaks, London, and New Delhi: Sage Publications.

Sokol-Katz, J., Dunham, R., & Zimmerman, R. (1997). Family Structure versus Parental

Attachment in Controlling Adolescent Deviant Behavior: A Social Control Model. *Adolescence, 32*, 199–215.

Sommers, I., Baskin, D. R., & Gagan, J. (1994). Getting Out of the Life: Crime Desistance by Female Street Offenders. *Deviant Behavior, 15*, 125–149.

Spiker, D., Ferguson, J., & Brooks-Gunn, J. (1992). Enhancing Maternal Interactive Behavior and Child Social Competence in Low Birth Weight, Premature Infants. *Child Development, 64*, 754–768.

Stark, E., & Flitcraft, A. (1996). *Women at Risk Domestic Violence and Women's Health.* Sage Publications.

Valliant, G. E. (1995). *The Natural History of Alcoholism Revisited.* Harvard University Press, Cambridge Massachusetts.

Wainrib R. B. (Ed.) (1992). *Gender Issues Across the Life Cycle.* Springer Publishing Company New York.

Walters, G. D. (1994). Drugs and Crime in Lifestyle Perspective. *Drugs, Health, and Social Policy Series, 1*. Sage Publications.

Warr, M. (1993). Parents, Peers and Delinquency. *Social Forces, 72*, 247–264.

West, D. J. (1973). Who Becomes Delinquent? *Second report of the Cambridge Study in Delinquent Development.* London, Heinemann Educational Center, Publishers for the Cambridge Institute of Criminology.

Wood, P., Pfefferbaum, B., & Arneklev, B. J. (1993). Risk-taking and self-control: Social psychological correlates of delinquency. *Journal of Crime and Justice, 16*, 111–130.

Wright, K. N., Ph.D., & Wright, K. E. MSW (1994). *Family Life, Delinquency, and Crime: A Policymaker's Guide, Research Summary.* Office of Juvenile Justice.

Yllo, K. A., & Straus, M. (1990). Patriarchy and Violence Against Wives: The Impact of Structural and Normative Factors. In: M. A. Straus & R. J. Gelles)Eds), *Physical Violence in American Families: Risk Factors and Adaptations to Violence in 8145 Families.* Transaction Publishers, New Brunswick and London, Chapter 22.

INDEX

Printed and bound by CPI Group (UK) Ltd, Croydon, CR0 4YY
16/09/2026
14963383-0003